FOOD POLICY IN MEXICO

The Search for Self-Sufficiency

Food Policy in Mexico
The Search for Self-Sufficiency

EDITED BY

James E. Austin AND Gustavo Esteva

CORNELL UNIVERSITY PRESS

————ITHACA AND LONDON————

First published 1987 by Cornell University Press.

International Standard Book Number (cloth) 0-8014-1962-X
International Standard Book Number (paper) 0-8014-9453-2

Library of Congress Catalog Card Number 86-19815
Printed in the United States of America
*Librarians: Library of Congress cataloging information
appears on the last page of the book.*

*The paper in this book is acid-free and meets the guidelines for
permanence and durability of the Committee on Production Guidelines
for Book Longevity of the Council on Library Resources.*

Contents

Preface

The seeds for this book were planted twelve years ago, when our paths crossed for the first time. We were riding similar currents of concern about rural development and food policy in Mexico and elsewhere. Over the years, as fellow observers, advisers, academics, and implementers as well as friends, we have struggled to understand the problems, processes, and possibilities of food systems. When the opportunity arose to mount a study of the recent experience of food policy formation and implementation in Mexico, particularly that of the Sistema Alimentario Mexicano (SAM—Mexican Food System), we accepted it with much enthusiasm, and with some trepidation: enthusiasm, because we perceived an opportunity to make a meaningful contribution to increased understanding of food policy, both in Mexico and elsewhere; trepidation, because of the difficulty of trying to shed additional light on the SAM experience, which had already been in the political and intellectual spotlight during the entire three years of its existence (1980–1982).

The undertaking was a major one, requiring strong institutional support and extensive intellectual collaboration from many colleagues inside and outside Mexico. A shared conviction that a scrutiny of the SAM experience was important to Mexico and to the international food community provided the cohesion and commitment essential to a collaborative effort.

Financial and material support was provided by Harvard University's Graduate School of Business Administration, the Sociedad Mexicana de Planificación (SMP—Mexican Planning Society), and El Comité Promotor de Investigaciones para el Desarrollo Rural (COPIDER—Committee to Promote Rural Development Research). The research was part of a global food policy project carried out in conjunction with the Harvard Business School's 75th Anniversary Research Colloquium Series. It also was part of the larger food and peasant research agendas of the SMP and

9

COPIDER. This fusion of interests, resources, and distinctive institutional strengths provided the basis for very effective international research collaboration. We deeply appreciate the sponsorship of this effort.

The contributing authors to this volume are extraordinary individuals. We were most fortunate to have their cooperation. Their distinct disciplinary perspectives and experiences as academics, or advisers, or policy makers, or managers have provided multiple lenses through which we can more richly and thoroughly examine the Mexican food policy experience. All of our contributors have been careful observers of food policy and of SAM. Many have distilled their insights from other, larger research projects to make their contribution to this collection. We express our gratitude to them for the efforts and sacrifices entailed in making this book a reality.

The large variety of perspectives—within a wide spectrum of ideologies—and the plurality of theoretical and methodological approaches do not relegate the book to superficial eclecticism. On the contrary, we believe they enrich the text with nuances, similarities, and differences that reflect, in their own special ways, the plurality of interests and trends of thought and action that shape food policies in the real world.

We also express our appreciation to the many individuals from public and private institutions in the food system who gave so generously of their time and insight in our two research workshops and in individual meetings. Our deep thanks also go to the administrative staffs of the Division of Research and Word Processing Department of the Harvard Business School, the Sociedad Mexicana de Planificación, and COPIDER for the superb assistance they provided throughout the life of this project. Finally, we are grateful to Frank Meissner and Ray Goldberg and to the Cornell University Press reviewers for their constructive critique of the draft manuscript; to Helen Frey Rochlin for her invaluable editing input; and to Leonor Corral for her skillful translation.

In reflecting upon this research effort, we are certain that we have learned much, we believe that the research has contributed to the current food policy formation process in Mexico, and we hope that it will increase the understanding of others in Mexico and elsewhere. With research, the end is always a new beginning. Our final hope is that this effort will spark and facilitate further food policy research by others. Deeper understanding is the basis for better policy.

JAMES E. AUSTIN AND GUSTAVO ESTEVA

Boston and Mexico, D.F.

PART I

INTRODUCTION

CHAPTER 1

The Path of Exploration

James E. Austin and Gustavo Esteva

A basic challenge facing Third World policy makers is meeting the people's food needs. Not only is this a social necessity; it is also a political imperative. Food shortages topple governments. The multitude of constraints surrounding this challenge—burgeoning populations, low agricultural productivity, low and skewed income levels, capital scarcity, shrinking foreign reserves and food-importing capacity—convert it into a seemingly overwhelming struggle. That politicians are very uncomfortable when their country depends on foreign suppliers for basic staples, especially when they have witnessed those suppliers withhold exports on political grounds, only further complicates the problem.

Given these circumstances, it is not surprising that policy makers view as highly desirable the goal of self-sufficiency in basic staples, a goal frequently stated yet seldom attained. It often clashes with economic theory and the law of "comparative advantage," yet it reflects a reality in the political economy of developing countries. The search for self-sufficiency looms large on the development agendas of most Third World nations, yet there are relatively few detailed studies of countries' experiences with this struggle.

A greater understanding of the key issues surrounding the goal of self-sufficiency is essential for public policy makers, food policy analysts, development specialists, and private decision makers involved in developing national food systems. It is important to explore the substance and rationale of strategies for food self-sufficiency, as well as the means by which they are implemented and their impact.

What is the substance of the self-sufficiency strategies?
There may be several paths toward self-sufficiency: increasing cultivable acreage by extending the agricultural land frontier; improving yields on existing croplands with growth-enhancing inputs such as special seed

varieties, agrochemicals, and irrigation; altering food access and consumption patterns to achieve a more equitable distribution of whatever is produced, thereby eliminating shortages due to income disparities. A strategy's substance is also shaped by definitions of self-sufficiency and by the relative roles assigned to peasants and modern producers.

Why are the self-sufficiency strategies chosen?
To fully understand a food policy, one must examine its genesis. Policy formation occurs in a bubbling cauldron in which political, economic, demographic, and cultural forces swirl tumultuously; one cannot assume that a self-sufficiency policy is an instant creation. Current events serve as catalysts, but historical forces can significantly condition outcomes.

How are self-sufficiency strategies implemented?
The leap from strategy to implementation is long and precarious. Many a pretty plan has failed to progress from word to deed. The food system is a complex, interrelated system of institutions. The reactions of government ministries and state-owned enterprises to a strategy are far from automatic; their dynamics must be understood. Similarly, the responses of farmers and private agribusiness enterprises to a strategy's stimuli also require analysis. Implementation is a dimension of the food policy analysis that has been relatively neglected by researchers.

What impact do the self-sufficiency strategies produce?
A basic issue is how to assess impact. Production increases are one measure, but it is not always clear whether an increase is due to the strategy or to other factors. One is also interested in the cost incurred in achieving the production gains. Another outcome variable is consumption effect but measuring changes in nutritional status demands special evaluation techniques. Other indirect impacts are on income and employment. A final concern is how other macro policies affect the outcome of a self-sufficiency policy. Documenting the impact of national self-sufficiency strategies is not enough; the approaches and problems in carrying out impact analyses are also of concern.

The Case of Mexico

This book addresses the questions above with an in-depth examination of one country's experience. Such analyses of food policy experiences are few, and Mexico provides a particulary rich case study. By understanding the Mexican experience, one can gain insight into these basic food policy questions. The richness of the Mexican case stems from that country's recent formulation and implementation of a food policy aimed at self-sufficiency. In 1980 the Mexican government announced its new program,

Sistema Alimentario Mexicano (SAM—Mexican Food System). Its strategy was to sow the country's oil revenues into agriculture in order to reap self-sufficiency in staples. "Comparative advantage" arguments for exporting oil and importing food were cast aside. Traditional peasant producers were placed at the heart of the strategy; their productivity was to be accelerated through the greatly increased provision of credit, crop insurance, improved seed, fertilizer, and crop support prices. Conceptually, the strategy distinguished itself from other countries' food policies in two ways: first, unlike traditional production-oriented approaches, it started from the consumption side, assessed nutritional needs, set consumption goals, and then derived the production targets; second, it employed a systems approach that led to an integrated strategy with action steps at each stage in the food system chain. Last, this was not a small pilot project but rather a massive national program carrying high political and economic priority.

The size and innovativeness of the strategy, and the importance of the underlying basic food policy issues, sparked intense interest and controversy within Mexico and in the international-development and academic communities. SAM was born in the fourth year of President López Portillo's administration and died with the 1983 entrance of the succeeding administration of President de la Madrid. From a researcher's perspective, this sequence begins to take on the fortuitous appearance of a controlled experiment: a precise beginning, clearly formulated objectives, observable procedures and results, and a precise ending. To a great extent these conditions are true and make the Mexico case study so valuable. However, the real world of policy making and implementation is messy. Retrospective policy analysis is clouded by imperfect information, uncontrolled variables, and the intertwining of past, present, and future. Thus we approach our task with the humility of analytic imperfection but also with the confidence that the exploration will add to a deeper understanding of the underlying food policy issues in developing countries. Although the book focuses predominantly on the 1980–1982 SAM period, it also traces Mexico's historical food policy roots as well as examines SAM's successors: Programa Nacional de Alimentación (PRONAL—National Food Program), which emerged in October of 1983, and Programa Nacional de Desarrollo Rural Integral (PRONADRI—National Integral Rural Development Program), which was announced in early 1984. SAM died, but in the sphere of food policy, reincarnation is possible.

The Analytic Map

Following this introductory chapter (Part I), the book is divided into the following four sections: Strategy, Implementation, Impact, and the Future. Studies on food policy tend to concentrate on the overall policy at a strategic level. We analyze this fundamental dimension in Part II in an examination of the historical evolution of food policy in Mexico culminating in the formulation of the SAM strategy; here we address issues of the *substance* and *choice* of self-sufficiency strategy. In Part III we examine the *implementation* of a self-sufficiency strategy. The implementation of strategies has received less coverage in the literature; the strategy that appears most suitable on paper may become less than desirable in reality. It is therefore crucial to examine the problems and achievements of the SAM implementation process. Accordingly, in Part III we analyze SAM's key institutions and inputs: state-owned enterprises, credit, seeds, peasant organizations, distribution systems, and private agribusiness entities. Data limitations prevent a precise measurement of SAM's *impact* on nutritional status or income distribution, or of its interplay with other macro policies. However, in Part IV we do discuss some of the problems in making these assessments and present approaches for approximating and understanding the effects; our aim in this section is to illustrate methodological issues rather than to provide firm empirical results of SAM. Mexico's food policy did not die with SAM. In Part V we explore the future of Mexico's new food policy, PRONAL. This allows us to return to the global dimension of strategy and to relate the SAM experience to the design and potential of PRONAL, so as to appreciate fully its intellectual and political heritage. The final part of the book attempts to draw some conclusions from the SAM experience and to relate them to the new food policy.

We now identify the authors' foci and the key questions addressed in each of the chapters.

In Chapter 2, Gustavo Esteva, who has studied the evolution of Mexico's food system for several decades, begins the strategy analysis by placing SAM within its historical context. A brief review of the country's food situation over the past four centuries and the various policy formulas used to approach it addresses the question, Was SAM a response to longstanding problems or to recent circumstances? Moreover, this historical analysis allows us to compare the evolving Mexican food approaches: To what extent were past policies based on balancing local production capacities and consumption needs and with what effects? How autonomous has the Mexican food system been?

This historical framework brings the analysis up to 1980, when SAM was designed, a subject that is dealt with in Chapter 3 by Mario Montanari. Montanari chaired the technical group that worked for Casio Luiselli, the promotor and coordinator of SAM, and he played a key role in conceptualizing the strategy. His analysis shows the challenges faced by SAM's architects and the theoretical and methodological approach they employed. Why did they adopt a systems approach? What was the rationale for focusing on peasants on rain-fed land and for using the various policy instruments?

Chapter 4 begins the Part III analysis of SAM's implementation. James Austin had closely followed the conceptual and institutional evolution of Mexican food policy for over ten years when, with his colleague Jonathan Fox, he undertook an analysis of Mexico's state enterprises within the food sector. State enterprises were the primary government instrument for carrying out SAM's strategy. Austin and Fox's analysis highlights some central aspects of SAM's implementation and focuses on some issues regarding state enterprises in general. What was their significance in implementation? As implementers, how did they respond to the shift in food policy produced by SAM? Which factors influenced the nature of their responses? What were the main problems with their actions?

During SAM, Raul Pessah was adviser to the director of the Banco Nacional de Crédito Rural, the primary conduit for rural credit. Since credit was a fundamental component of the strategy, Chapter 5's detailed and, indeed, unique analysis of agricultural lending policy allows us to understand both SAM's implementation problems and the response capacity of a key state enterprise to its guidelines and demands. Did the volume and allocation of official credit change? Did the credit flow more to agriculture or to livestock? To irrigated or to rain-fed land? To basic staples or to other crops? Was the loan size adequate to acquire SAM's technological package? Was the traditional geographic concentration of credit lessened? How did the bank's organizational structure and procedures change in response to the SAM strategy?

The accelerated use of improved seeds was central to achieving the productivity gains envisioned by SAM. To what extent was the Mexican seed industry able to supply this increased demand? What were the nature and effects of the interaction between the public-sector seed institutions and the private-sector companies, local and transnational? What were the implications of peasant adoption of the improved varieties? David Barkin, a well-known analyst of the Mexican countryside and head of an interdisciplinary research team that recently completed a major study of the production and use of improved seeds in Mexico addresses these questions in Chapter 6.

One of SAM's most innovative aspects was its clear commitment to the peasantry. A critical implementation issue was how to develop and disseminate the new technology to these traditional producers. Gustavo Viniegra Gonzalez, an outstanding specialist in the field of rural technology, approaches this subject in Chapter 7 by examining five empirical experiences related to the adoption of SAM's technological package. Rodrigo Medellin has worked with peasant organizations for many years, and he uses a participatory research method to present, in Chapter 8, the peasants' perspective on the implemenation of SAM. What is the peasants' capacity to respond to the state stimuli? What role can they play in a process of transformation such as SAM? Viniegra Gonzalez approaches the peasants' reaction as the *object* of policy, while in Medellin's study they are seen as the *subject* of the process.

SAM also gave strategic importance to food distribution. All the links in the supply chain were considered in the design of the strategy. Some aspects of the implementation of food distribution are dealt with in Chapter 4, where state enterprises are considered. In Chapter 9, Cynthia Hewitt de Alcantara studies the problem of supplying in Mexico City, where one-fifth of the population is concentrated and where the springs and mechanisms that influence the national food-supply system converge. How is Mexico City supplied? What are the problems in this supply system? To what extent did SAM address these problems? Hewitt de Alcantara has carried out a research project on this issue as part of the Food Systems and Society project of the U.N. Research Institute for Social Development.

The section on implementation begins with a chapter on public enterprises; it ends, with Chapter 10, with the experience of private enterprise, another key agent in SAM's implementation. Carlos Sequeira, an experienced analyst of the role of private agribusiness in Mexico and other countries, examines it with respect to SAM. What was its contribution to the strategy's implementation? What were the benefits it derived? What were its perceptions of SAM?

Chapter 11 begins the Part IV examination of the impact of SAM by looking at production and costs. On the production side, the questions are these: How much did production increase? What increased? Were the self-sufficiency targets reached? What were the sources of the increases: expanded area? improved fields? good weather? Who produced the gains: the peasants or the commercial growers? On the cost side, there is the issue of the size of SAM's price tag. What costs are actually attributable to SAM? How do these outlays compare to expenditures in other sectors? Finally, how do SAM's costs compare with the value of the production and consumption gains? These are the questions that Armando Andrade and Nicole Blanc, researchers from the Fondo de Cultura Campesina, try to deal with, in spite of the well-known limitations and deficiencies of the available information.

In Chapter 12, Jacobo Schatan explores the effects of SAM on consumption and nutrition. The lack of data on nutritional status prevents a conclusive evaluation. However, Schatan, a U.N. expert on food issues and a member of the U.N. Research Institute for Social Development research project Food Systems and Society, critically examines SAM's approach to nutrition. What was the nature of the nutritional problem? How valid was SAM's conceptual and methodological approach to nutritional improvement? What was the likely impact on consumption?

Macro price policies greatly affect performance of any food policy. Oil-exporting countries have been subject to considerable pressure in this regard. What have been the effects on Mexican agriculture? What are the ties between macro prices, energy, and food-system performance and the implications for SAM? In Chapter 13, C. Peter Timmer, a pioneer food policy analyst, outlines a model that examines these issues, applying 1960–1980 data from Mexico and other countries. This analysis provides a frame of reference from which to consider SAM's ties with macro prices and structural change.

Bill Gibson, Nora Lustig, and Lance Taylor worked with the team that formulated and coordinated SAM's macroeconomic analysis. In Chapter 14, they explore, with the use of the econometric model they built in the course of their activities with SAM, the projected impact of the strategy on the structure of income distribution. To what extent did SAM redistribute income and to whom? What distributive effects would changes in the SAM strategy have?

In October 1983, PRONAL was formally announced. This is one of the two Special Programs of Mexico's National Development Plan for the 1983–1988 period. Officially, PRONAL replaced SAM. Part V looks at future food policy in Mexico. Chapter 15, consisting of three sections, begins by presenting the official summary of PRONAL, along with its formal introduction by President Miguel de la Madrid. In the next section we see the strategic thrusts of the new food policy in Mexico. This is described by María de los Angeles Moreno, undersecretary of Programs for Social and Rural Development of the Ministry of Planning and Budget, who spearheaded the development of PRONAL and who is technical secretary to the national food commission in charge of implementing it. The third section is by Clara Jusidman, who was the director of the Center for Integrated Rural Development Research of Mexico's Ministry of Planning and Budget and participated in the technical design of PRONAL. Jusidman analyzes the extent to which and conditions in which the new administration learned from SAM's experience and discusses the implementation problems facing PRONAL.

Finally, Esteva and Austin attempt in Chapter 16 to integrate the answers to the questions posed in the foregoing chapters and to distill the lessons from SAM. They also examine the extent to which SAM's legacy was incorporated in PRONAL and highlight the challenges facing Mexico's future food policy.

PART II

STRATEGY

CHAPTER 2

Food Needs and Capacities:
Four Centuries of Conflict

Gustavo Esteva

SAM was the first serious attempt to implement a food policy in Mexico, but it never reached three years of age. Even before its corpse was cold a new national food program, PRONAL, was created in its place. It would be tempting to understand SAM's creation, rise, and fall as a mere administrative happening. It was born unexpectedly, within an advisory office of one of Mexico's presidents; it died suddenly in another, under a different president. It rose through official channels onto the public stage, to become a fad, but it never seemed to leave the bureaucratic sphere that appeared to be its natural milieu. SAM was indeed an administrative experience, and we examine it as such in this book. But it was many other things as well, and to analyze these it is necessary to look at the past. It is true that SAM was a contemporary phenomenon whose origin can be traced, at most, to the immediate postwar period. Such a limited horizon, however, would leave many of the roots of the strategy unexplained.

It is increasingly acknowledged, though not always with ease, that the needs of a group or a society are an expression of its cultural patterns. There are social and political filters or processes that define these needs and that must be understood in order to satisfy them. This is particularly clear in the case of food, and Elias Canetti has succinctly restated this proposition: "Everything that is eaten is an object of power." A government's food policy, however, should not concern itself with feeding people, but rather with protecting vernacular production. The term *vernacular* acquired meaning with the practical need of escaping from the ideologies of development and underdevelment. *Vernacular* is a term of Latin origin that is used mainly to refer to language, the language acquired without the help of a teacher (the mother tongue). It was used in Rome to refer to any value generated at home—home-made—for consumption at home. This analysis of Mexico's food policy attempts to examine the historical evolution of official efforts to enhance or destroy its

23

people's ability, individually and collectively, to produce their material needs, that is, their vernacular ability and autonomy.

One school of thought on world food problems argues that all countries have sufficient resources to feed their populations (Collins and Lappé, 1977). Mexico is a good illustration of this hypothesis. Many analysts attribute the country's food problems, not to lack of resources or abilities, but rather to the prevailing pattern of production and distribution, which tends to concentrate productive resources in the hands of a few who have used them inefficiently and unfairly. Other analysts attribute the difficulties to the "population explosion," arguing that the number of people exceeds available resources or the capacity to exploit them. In the present framework, I do not dismiss the maladjustments caused by an accelerated growth of the population, and I admit that, during certain periods, such growth has seriously threatened the country's food system. I reject, however, Malthusian or neoMalthusian simplifications that explain deprivations and crises on the basis of demography or resources. In my own view, the most important factor is the way the peasant food system is inserted into the national system—and that, in turn, into the international economy. In other words, these crises are inversely related to the autonomy of the peasants and of the Mexican food system, that is, to its ability to satisfy—locally, regionally, and nationally—the people's nutritional needs. One formulation of this hypothesis, applicable to the past few decades, focuses on the process of internationalization of capital (Barkin and Suárez, 1982; Esteva, 1983).

As I point out subsequently, the loss of an autonomous Mexican food-producing capacity has periodically caused serious problems and can be attributed to the decay of local and regional food systems dominated by peasants. The demands imposed on them have exceeded their resources, while their potential to be self-sufficient and their capacity to change are denigrated. With support, they could have increased their contribution to feeding the rest of their society without neglecting their own needs; Viniegra Gonzalez's and Medellin's analyses in Chapter 7 and 8 show that peasant systems respond to economic incentives better than is conventionally accepted.

Before 1980, government food-related actions were directed to two separate dimensions: (1) agrarian reform and agricultural production, and (2) supply to the cities. The former included the ownership and exploitation of resources for food production. Weaknesses of policy, partly due to the failure to solve the problem of land tenure, created the need for measures directed to stimulate supply. From the start, these took the form of official regulation of the market, and they were gradually extended to encompass a broader field of action. My hypothesis suggests that both problems—agrarian reform and agricultural production, and supplying the cities—remained unresolved because the transformations in the country's food demands and productive structure as the country was integrated into the international market impeded the development of adequate linkages between rural production and urban supply.

In the Beginning There Was Malnutrition

Pre-Columbian residents of what is now Mexico enjoyed a dynamic, self-sufficient food system, with surpluses that allowed them to face natural adversities. Their population and organizations expanded and prospered, reaching somewhere between 7 and 25 million by the time the Spaniards arrived at the beginning of the sixteenth century. One hundred years later, only one million natives remained. There were many causes for this catastrophe, but the main one was lack of food. There is evidence of a very high correlation between epidemics and famines (Florescano, 1979), and many deaths were due to punitive expeditions against hunger-driven revolts. How did the food system become inadequate?

Of course, additional needs appeared. The Spaniards' own consumption did not represent too burdensome a demand, especially during the first decades, because their numbers were small. A more significant fact was the extensive use of the land for cattle raising. Demand for meat and leather increased, for both domestic use and export. The new production of wheat and sugar, partially exported, also required a sizable volume of resources. The emergence of mining was perhaps the factor that pushed the food system beyond its capacity. In order for the mines to operate, indigenous communities had to provide the men to work them. This weakened community productive capacity, while increasing demand—to feed workers, Spanish mine managers, and slaves in the mines. This same pattern prevailed as growing number of workers were drafted for the construction of public buildings and for many other activities (Israel, 1975).

All these additional requirements were imposed on indigenous communities without a change in their productive systems. Consequently, when surpluses were depleted they were forced to organize the reduction of food intake in the whole community. This triggered the vicious cycle of population destruction through epidemics and famines. The end result was not so much lack of food as lack of men.

New Spain's first century was, in short, one of hunger and scarcity. Except for large cattle properties (subject to the Spanish system of cattle raisers' unions) and for the part of the land devoted to wheat and sugarcane (also subject to other Spanish forms of labor organization), the native food production system was maintained, though deprived of resources and submitted to new definitions of needs that, evidently, exceeded its productive capacity (Frank, 1976).

The Colonial Cycle

The wealth of New Spain must have seemed unlimited to the astonished eyes of the first Spaniards. But, in fact, by taking it without any attempt to replenish it, they almost destroyed it. The lesson of these first decades of wild colonization was clear: taking over wealth without concern for anything else meant running the risk of not leaving anything or, rather, not leaving *anybody* to carry on the activities required for its appropriation. It became necessary to consider the social reproduction of workers. Thus, at the turn of the seventeenth century, there evolved the institutions that subsequently characterized the colony, the *hacienda* (large estate) and the indigenous community.

Spain's recognition of some property rights of the indigenous peoples was decisive. The original deeds, issued by the Spanish Crown, are still in force today: according to Mexico's constitution, they are proof of inalienable ownership rights.

Nevertheless, in the indigenous communities, this ownership was challenged daily as the Spaniards developed their ranches and haciendas. No doubt their eagerness to possess land was important, but still more significant was their need for labor. When slavery and the tribute system decayed as foundations for labor organization, the *hacendado* (estate owner) required an alternative process of recruitment. The best way was to "free" the natives from their ties to their ancestral community land. The natives were forced to work for the estate owners, since new members of the community could no longer settle on idle lands. They were obliged to pay in money the tribute that had been paid in kind or with labor. As they accumulated debts, these would later bind them definitively to the haciendas, becoming *peones "acasillados"* (tied through indebtedness).

By the eighteenth century, haciendas dominated the countryside. They covered 70 million hectares, against only 18 million belonging to communities. Their power was unquestionable, both regionally and locally, and in the cities the master of the hacienda easily ruled the other social classes.

Owing to a large variety of factors, haciendas and communities tended to develop as relatively autonomous units of production and consumption. With time, the autonomy of the communities and haciendas was weakened by their exposure to market instability, and the latter tended to reduce the area sown and to isolate themselves by establishing an expensive storage system (barns and silos). Consequently, society became increasingly vulnerable to the demographic catastrophes (epidemics and famines) that occurred repeatedly during the eighteenth century, though this did not prevent the sustained population growth that characterized the period. The colonial period ended just as it had started, with hunger.

Independence Is Pending

Mexico, like all other Hispanic countries, inherited the need for a central unifying political structure, but as Mexico became independent (1810–1821) the country still lacked the economic, social, and political structure around which it could give some sense of reality to the national ideal it seemed to be striving for. During the first fifty years of independent life, in the midst of constant civil war, Mexico lost half its territory, suffered foreign interventions, and was divided into a thousand pieces, losing both the appearance and the substance of a nation. The exercise of autonomy was localized: deprived of a center, or with a weak center that was alien and distant, decentralization was no democratic dream but a nightmare of disintegration (Otero, 1964).

The food situation clearly reflected the prevailing conditions. The indigenous community's only hope was self-sufficiency. The hacienda entrenched itself in its barns and operated in autarky, but with no real possibility of expanding. The small agricultural property, which had waited so long to appear, found its opportunity in this chaotic situation in the form of self-sufficient and highly flexible farms that tried to take over the spaces left open by the other forces. The Catholic church was one of the very few colonial institutions that resisted the movement toward dispersion. It adopted its own forms of labor organization or submitted to others that would extend its spiritual domination to concrete economic patterns of control.

Mexicans' first attempts at independent life were not exactly successful, but the population adapted itself to the local pattern of production and consumption more successfully than it had to colonial rule. When local structures experienced less pressure from the center, they were able to display their vitality and survive under exceptionally adverse conditions. During this period there was no trace of massive famine.

Willingly or Unwillingly

The arrival of Benito Juárez to power, during the so-called Restored Republic (República Restaurada, 1867–1876), laid the foundations for a democratic and liberal organization of Mexican society. A whole generation of liberal intellectuals and politicians, those who had contributed to shaping the 1857 constitution, undertook the task of building the country with a new republican structure.

The liberals promoted all types of reforms. To break the prevailing chains of land ownership and to build a republic of small producers, they began by dismantling the church's large estates and the indigenous communities' collective property. They thought this would free the Indians and the hacienda peons from all forms of oppression and thus pave the way for small- and medium-sized agricultural properties. They trusted that this form of labor organization would lay the foundations for a new, free, prosperous, and democratic society following the U.S. model of the time, which was homesteading on family farms.

Rather than producing the expected results, this strategy fostered the plundering of communities. The abrupt occupation of peasant lands by individuals and corporations that would undertake productive development provoked numerous rebellions that were severely repressed. The church's reaction and the disorder produced by it all contributed to further disorganization. In 1876 the revolution of Ayutla abruptly ended this project and enthroned Porfirio Díaz as dictator.

When the dictatorship started, agriculture was operating under highly varied production systems, but in general it was linked to domestic consumption and barely connected to foreign demand. Small units, governed by different land tenure regimes, tried to survive in virtual autarky, producing only for their own consumption. Juárez's reforms, however, triggered profound and thorough changes in the land tenure system, promoting the emergence of private property. After the initial convulsions—which destabilized Juárez's regime and created the conditions for the movement that overthrew him—the effects of his reformations were felt on productive structures: the division of labor was emphasized, autarky was weakened, and there were advances toward the creation of a national market. This was the same road chosen by Porfirio Díaz. True, it was not until the end of three decades of dictatorship that he decided to eliminate the taxes that had blocked the free transit of commodities and contributed to compartmentalizing production. Nevertheless, from the very beginning he opened agriculture to capital and to foreign markets, exposing it to the rigors of free competition and stimulating entrepreneurial initiative where it seemed to be practically nonexistent.

To a great extent, production was based on the dismantling of indigenous communities whose lands could be legally appropriated through simple administrative procedures (Cosío Silva, 1965). New laws created the surveying companies (*compañías deslindadoras*) that encouraged foreign migration to promote colonization of idle land. Between 1877 and 1910 these surveying companies helped deliver over 40,000 titles, covering almost 40 million hectares. Since estate owners could avoid the land-clearing action of the companies, these activities encroached particularly on indigenous communities and small properties that were absorbed together with the idle land. In just two decades (1876–1894), one-fifth of Mexico's land ended up in the hands of fifty owners, and the number of haciendas trebled.

The results clearly reflect the orientation (Meyer, 1973; Mejía Fernández, 1979). Between 1877 and 1910, agriculture grew at an average annual rate of only 0.65%, but agricultural production for foreign markets increased by 6% per year, implying a gradual reduction of production for domestic consumption. The overall productive structure was substantially altered. Production of maize, Mexico's main foodstuff, decreased from 52% of total agricultural output in 1877 to 33% in 1910, while export products increased from 4 to 20%. In fact, maize production fell from 2.7 million tons in 1877 to only 1.4 milion in 1894. It recovered in 1897, when import and transport duties were annulled. In any case, per capita production of maize fell from 282 kilograms in 1877 to 144 kilograms in 1910. Similarly, in 1910, production levels of wheat, barley, beans, and chilies were the same as they had been in 1877. Maize and wheat imports grew constantly.

By 1910, 12 million people—of the 15 million that constituted the total population—worked in agriculture. Around 8000 haciendas, in the hands of a smaller number of owners, occupied 113 million hectares with 4500 managers, 300,000 tenants, and 3 million indentured peons and share-croppers (*aparceros*). There were 50,000 ranchers, with 10 million hectares, and 110,000 small owners with 1.4 million hectares. Around 150,000 indigenous communal landholders occupied 6 million hectares (12 million less than in 1810) and, together with their families, totaled one million people (50% of the number of Indians in the country). Half of the population lived in 57,000 towns directly controlled by haciendas and ranches, while the other half lived in 13,000 "free towns" that suffered several forms of exploitation: one of these was temporary work as peons on the haciendas. Less than 1% of the population owned over 90% of the land, and over 90% of the rural population lacked any access to it.

Agricultural, cattle-raising, and agroindustrial haciendas were extensive rural commercial operations; there was one with 7 million hectares in the hands of a single owner. Hacienda owners, both Mexicans and foreigners, lived in luxury in the provincial capitals, in Mexico City, in the United States, or in Europe. They left their administrators, foremen, and accountants to operate the hacienda, where labor was remunerated not with wages but with a small parcel "granted" to the workers by the *hacendado*. "Free" workers, who lacked housing, a plot, and a safe job, already accounted for a significant share of the rural population.

Food production, as indicated before, advanced painfully slowly and was always insufficient to meet the needs of the population. In the countryside, where three-fourths of the people lived, mechanisms to regulate production and supplies were integrated into the productive system: the indigenous community, with its traditional institutions to allot resources and distribute food, and the haciendas, whose barns were the centers for distribution and reserves. In the cities, where one-fith of the population lived, a relatively modern commercial system, mostly foreign, was created. At the turn of the twentieth century, for example, in Mexico City there were 212 registered traders, of whom only 40 were Mexican. Consistent with liberal principles, the government had very little influence on commercial operations proper, except through import protection. When conditions became unbearable, the government reduced this protection and authorized the free import of cereals to reduce scarcity and excessive price increases. There were also special support programs to take care of specific situations. At times, for example, the lowering of taxes on flour imports were used to subsidize imported maize or to distribute it to the poor (Esteva, 1979).

The case of cattle illustrates the overall impact of the food system. Under Porfirio Díaz, cattle raising was extensive and inefficient. It was more an activity for prestige than profit, with emphasis on the size of the ranches and on the number of head. The business, if any, was fundamentally to produce lard and leather, particularly for export. In a country with people dying of hunger, milk was a mere by-product and was often wasted. The irrationality of this was understood even at the time, as Fernando Pimentel y Fagoaga demonstrated when he declared his preference for cows more civilized than the Mexican ones.[1]

In summary, the major trends during the 1877–1910 period were as follows: local autarky and production of staples decreased, while land concentration, export production, staples imports and foreigner production, commercialization, and financing all increased.

The liberal ideals of the previous era were under attack from the monopolistic structure of agrarian property on which the dictatorship was based. Large landowners could resist the pressures from the land surveying companies and from foreign markets while controlling the internal market by resorting to two mechanisms: the old protectionist tariff policy and the new currency devaluation. The stability of the dictatorship gave capital, trade, and credit all sorts of safety, but it could not inject them with dynamism, risk-taking capacity, or a modern structure.

In spite of the massive inflow of capital and a permissive attitude toward foreign market signals, agriculture was maintained, not so paradoxically, in a somehow artificial way. In fact, it could not live without tariff protection and currency depreciation. The large subsidy this implied

1. Don Fernando said: "Our cattle, raised under the sun, with fresh air and in the open air, used to being ill-fed, to having no shelter and miraculously to survive, thanks to their great resistance developed through years of a crude and difficult life, cannot be compared with foreign cattle. They are educated, civilized, and perfected by strain and by training. When an American or European cow is fed with *nopal* leaves, weeds of the Santa Anita canal, sleeps in open corrals, and grazes on dry weeds, either its productive capacity will decrease or it will inevitably die. Cattle raising is not, cannot be, the job of the first man who has a ranch or money to buy fine cattle. It must be a job of patience, perseverance, observation, and meticulous experimentation. Then, just as the cow needs care, special attention is required for the offspring. An ill-fed and small calf, living far from its mother, in carelessness and dirt, cannot become a superb bull suitable for breeding. The most it can do is grow a few inches, if not die of epizootia. It will become a plowing bull and breed calves as thin as itself. Placing a magnificent bull in a stable will not greatly change the quality of calves, as long as the present stabling system prevails. And only God knows when will it disappear" (Cossio Silva 1974: 14.3).

did not contribute to making agriculture capable of handling the domestic needs for raw materials and food or to operating on competitive terms in foreign markets (Cosío Silva, 1965).

Liberalism Learns Social Revolution

In 1910, the poor peasants who one hundred years earlier had made independence possible again tried to gain their liberation through a popular insurrection. As main protagonists of a large social movement, peasant armies suddenly became the dominating force (Womack, 1968; Wolf, 1969).

Between 1910 and 1920, 1 to 2 million people died, around 10% of the population. The 1910 standards of living and of production were not regained until 1920. Violence aggravated regional and sectoral inequality. The number of people working in agriculture increased, but their production and income decreased. Oil and minerals production, in the hands of Americans and the British, grew almost continuously and, at times, at spectacular rates; in 1920 these became the main export products. Both industry and trade, and the cities, came out relatively stronger.

In 1917 the oppressive and exploitative system of the Porfirio Díaz dictatorship was broken. The 1917 constitution reflected the country's new economic and sociopolitical position, combining Juárez's liberal spirit of 1857 with a social orientation of the state that attempted to safeguard popular rights and build a fair society through free education, workers' rights, and agrarian reform. The latter was explicitly based on the sovereign principle expressed in Article 27 of the constitution: "Ownership of land and waters within the limits of the national territory originally belong to the nation, which has had, and has the right to transfer their domain to private individuals, constituting private property.... At all times, the nation will hold the right to impose on private property the rules dictated by public interests, and to regulate the utilization of natural elements susceptible of appropriation, conducive to an equitable distribution of public wealth and to its conservation."

The *hacendados* were unable to face the revolutionary movement in a united fashion, but they survived by their ability to adapt to the new conditions. They took advantage of the conflicts among revolutionary groups and, in many cases, fostered and enhanced them. While they rapidly created organic links with the new power structure, peasants were unable to take command of the process in which they were the main armed protagonists.

From 1916 to 1920, "legal grants" were made to 334 *ejidos* (lands legally used without personal ownership)[2] covering a total area of 382,000 hec-

2. *Ejido* is an ancient form of land tenure whose reconstitution was demanded by peasants during the revolution and which at present is a specific form of land tenure, stipulated by

tares. Each of the 77,000 members of the *ejidos* received less than 5 hectares. In the meantime, an active discussion continued about the lands under "temporary occupation": 3 to 4 million hectares that peasants had occupied directly. To put these figures into perspective after ten years of revolution, let us recall that haciendas in 1910 held 113 million hectares.

From 1920, instead of the radical agrarian reform provided implicitly in Article 27 of the constitution, the development of small- and medium-sized parcels was emphasized, and the subject of eliminating the *hacienda* was evaded (Dulles, 1961; Molina Enríquez, 1976; Reyes Osorio, 1978).

Between 1917 and 1934 close to 11 million hectares were distributed to around 6000 *ejidos* and almost one million *ejidatarios,* each getting a little over 10 hectares. Rural development was stimulated, from the mid-1920s, through the actions of the National Irrigation and Roads Commissions, the National Agricultural Credit Bank, and four regional *ejidal* banks, complemented by the support given to rural education.

Between 1920 and 1930 food production remained almost the same. Maize and bean production decreased substantially: maize, from 2.3 million tons in 1920 to only 1.3 million in 1930; beans, from 169,000 tons to 119,000. But production of coffee increased, as well as that of sugarcane, tobacco, and cotton; these were each oriented toward foreign markets, which had an overwhelming influence in determining producers' decisions.

As for the regulation of supply, very little changed during this period. The passage of the revolutionaries through the haciendas was always accompanied by looting of barns that were opened to them. The idea of a mechanism to regulate prices in the countryside started with the revolution. In the constitutional draft of 1913, for example, Vice President Pino Suárez mentioned the intention of finding a better formula "to defend small producers and prevent their income from being cut by monopolies and speculators." This idea, however, did not materialize in a concrete institution at that time. Except for periodic moves by the government's agricultural bank, very little was done to control supply, either in the countryside or in the cities.

The sort of modernization sought by revolutionary leaders of the 1920s is well documented (Córdova, 1973). They expected much from the breaking up of large properties. "The job of any truly nationalist government,"

law, in which the individual right to the exploitation of a specific plot of land is recognized. The right is not equivalent to private property, since it is restricted to direct exploitation of the plot of land. Legally, the *ejidatario* cannot sell, rent, or mortgage his plot of land. Inheritance is regulated by certain legal stipulations that may prevail over the explicit will of the *ejidatario*. The organization of production in the *ejido* may take different forms within the limits and restrictions laid down by the law, on the basis of *ejidatarios* decisions. It must satisfy different formal requirements in order to acquire legal representation.

said President Calles, "must be, first and foremost, to create small properties, turning peasants into owners of land on which they can work."

Calles did not forget the role of the state. On one hand, when turning
back to the individual form of land tenure, provisions had to be made so
that large properties would not be re-created through commercial speculation. On the other hand, the state had to facilitate peasant labor and fight
for a rational organization of crops. Calles believed in authentic cooperation between the state and the peasants. He wanted to "organize rationally
the development of crops and encourage the utilization of agricultural industries ... by organizing credit and agricultural cooperation." He attempted to give shape to a general plan of rural education that would be
part of a larger plan to cover the whole of the economy. In 1928 he made a
decisive step toward the institutionalization of the revolutionary process
by promoting the agglutination of all political factions with the creation of
the National Revolutionary Party (Partido Nacional Revolucionario).
Here, nevertheless, he only took a few weak steps before leaving office.

Attempting Another Alternative

At the beginning of the 1930s, a double force rose against the landhold-
ing sector that had gained strength in the 1920s: a new peasant dynamic,
moved by the antiagrarian inclinations of the Calles administration, and
the industrial and commercial structures of economic and political power
that advanced at the pace of capitalist expansion. Mexico was feeling the
effects of the Great Depression and of international events that preceded
World War II. In order to provide a political solution to the country's internal conflicts, the government adopted a new radical framework, the six-
year plan of 1934, and put into practice coherent mass political activities
in which popular mobilization was the mainstay for official efforts to
transform society. The plan was rejected by conservative groups. Ramón
Beteta's defense of 1935 reflected the general attitude of the revolutionaries of the time, who thought it possible for the country to create a *different* destiny from that of other people. "By watching the latest crisis of the
capitalist world, we believe we could use the advantages of the industrial
era without suffering from its well-known defects. ... We have dreamed of
a Mexico made up of *ejidos* and small industrial communities, electrified
and healthy, where goods are produced to meet the needs of the population, where machinery is used to alleviate man from heavy chores, and not
for the so-called over-production" (Simpson, 1937).

Between 1934 and 1940, General Lázaro Cárdenas's administration distributed 20 million hectares to 11,000 *ejidos,* that is, 750,000 peasants
receiving an average 25.8 hectares. Thus, in a period of six years, the
government distributed almost twice as much land as during the previous

eighteen years, although, in general, it was of inferior quality (Esteva, 1983).

With Cárdenas, the *ejido* became the pillar of the nation's argicultural economy; it occupied half the arable land, from then until now. One decisive step in the new direction was taken in 1935, when the large agroindustrial corporations of big landowners were expropriated. These had not been touched before for fear of affecting production. These corporations were delivered, in the form of cooperative *ejidos,* to workers and peons that had organized combative unions and that demanded expropriation by government (Ashby, 1967). Under Cárdenas, small independent producers were encouraged and hacienda owners were placed on the defensive. This was possible because of Cárdena's support of peasant mobilization and, especially, of the collective organization of production. For him, the relationship between the peasants and the state was defined as a reciprocal commitment of social transformation.

Agricultural production, which had been relatively stagnant in the 1920s, grew at greater speed during the 1930s. Though almost all items recorded increases, the highest were for products for the domestic market. The food pattern was defined within the framework of the six-year plan, which viewed the state as a sort of arbitrator and tutor, to maintain essential order and economic coordination among industrialists, traders, and consumers. The plan recognized "that the improvement of the standard of living of the Mexican people not only requires a continuous and vigilant defense of the workers' wages, but also the maintenance of prices at convenient levels. To this end, distribution channels should be established, adding but a minimum cost to that of production, and eliminating the greatest possible number of intermediaries. Consequently, the cooperative organization of consumers will be encouraged, but since it cannot be fully established in the short term, the State shall regulate domestic trade to ensure that it will meet its function."

In 1934, the first state regulating institution was founded: the National Warehouse Depository (Almacenes Nacionales de Depósito, S.A.). Its creation and initial orientation corresponded to that of the first six-year plan. In June 1937, the Regulatory Committee for the Wheat Market (Comité Regulador del Mercado del Trigo) was institutionalized to regulate wheat prices and the Mexican Export and Import Company (Compañia Exportadora e Importadora Mexicana, S.A.—CEIMSA) was created as a state trading agency; the latter was to gain great importance in the regulation of the market through foreign trade. In July 1938, the Regulatory Committee for the Staples Market (Comité Regulador del Mercado de Subsistencias) was created to "support legal sales prices of products set by public power, through the purchase and sale of same in the market." With this committee, a wide variety of regulating activities were started; these served as the basis for public action in the field during the following four decades (Esteva, 1979).

Table 2.1. Land pattern changes, 1910–1940

	1910	1940
Haciendas		
Area (million ha)	113	80
Number	8,431	9,697
Average size (ha)	13,403	8,250
Ejidos		
Area (million ha)	0	30
Number	0	1.7
Average size (ha)	0	17.6
Ranches		
Area (million ha)	9.8	19.6
Number	48,633	280,639
Average size (ha)	200	70
Small farms		
Area (million ha)	1.4	4.6
Number	109,378	928,539
Average size (ha)	12.8	5
Land access		
Share of rural population with access	8.7%	6.7%[a]

[a]Half the active rural population, 1.9 million people, were agricultural laborers, around 700,000 of whom were *ejidal* or small farmers.

Source: Silva Herzog, 1934; Simpson, 1937; Dulles, 1961; Martínez Ríos, 1970; Molina Enríquez, 1976; Reyes Osorio, 1978; Esteva, 1983.

The net results of the agrarian reform efforts between 1910 and 1940 are shown in Table 2.1. Basically there was a major redistribution of hacienda lands along with the opening of new lands to create the *ejidos*. However, the haciendas' 80 million hectares were still one and half times greater than the *ejidos'* 30 million hectares. The numbers of ranches and small farms also increased tremendously, but average farm size dropped, which suggests fragmentation of land. Land access rose dramatically, responding to the original calls of the Mexican revolution, but increasing numbers of farmers were unable to make a full-time living on their small plots and had to work off-farm as wage laborers.

Despite its efforts, Cárdenas's administration did not develop an integrated food policy. The internal and external events of the late 1930s and early 1940s created the conditions for change without, however, allowing time for its consolidation.

Hunger on an Industrial Scale

As soon as World War II ended, Mexico decided to gamble on industrialization. To this end, it needed adequate supplies of cheap food, foreign exchange, accumulation of funds, and workers, and all this was to be gained through agriculture. The government's model gave responsibility for agricultural production to a group of high-productivity enclaves, to which it channeled most agricultural development resources. These enclaves, next to the urban centers where industry and services were concentrated, were to absorb peasant labor, whose parcels would disappear as more dynamic agents occupied their places or as their scarce resources simply decayed (Hewitt de Alcantara, 1978).

The model began to be applied with the extension of the agricultural frontier, especially of irrigated land. From 1950 to 1970 the cultivated area expanded at a rate of 1.5% per year, mainly by state action. From 1950, 70% of federal government resources were destined for agricultural development, and this was used for the construction of large irrigation works. The enclaves were freed from the pressure of landless peasants by giving the latter other, less-productive land.[3] It is estimated that, by 1970, 30 million hectares had been "distributed" on paper (Martínez Ríos, 1970; Molina Enríquez, 1976).

One fundamental tool of this policy was credit, which was concentrated in a few areas and destined for commercial crops: wheat, cotton, sorghum, strawberries, tomatoes, oilseeds, and others. A study made at the end of the period revealed that 25% of the producers did not receive any credit at all, and 50% obtained it from "noninstitutional" sources, meaning the local moneylender (Esteva, 1979; Echenique, 1980).

Special attention was given to the production of modern inputs—fertilizers, improved seeds, agrochemicals—and to mechanization. The fertilized area increased from 4.8% of the total harvested area in 1950 to almost 40% in 1970. This input concentrated on a few crops, where it became the main source of yield increases. A similar pattern, including subsidies for the benefit of commercial areas, was followed with improved seeds, other inputs, and mechanization.

The regulatory intervention of the state was based on the experience gained during the 1930s. In 1953 rural support prices were established in the whole country; but these prices were received only by larger commercial growers, because the areas of peasant economy tended to operate with a system of intermediaries who sold to the state at the prices set but paid much less to peasants.

From 1943, agricultural science and technology were also at the service of the model. Genetic research was carried out initially at Mexico's Ministry of Agriculture's Office of Special Studies and subsequently at the International Center for the Improvement of Maize and Wheat (CIMMYT),

3. Of the total area delivered to *ejidatarios,* 1.5% was irrigated land from 1947 to 1952, 1.2% from 1953 to 1958, 0.8% from 1959 to 1964, and 0.5% from 1965 to 1968. During these same periods, percentages for nonarable land distributed to *ejidatarios* were 78.8, 74, 81, and 91.3% (Esteva, 1983).

both financed by the Rockefeller Foundation. The Green Revolution was set in motion, with its "miracle seeds" operating as the vanguard of an integrated model that could only be developed in these enclaves or that, seen from the other side, was conceived as a technological model appropriate for their artificial conditions, rather than for those prevailing in the rest of the country. (Barkin's analysis in Chapter 6 explores the role of improved seeds in the SAM strategy.)

Thus things were arranged for the development of commercial agriculture. It settled on works of large infrastructure created by the state (which also took care of providing water at subsidized prices and of maintaining the works), in areas that required no large investments for productive improvements; commercial agriculture enjoyed the double incentive of low labor costs and high support prices. It also enjoyed cheap credit, modern subsidized inputs, support for mechanization, reasonably efficient technical assistance, and the advances of agricultural research.

In 1950, the picture clearly reflected the changes that took place ten years after the Cárdenas efforts (Cline, 1962; Hewitt de Alcantara, 1978). In authentic small properties, in recently established "colonies," and in large landholdings (*latifundios*), a strong group of landed entrepreneurs took a firm step forward: 10.5% of producers, private owners of over 5 hectares, accounted for 54% of the total value of production in 1950. Next to them, a certain number of collective *ejidos* survived, though facing increasing difficulties. Scattered throughout the country, the parceled *ejidos* multiplied, as well as the small farm properties: private owners of less than 5 hectares accounted for 35.1% of producers and contributed only 8.7% of production, while figures for *ejidatarios* were 54.5 and 37.3%, respectively. Seasonal workers (400,000 according to the census) and agricultural workers (1,793,000 according to estimates) worked in well-established fields or in agroindustries for the development of coffee and sugarcane. Numerous indigenous communities barely survived, left aside and exploited as the real pariahs of the countryside. For them, the preservation of their social structure constituted a very weak defense against the commercial penetration of the modern sector of the economy.

The initial results of the government's policies encouraged optimism and strengthened the decision to continue. From 1952 to 1956 the volume of agricultural production increased at a rate of 6.5% per year. The harvested area grew by 1.9% per year during the period, and yields, by 3.8%. During the following ten years, these rates fell by almost half, but it was still possible to sustain production increases that allowed for self-sufficiency in basic food production. This was later exceeded with surpluses of these products, which were added to the commercial exports (Esteva, 1983).

Frequent peasant protests in the late 1950s were an early signal that things were not going as smoothly as they seemed. Nobody, however, seemed to pay very much attention to this. Beginning in 1965, symptoms became very evident: "hunger caravans," rural guerrilla fighting (localized but constant), a decrease in the rate of growth of production. By 1970 it became clear that, though agricultural progress had not been entirely an illusion, something was wrong with the model.

Agriculture, it is true, had strongly contributed to finance industrial development, through both the net transfer of resources and the sustained inflow of foreign exchange. But the effort left it exhausted. Though the system of trade relations maintained considerable volumes flowing to the cities, a countertrend evolved, compensating for the general decay of production and living conditions in the countryside. It was not only necessary to recover the pace of public investment in the sector, whose share of the total had decreased, but it became urgent to multiply the "welfare" expense for rural communities, which had reached unbearably low levels of employment and income.

The modern sector, backbone of agricultural development, showed indications of lacking the necessary internal dynamism, and it became more evident that it depended on the extraction of resources from traditional agriculture. In other words, real enclaves capable of developing autonomous strength had never been constituted. Productive areas, the pride of their creators, were fundamentally a large funnel that provided cheap and subsidized labor and resources. The mechanism that had been used to create them started to operate as a brake. Instead of being oriented to the optimal utilization of resources, assuming risks for the achievement of maximum gains and extending investments in the countryside, modern farmers now tended to prefer suboptimal exploitation—requiring fewer technical and market risks—and started to channel their gains into several forms of conspicuous consumption and to sectors other than agriculture. Expanding or even maintaining the strength of these enclaves no longer seemed possible. Because of the increasing unit cost of works of large infrastructure, they could only be developed at the expense of other essential public investments with no advantage at all.

The whole sector exhibited substantial weaknesses (Hewitt de Alcantara, 1978; Esteva, 1983); for example, total production, which from 1960 to 1965 still grew at the rate of 8.2% (especially because of increases in irrigated area of 3.3 million hectares and because of technological changes in commercial agriculture), grew only at a rate of 1.8% per year from 1965 to 1970. Exports (a fundamental pillar of the model), which had increased in value by 10.1% yearly from 1960 to 1965, grew only by 0.9% during the second half of the decade.

What had happened, meanwhile, to the main protagonists of this story? By 1960 a little over 3% of the properties held 43% of arable land and yielded 54% of total agricultural product. In contrast, half the plots controlled one-eighth of the land and contributed around 4% of the product, and 84.1% of the plots produced only 21.3% of the total value of production. In 1970 the structure was similar.

Wage labor, which from 1950 to 1960 apparently increased to reach 2.2 million in 1960, grew at 1.5% per year during the following decade, to 2.6 million in 1970.

The general situation of *ejidatarios,* small landholders, common landholders, and temporary or permanent workers showed a clear deterioration in 1970. Employment is a good illustration. Of the total agricultural labor force in 1970, 36% barely survived with their parcels and had to work temporarily as day laborers or in other activities. Another 58% were landless or peasants who received a minimal share of their income from their plots. These constituted a large population inserted in the agricultural sector, but actually marginal to it. In 60% of maize-producing plots and 33% of bean-producing land, production was exclusively destined to peasant household consumption. Of the total land, 20% did not even produce maize or beans in sufficient amounts for the consumption of peasant families, who were forced to purchase the balance of their needs with income derived from other sources. The other sources of peasant income were also in open deterioration by 1970: the production and sale of crafts, the exploitation of forests, small-scale cattle raising, and many other complementary activities for peasants were in crisis. As for employment as temporary or permanent day workers, not only was the prevailing wage below the rural minimum, but for some groups—like those of sugar harvesters—working conditions were completely subhuman.

Thus it can be stated that the government's policies brought the large mass of traditional peasant agriculturalists to the brink of extinction, but it failed to achieve the expected counterpart: the construction of a modern agricultural sector capable of feeding the population, of generating foreign exchange, and of productively absorbing peasants.

Stagnation

With hindsight, we can see that the model was already dead by then. There was no hope of revival, nor were there conditions for change. No one saw the problem clearly, and many still benefited from the government's policies. Consequently, there was a long, costly agony that aggravated things further. For a period of ten years, the government attempted to patch up the model and apply palliatives.

From 1971 to 1979, the activities of the sector advanced, but with difficulties. The value of total production went from 53.2 billion pesos in 1970 to 68.8 billion (1970 pesos) in 1979, a rate of only 3.3% per year, almost the same as that of population increase. The total harvested area showed no increase: around 15 million hectares during the period. The area for the four basic foodstuffs (maize, wheat, beans, and rice) went down from 10.2 million hectares in 1970 to only 7.9 million in 1979. The area for maize fell from 7.4 million to 5.5 million during the period. Annual production of these main products barely changed: from 12.6 million tons in 1970–1972 it went to 13.5 million in 1977–1979 and, in fact, it dropped to only 11.9 million for the latter year.

During the first part of the period, the contribution of agriculture in terms of foreign exchange still appeared to be a justification for the policy being followed. In spite of the constant increase in food imports, which reached 22.2 million tons during the decade, the agricultural trade balance continued to be favorable. In 1978, when massive oil exports had already started, the contribution of agriculture in terms of foreign exchange was still higher. Until the oil boom, the structure of exports, which since 1959 was 70% agricultural raw materials and foodstuffs, was maintained. By the end of the decade, however, the agricultural balance became negative. Though exports were kept at a high level, imports exceeded them.

Efforts to prolong the agony were made on all fronts. Public investment in the sector recovered their historical levels, as a share of the total, around 20% throughout the decade. Some efforts were made in infrastructure, especially large and small irrigation works. Support price policies were reoriented: after a decade of stagnation, they were increased in 1973. Differential systems of support to producers were implemented to protect the peasant economy, with the use of rural warehouses, consumer credit, outlets and sale centers for basic foodstuffs, fertilizers, implements, and many other measures. Credit was used again as a development tool, production of fertilizers and seeds increased more rapidly, and large programs to improve the living conditions of the rural population were implemented.

During this decade, the government's food-marketing agency, CONASUPO, expanded dramatically (Esteva, 1979). Between 1971 and 1975, CONASUPO grew both quantitatively and qualitatively: its budget grew tenfold, and subsidies applied through it quadrupled. The number of affiliates increased from four to nineteen. It diversified its programs and operated a network of around 10,000 retail outlets, besides controlling—as a state monopoly—all foreign trade in basic foodstuffs, operating the main storage systems of the country, and having a large variety of regulatory instruments and supply activities.

None of this, which in a few lines covers efforts that were more than spectacular, was enough. The model had become seriously bankrupt. Many felt this, though it was not expressed numerically nor known until the announcement of SAM on March 18, 1980. It was said then, with data collected shortly before by the president's advisers, that half of the population had a food intake level that was clearly insufficient and inadequate. One-fourth of the Mexican population lived in that long process of agony and lack of physical and mental development that is known as severe malnutrition. The country had a serious food dependence; it was forced to import basic foodstuffs at a rate of 1 million tons per month during the days of food power, when the idea that food is a political weapon for international negotiations was proclaimed and practiced. A linear projection showed that if current trends did not change, by 1990 the country would have to divert two-thirds of its foreign exchange earnings from oil—then on an upward trend—to import food. The figure, according to trends seen during the 1970s, would reach 100% by the year 2000. In other words, all the oil exports of the fifth largest oil-producing country in the world would have to be used to feed its population.

By 1979 things started to heat up. Relatively uncertain statistics, based on the analysis of daily reports in newspapers, estimated that, by the end of that year and the first months of 1980, peasants were taking over one official bank by assault every six days, one municipal building every nine, an agrarian federal agency or agricultural office every fourteen. Attacks on warehouses and highway takeovers returned. Cities were plagued by their own conflicts and by unexpected migration. Things were really getting hot. And then, SAM arrived.

Historical Lessons

Before going on to SAM's development, we can extract some lessons from our historical analysis. A first, rather obvious lesson is that one cannot expect the impossible; there must be a relationship between food needs and capacities, both quantitative and qualitative. If needs are defined without considering capacities, the model will fail, inevitably. We should have known this since the sixteenth century, but we ignored it until the past decades.

This issue has two aggravating factors:

The countryside is always asked to give more than it gets. The result is that it will not; it cannot do so for long; it explodes. Producers are periodically required to provide more food and raw materials: this was the case under the Spanish Crown, and under all subsequent systems. At the same time, the countryside is deprived of resources. An increasing share of what it owns is extracted and diverted to other activities: to mines under colonial rule; to industrialization in the past decades.

The countryside is expected to provide not so much foodstuffs as wealth or gain. People and resources of the countryside were, for the Spaniards, merely a primary source of wealth, accumulation, and power. They should have been a source of democracy, according to Benito Juárez. For postwar governments, they were the building materials for industrialization and for the construction of a modern society. Periodically, there is the rediscovery that the purpose of the countryside is to produce food. But this awareness generally comes too late to avoid a high economic, social, and sometimes political price.

A second lesson that should be learned from historical experience is also quite elementary: Those who are most interested in food production are precisely those that produce food; therefore, they should be entrusted with the task. Food consumed by the population in Mexico is mostly produced by peasants, who hold the dual status of producers and consumers of food. This, to a great extent, defines their existence. Therefore, they should be given the task of producing food. They can do it, they want to do it, and they must.

Such a lesson has important policy implications. Mexico's history reflects a constant conflict between a wide range of forces, internal and external, and peasant producers. Peasants are deprived, through thousands of mechanisms, of their capacity to act. They see their resources taken away or destroyed. They are historically, economically, and politically pushed aside. Once in a while, when things reach an extreme, the country turns to them to produce what is required. Or they mobilize, rise up, and make others take them into account. Again, however, this happens too late, and at a very high price.

Since due priority has not been given to food or to the use of available resources to produce it, since the food model has been exposed to determinations that are alien to producers and consumers, and, above all, since there is no recognition that the peasants must be the main protagonists of food production, the food patterns developed throughout Mexico's history have tended to dismantle, rather than protect, autonomous production capacity. Because for centuries no one has spoken of autarky to refer to these issues, the autonomy that is at stake refers to a *form of insertion* into the international patterns of production and consumption of food. The road toward food sovereignty implies an advance toward a self-determination with which it may be possible to create autonomously the production patterns and consumption standards that will allow for international trade to become an *additional* opportunity for improvement, rather than a threat flooded with expectations and frustrations.

Policy makers actively discussed these subjects at the beginning of the 1970s. Between 1966 and 1969, there were marketable surpluses of some basic food products; in 1970 there was a shortage. The deficit was attributed to bad weather. "And now they will even blame me for bad weather," said the president at that time, showing, according to some analysts, that he had understood nothing. At the beginning of 1971, under a new administration, there were again some exports, which gave rise to intense discussion over the policy to be followed. It was said that such "surpluses" were due to the policy of support prices and other tools that were inefficient subsidies to producers. It was even said that foreign consumers were thereby being subsidized. The doctrine of "comparative advantage" was put forth, as it would be later in the decade, to keep support prices down and to open agriculture to the forces, competition, and signals of the international market.

The academic diagnosis supported that position. One economist who represented the dominant trend and whose estimates were the basis for the policy being adopted participated in a so-called forecasting exercise and wrote:

> Mexico no longer imports agricultural products. On the contrary, the absolute level of agricultural production even allows us to export and to maintain Mexico's share of the international market.... If the country's agricultural development continues during the coming decade at a level of 7% or 8%, there are bases to predict that we will continue to have a self-sufficient agriculture.... The agricultural problem of the coming decade is not on the production side: it is inscribed within the framework of the general problems of the country of achieving a better distribution of income. (Rodríguez Cisneros, 1979)

It is not very elegant to highlight the clumsy diagnosis of such a prediction precisely at the beginning of the decade of disaster, when Mexico imported almost one-fourth of its basic needs and faced the most serious production problems, in other words, when its agricultural sector not only stopped growing but even had a negative evolution. It serves, however, to highlight the intellectual context that prevailed at the time among policy makers. These phrases reflect, moreover, a definition still shared by many and that seems to be in open contradiction to historical experience. For Rodríguez Cisneros, as for many theoreticians and politicians of that time and of today, there are two problems, not one. On one hand, there is the issue of production, which can be considered as solved, according to Rodríguez Cisneros. On the other hand, there is the problem of welfare and justice. But it is a different problem: "There is no need for a more accurate estimation," says Rodríguez Cisneros, "to realize the urgent need to increase peasant income."

In the midst of these discussions, in 1972, Mexico officially adopted the goal of self-sufficiency within the framework of the national program of the sector for that year. This was done within the Coordinating Commission of the Agriculture Sector (Comisión Coordinadora del Sector Agropecuario), created to face these problems in a concerted fashion. This was the first postwar administration to adopt a policy and political definition in a high-level official document for food objectives proper. Of course there were some previous approximations, like the second six-year plan (1941–1946), the National Investment Program (1953–1958), the Immediate Action Plan (1962–1964), and the National Economic and Social Development Plan (1966–1970). The 1972 definition established an important difference: it separated the objectives of producing basic foodstuffs for internal consumption and for export goods, and it gave top priority to the former. But institutional inertia and market signals inevitably made the latter prevail. The 1972 definition also suggested, in embryonic form, the possibility of proposing a unified food policy and not separating, as in the past, agrarian issues from those of supply, or production issues from welfare. By 1972 there was a suspicion that these were all a single problem, one that had to be understood as a global challenge and for which global responses had to be found.

The concept of self-sufficiency developed in 1972 included a margin of flexibility that recognized variable limits to annual harvest results: 5% for maize and beans and 10–25% for other products.

Cooking up SAM

All these issues were debated constantly during the 1970s. They still are. In words and actions, in academia and in public forums, the process of constructing a policy for meeting food needs is a matter that attracts great attention and intense discussion, particularly to explain local-regional-national-international dynamics. Peasants have been playing an increasingly active role in this discussion. From 1971 to 1976 their mobilization was less dispersed, violent, and anarchic than during the late 1960s.

Some of their demands began to get a response in 1975, during the administration of President Echeverria Alvarez. On November 20, 1976, ten days before leaving office, the president delivered to peasants around 40,000 hectares of irrigated land in the valleys of Yaqui and Mayo, one of the most modern productive areas of the northwest. He thus took action against the heart of the powerful commercial agriculture created during the postwar period, challenging the constellation of forces that had promoted and sustained it. The reaction was swift. It was part of one of the worst crises for the Mexican state of the century. The agrarian problem became, for the new administration, a crisis issue. To continue along the lines of this expropriation would mean challenging the most powerful economic interests of the country. To change it would imply the abandonment of peasant producers.

Four days before that historic date, on November 16, 1976, two militants of the pro-peasant forces of rural development had attended a special meeting with the president-elect of Mexico, José López Portillo, who would take power two weeks later. Also present was the famous "group of twenty five," which in the brief interim after the election had helped him formulate his program. Most of the members of the group had already been appointed ministers of state or directors of decentralized organizations in the new administration. That fact was secret and, of course, unknown to the two stubborn and untimely speakers for the peasant cause. After witnessing an unequal battle in which efficiency, "comparative advantage," economic realism, political stability, and many other things were thrown over the heads of the weary pair, the president-elect stood up and, with a sorrowful smile, said, "In any period of rapid changes there is a group that always gets the worst part. This time again, it will be the peasants."

The fate of many was being sealed: first and foremost, that of the peasants, who for the following three years would face an adverse policy within a food model that practically condemned them to disappear and that institutionalized hunger and malnutrition; second, that of the country, which lived in the dream world of the oil boom; and, third, that of one group of advisers to the president who were not part of the "group of twenty five," but who had accompanied him during his political campaign and were very close to him. They were young, but they were also professional and devoted. They had come close enough to a sound diagnosis of the food problem and, since then, had showed some inclination toward the peasants as protagonists of rural development. Apparently they were trying to learn from historical lessons, and they did so in very modern terms—with advanced degrees that some had just completed abroad. It was thought that they would occupy high positions under the new administration. Perhaps they themselves thought it, too, but their fate was different. They stayed close, very close to the new president. Physically they were installed in the presidential estate, with their offices in the garden. But they occupied no executive positions. They were appointed advisers to the president. Among other things, they were asked to continue studying the food issue.

They did so. Three years of additional studies allowed them to produce significant amounts of papers, analyses, graphs ... and concrete knowledge. They learned more of food issues. They conceived some concrete ideas that could shape a new policy. They thought about it, time and again, perhaps losing their patience because they were never heard sufficiently. But in January 1980, President López Portillo called them to his office. He wanted a new food policy, and soon. The grill was too hot for the pancakes. The very poor harvest of 1979 had demonstrated the limitations of the policy and, consequently, wild Mexico was waking up.

A few weeks later, SAM was born.

Bibliography

Anderson, Charles W., and William P. Glade, Jr. 1963. *The Political Economy of Mexico: Two Studies.* Madison: University of Wisconsin Press.

Ashby, Joe C. 1967. *Organized Labor and the Mexican Revolution under Cardenas.* Chapel Hill: University of North Carolina Press.

Barkin, David. 1978. *Desarrollo regional y reorganización campesina.* Mexico City: Nuevo Imagen.

Barkin, David, and R. Suárez. 1982. *El fin de la autosuficiencia alimentaria.* Mexico City: Nuevo Imagen.

Bartra, Armando. 1975. *Notas sobre la cuestión campesina (Mexico 1970–1976)*. Mexico City: Siglo XXI Editores.

Brandenburg, Frank. 1974. *The Making of Modern Mexico*. Englewood Cliffs, N.J.: Prentice-Hall.

Cline, Howard F. 1962. *Mexico: Revolution to Evolution, 1940–1960*. London: Oxford University Press.

Collins, Joseph, and Frances Moore Lappé. 1977. *Food First: Beyond the Myth of Scarcity*. Boston: Houghton Mifflin.

Córdova, Arnaldo. 1972. *La Formación del Poder Político en México*. Mexico City: Ediciones Era.

Cosio Silva, Luis. 1965. "La agricultura." In *Historia moderna de México, El Porfiriato, vida económica*. Mexico City: Editorial Hermes.

Cosio Silva, Luis. 1974. "La Ganadería." In *Historia Moderna de Mexico: El Porfiriato*, ed. Daniel Cosio Villegas, vol. 1: *La vida económica*. Mexico City: Hermes de México.

Dulles, John W. F. 1961. *Yesterday in Mexico: A Chronicle of the Revolution, 1919–1936*. Austin: University of Texas Press.

Echenique, J. 1980. *Crédito y desarrollo agrícola en México 1940–1978*. Mexico City: Nuevo Imagen.

Esteva, Gustavo. 1965. "Las transnacionales y el taco." *Documentos de trabajo para el desarrollo agroindustrial* no. 1. Mexico City: SARH, DGDA.

_____. 1979. "La Experiencia de la Intervención Estatal Reguladora en la Comercialización Agropecuaria de 1970 a 1976." In *Mercado y Dependencia*. Mexico City: Nuevo Imagen—CISINAH.

_____. 1983. *The Struggle for Rural Mexico*. South Hadley, Mass: Bergin & Garvey.

Florescano, Enrique. 1979. *Origen y desarrollo de los Problemas Agrarios de México (1500–1821)*. Mexico City: Ediciones Era.

Frank, A. G. 1976. *La agricultura mexicana, 1521–1630*. Mexico City: Comité de Publicaciones de los Alumnos de la Escuela Nacional De Antropología e Historia.

Gomex, J. F. 1970. *El Movimiento Campesino en México*. Mexico City: Editorial Campesina.

Hewitt de Alcantara, Cynthia. 1978. *La modernización de la Agricultura Mexicana; 1940–1970*. Mexico City: Siglo XXI Editores.

Huizer, G. 1979. *La lucha campesina en México*. Mexico City: Centro de Investigaciones Agrarias.

Israel, J. I. 1975. *Race, Class and Politics in Colonial Mexico*. Oxford, England: Oxford University Press.

Martinez Rios, Jorge. 1970. *Tenencia de la Tierra y Desarrollo Agrario en México*. Mexico City: UNAM

Mejia Fernández, Miguel. 1979. *Política Agraria en México en el siglo XIX*. Mexico City: Siglo XXI Editores.

Meyer, Jean. 1973. *Problemas Campesinos y Revueltas Agrarias (1821–1910)*. Mexico City: Sepsetentas.

Molina Enriquez, Andrés. 1976. *La Revolución Agraria en México*. Mexico City: LER.

Otero, Mariano. 1964. *Ensayo sobre el Verdadero Estado de la Cuestión Social y Política que se Agita en la República Mexicana.* Mexico City: Instituto Nacional de la Juventud Mexicana.

Restrepo, I., and J. Sanchez. 1972. *La reforma agraria en cuatro regiones.* Mexico City: SS.

Reyes Osorio, Sergio. 1978. *Estructura agraria y desarrollo agrícola en México.* Mexico City: FCE.

Reyes Osorio, S., et al. 1969. *Reforma agraria, tres ensayos.* Mexico City: CENAPRO.

Rodriguez Cisneros, Manuel. 1979. "Agricultura." In P. González and E. Florescano, eds., *Mexico hoy.* Mexico City: Siglo XXI Editores.

Silva Herzog, Jesus, 1934. *La reforma agraria en México y en algunos otros paises.* Mexico City: Author's edition.

_____. 1960. *El Mexicano y su morada.* Mexico City: Cuadernos Americanos.

_____. 1960. *Trayectoria ideológica de la Revolución Mexicana.* Mexico: Cuadernos Americanos.

Simpson, Eyler N. 1937. *The Ejido: Mexico's Way Out.* Chapel Hill: University of North Carolina Press.

Stavenhagen, R. 1971. *Las clases sociales en las sociedades agrarias.* Mexico City: Siglo XXI Editores.

Tannebaum, Frank. 1966. *Peace by Revolution: Mexico after 1910.* New York: Columbia University Press.

Vernon, Raymond. 1963. *The Dilemma of Mexico's Development: The Roles of the Private and Public Sectors.* Cambridge: Harvard University Press.

Whetten, Nathan L. 1948. *Rural Mexico.* Chicago: University of Chicago Press.

Wolf, Eric R. 1969. *Peasant Wars of the Twentieth Century.* New York: Harper and Row.

Womack, John, Jr. 1968. *Zapata and the Mexican Revolution.* New York: Knopf.

Yates, Paul Lamartine. 1978. *El Campo Mexicano.* Mexico City: Ediciones El Caballito.

CHAPTER 3

The Conception of SAM

Mario Montanari

SAM was announced on March 18, 1980. It pushed the nation toward a more just and independent development path. The challenge of resolving the country's food problem forced the new Mexican food system to confront social inequality and national dependence. The low production and income of the more than 40% of the population earning its living in the food sector and the malnutrition of almost half of all Mexicans were, and still are, the most dramatic expressions of the food situation. But the limited national production of basic grains of the previous several years—closely associated with changes in the crop patterns and the internationalization of consumption habits under the influence of transnational corporations—and the sustained increase in imports of production inputs to the food chain were leading to growing foreign dependence.

The Mexican government recognized the food situation as a national problem and opted for food self-sufficiency. Because of its economic, social, political, and cultural interrelations, the solution demanded a multisectoral and interdisciplinary perspective and an all-encompassing strategy. The SAM decision meant a reconsideration of the role of agriculture and the beginning of a fundamental reorientation of agricultural policy. Transcending sectoral divisions, SAM was a strategy for organizing and directing actions and resources toward production, commercialization, industrial transformation, distribution, and consumption of food. Taking the peasant economy as one of its principal bases, SAM assumed that a state-peasant alliance would play a central role in solving the food problem. Born during a period of strong undervaluation of the dollar and sustained foreign exchange earnings based on oil exports, the strategy struggled against the temptation to follow the dictates of "comparative advantage." The measures gained ground not only because of the political will behind them, but also because of the vacuum in existing agricultural policy. Nevertheless, the existing institutional apparatus was difficult to adapt to the new conception.

The agricultural crisis in which SAM was proposed showed the non-viability of the earlier development path, a path that led to poverty for peasant producers, the loss of self-sufficiency in basic grains, the deformation of the agrofood development path, and a lack of fit between the different links in the food chain.

In the pages that follow, we examine the attempt at an alternative path, SAM, a path in which scientific and political tasks came together to address economic and social problems.

The Food Strategy Approach

Given the prevailing food situation—nutritional deficits, low production, adverse foreign trade relations—it was not enough to consider a new agricultural policy, a particular commercial policy, or even an aid policy for the most underpriviliged. The situation required an overall approach to the food problem, one that considered its causal interrelations.

An important early finding of development theory shows the inadequacy of sector-focused economic policies. In Mexican agriculture, the programs that sought production increases alone had obtained only transitional and partial successes and were unable to modify the fundamental conditions that prevented sustained production increases. The main problem was that imperfections and unequal power in the market channels tended to extract much of the profits generated from any increased production by the small farmers. The country needed analytic approaches that considered the agricultural problem in relation to the rest of the stages of food production and related activities, as well as new, integrative conceptions to direct economic policy. The experiences of underdeveloped countries and a glance at the international context made it clear that an "agricultural rationality" did not exist; rather, the behavior of the sector had been determined by its relations to agroindustrial demand. These relations are the ones that define the necessary technical and political parameters for food policy.

From the beginning, the team that would later give birth to SAM searched for a new approach to the problems of nutritional and food deficits. The limits of state institutions seemed to become more acute with the deepening of the food crisis; furthermore, agricultural producers' organizations were weakened economically, socially, and politically, beginning with the end of land redistribution and the rupture of the state-peasant alliance. The essential requirement for the food strategy's success was a design based on an understanding of the links among the phases of the food chain, from primary production to consumption, passing through commercialization, industrial transformation, and distribution.

Based on these premises, the SAM analysis adopted a systems approach whose global point of view allowed the integration of all elements (actors and functions) and relations that make up the food chain. It also recognized that different food systems have evolved throughout Mexico's history, giving rise to modern commercial producers as well as self-employed workers and traditional family production units. This heterogeneity required distinct strategies for each producer segment.

The systems approach operated on at least four levels to avoid the limitations of sectoral strategies. The first level concerned the national food system and its relation to the rest of the national and international economic system. The second level referred to the chain's subsystems by stage: production, transformation, commercialization, distribution, and consumption of food, as well as the para-food or auxiliary subsystems in the food chain. The third level dealt with the existing commodity systems, such as basic grains, meat, dairy, fruit, and fish, as well as some of the industrial product systems, such as fertilizers and farm machinery. At the fourth level, diagnosis and policy came together, giving priority to the spatial dimension of food systems. Special attention was given to the relations among the hungry and malnourished sectors, the professionals, public officials and consumers related to the food problem, and food policy decision making; in other words, the links between the state and the different social sectors involved. These groups were consulted during the phases of problem diagnosis and policy formulation.

The relation between the new strategy's target population and the governmental apparatus would be the basis for the development of the social and political force needed to assure the viability of the new food project. The intention was to respond to what had seemed to be the main weakness of previous projects: the inability to create a multiclass social subject, and its institutional counterpart, that were strong and articulate enough to overcome the resistance of those opposed to the alternative food strategy, to attract uncommitted social forces, and to bring together the receptive and interested ones. This combination of social forces could by no means be achieved by decree or governmental decision. Rather, state policy had to create the conditions to bring together the different, and sometimes conflicting, interests of these groups. It also had to make the food problem a political problem; that is, to see to it that food was perceived as a problem concerning society as a whole, as something that affected the possibility to reproduce the social system, the balance of power relations, and national development.

The Basic Outline of the Strategy

SAM's food self-sufficiency strategy focused on three things: (1) the nation's real capacity to produce the food required by its inhabitants, (2) the needs that were expressed in a basic recommended food basket, and (3) the strategic path to be followed by the social subject in order to modify the food structure and its power relations.

The new strategy implied that agriculture would no longer be subsidiary to industrial growth. A food chain that contributed almost 30% to the GDP and employed 47% of the economically active population could lend dynamism to the economy in general through high self-sustained growth. The new strategy also foresaw the achievement of self-sufficiency through the productive potential of rain-fed agriculture and fishing. SAM sought to have primary production play a fundamental role in a balanced relationship with the rest of the food chain, oriented toward a model of consumption adapted to the needs of Mexico and its population. By assigning resources in accordance with the national priority of reaching basic food self-sufficiency, the strategy set out by the Mexican state was linked to the goal of withdrawing from external dependence, reaffirming national sovereignty, and redefining the terms of the relations between the national and international systems.

SAM was born with a rejection of the option traditionally recommended by economic orthodoxy—commercial agriculture—and opted instead for the alternative that emphasized the structural nature of the crisis. The rejection was based on two assumptions. First, the strictly productivist, short-term approach would only deepen the already abysmal differences between the irrigated, market- and export-oriented agriculture and the rain-fed peasant agriculture that produced basically maize and beans. Second, the productive capacity of peasant agriculture was—and still is—underutilized. At the same time, the peasants in the rain-fed areas were the poorest group in Mexico, and the ones that generally showed the highest nutritional deficit. The definition of this balance of forces was of fundamental importance for designing the agricultural policy that was to be incorporated into the broader concept of food policy: unequal treatment for unequals. Knowing that malnutrition is largely a result of poverty, the food policy assumed certain actions that would increase the standard of living of the poor. The productive reactivation of the lowest income groups would permit an increase in their income and consumption.

The reorientation of agriculture and fishing toward the preferential production of basic foodstuffs was considered one way to ensure the population's access to food. At the same time, this orientation was to foster a degree of regional self-sufficiency and, in that sense, be a fundamental contribution to the important process of national economic and political decentralization. The goal was to attack poverty by generating more jobs, expanding the internal market, improving income distribution, and increasing the use of underutilized productive resources. This production-income-consumption strategy formed SAM's core. Consistent with this strategy were recommendations about the types of producers that could best provide the food, the quantities required of them, and the consumption pattern of the population, current and desired.

The starting point was to know the nutritional situation of the population. Data were collected through a national survey, and the results were compared with those obtained during the previous twenty years. The country's nutritional profile showed that between 1959 and 1979 there was a decrease in caloric consumption in the central, southern and southeastern regions. In rural areas, the average corn consumption decreased, while average bean, wheat, egg, milk, and cooking-fat consumption increased in relative terms. In urban areas, changes were less pronounced, but there was a tendency to incorporate industrialized products of low nutritional quality into the diet. Although an estimated 21 million people suffer some protein or caloric deficiency, approximately 9.5 million people or nearly 42% of the rural population suffered from caloric deficits of 25 to 40% below the Mexican minimum standard of 2750 calories per person per day. Another million were estimated to consume below 2000 calories daily in Mexico City.

The survey supplied consumption data from which the Basic Recommended basket (Canasto Básico Recomendado—CBR), an expression of the quantitative and qualititative goals for covering nutritional needs, was developed (see Chapter 12 for a critique of this methodology). The targets were set at 2750 calories and 80 grams of protein. Through this nutritional survey, it was also possible to define the population group with the most severe caloric and protein underconsumption problems: 19 million inhabitants in total, 13 million of whom were in rural areas. They became the priority target population.

The CBR, by specifying the population's food needs—for whom and what—and by going beyond effective demand as the basis for resource allocation, became an invaluable planning instrument for guiding the fulfillment of nutritional needs and for reorienting consumption. The distortion of consumption patterns had given rise to an increasing demand for foods that were considerably more expensive per unit of protein and calories. The consumption policy had two main objectives: to make the CBR accessible to the target population and to reshape consumption habits that had been negatively influenced by the mass media. Public-sector funds were increased in order to make the CBR economically accessible to the low-income target group. Subsidies for 1980 were estimated at 85 billion pesos, which was still less than the gasoline subsidy. A network of state-run stores and transportation services for deeper penetration into rural areas was developed with the dual purpose of receiving production and supplying basic products. Distribution and commercialization were reorganized to reverse their traditional use as mechanisms for extracting the producers' surplus; it was a great challenge to turn peasant participation in the channels of commercialization into a way of retaining their surplus as a means of capitalization. Nutrition education campaigns focused on offsetting distortions in consumption habits in order to make better use of the target population's scarce resources.

Targets were set to meet the projected CBR-based consumption needs by national production, replacing all imports for the target crops. The primary goal was self-sufficiency by 1982 in the country's two basic staples: 13,050,000 tons for maize and 1,492,000 tons for beans. The annual growth rates, relative to the averages for the 1976–1978 period, were to be 6.2% for maize and 13.2% for beans. For maize this growth was to be achieved by planting 1.5% more area (85% of which would be rain-fed) and increasing yields by 4.4%. For beans the increased would come from a 9.2% expansion in area (85% rain-fed) and yield improvements of 3.7% for irrigated land and 2.9% for rain-fed land. Self-sufficiency by 1985 was also targeted for rice, wheat, sesame, safflower, soybeans, and sorghum.

But why seek self-sufficiency in basic foods by relying on rain-fed agriculture? Some held that such an objective could be achieved with the nation's commercial, irrigated agriculture: indeed, some irrigated districts of the northeast were among the most modern and highly productive in the world. SAM's strategy propounded self-sufficiency through rain-fed agriculture because its goal was, not merely to increase the quantity of food available, but also to make it accessible to the population. The emphasis given to the revitalization of rain-fed agriculture is a clear reflection of SAM's integrated systems approach. The point of assigning the role of basic food supplier to rain-fed agriculture was that it was precisely the peasants from the *altiplano* and southern Mexico who formed the group with the greatest nutritional deficits. Almost 70% of the target population lived in rural zones; these were the subsistence producers who were pushed to the margin of survival by agriculture's modernization and forced to survive on ever-smaller parcels of land, insufficient to satisfy their own consumption.

The previous agricultural policy—one that accentuated social imbalances, one that even within its own logic was incapable of supplying basic foods—was deemed uncorrectable through minor changes. SAM's strategy promoted a basic reordering of the pattern of cultivation and a fundamental change in the criteria for delivering support. It clearly distinguished, and gave priority to, the target subject, attempting to develop an intersectoral relation that would restrain the traditional extraction of resources from peasant agriculture. In this way it hoped to reshape the economic relation between city and countryside.

SAM based its strategy on the following instruments: credit, production technology, "shared-risk" crop insurance, support prices, the rationalization of the relations between agriculture and livestock, and peasant organization. Technological change, which began by recognizing local technologies, included the components needed to increase productivity at a lower cost and in less time. In the rain-fed lands in Mexico, introduction of new or improved fertilizer, changes in sowing density, and massive use of improved seeds and adequate tools were highly effective when used within a framework reinforcing the distinct ecosystems.[1] While these methods were simple and well known by most peasants, adopting them required credit and was risky for subsistence-level farmers; changes in regular practices could jeopardize their survival. The "shared-risk" mechanism was a way of making it possible for peasants to make these changes. The state's coresponsibility meant the creation of incentives to acquire certain inputs, the technical assistance and follow-up for the production process, and, in the extreme case of disaster, economic support to assure the peasant's family consumption. The peasants, for their part, committed themselves to adopting the proposed innovations and the measures agreed to in the production plan, adapted to local and regional conditions. In effect, credit enabled the peasants to acquire the modern production technology, and expanded crop insurance reduced the consumption risk involved in the change of production technology.

According to SAM studies, this labor-intensive technological adoption process, if used for a reasonable period of time, would enhance production and consumption, thereby breaking the apparent conflict between investment and consumption. At the same time, considerable efforts were made to bring the productive sector's technological demands to the scientific-technological community in order to establish the relationship and feedback necessary for creating an autonomous technological capacity.

The initial subsidies for the technological package were to promote a process capable of breaking the vicious cycle of low prices–low production–low income–decapitalization. SAM sought, with simple, accessible, improved techniques, to enable the peasants to increase their production and, perhaps even more important, to retain their surplus through strengthened peasant organizations and participation in the phases of commercialization and industrial transformation. To reach the goal of enabling the primary producers to retain a greater part of their surplus, it was necessary to change the support price policy to make peasant efforts more remunerative. This policy was conceived as a basic instrument for reversing the unfavorable terms of trade between agriculture and the rest of the economy, to develop a basis for the capitalization of agriculture.

In Mexico's inflationary economy, however, the effect of agricultural price increases on urban workers' wages warranted special care. So that increased peasant income would be compatible with access to food for the urban population, the criteria for subsidies were reformulated. SAM's subsidy policy was oriented to the peasant producer, first as a temporary expedient in the absence of a self-sustaining process of capitalization and, second, for reasons of equity, to increase the consumption of those rural

1. This has been demonstrated by research conducted at the Colegio de Posgraduados de Chapingo and the Instituto Nacional de Investigaciones Agrícolas.

and urban populations suffering the most serious nutritional deficits. The subsidies allowed higher prices to the farmers without raising urban prices.

Permanently increasing agricultural production meant opening the still-considerable agricultural frontier. Underutilized by extensive cattle ranching, it needed to be freed up for the cultivation of basic grains and alternative feed crops. The full implementation of this policy would be in the medium rather than the short run. As it gathered sufficient force, it would have to confront the problem of the distribution of lands currently used for livestock, the basic nucleus of large landholdings in Mexico.

This reactivation of crop production expected in the first two years was to provide the country with enough basic grains to permit the transition to a second phase. Then certain production systems would be chosen for special emphasis, so that the nation would enter the 1990s, not only with self-sufficiency, but also with the capacity for a positive agrofood balance of payments of approximately 2 billion dollars annually. Mexico would only import milk and oilseeds.

In spite of the rapid growth of production, the food industry had been unable to supply the demand for processed products, mainly basic foods. This created an underutilized economic space that could be filled by peasant agroindustries, initially oriented to the target population, which could in the medium term be the basis for reorienting the distorted growth pattern of the food industry.

The link between the production support measures and the explicit objectives of improved consumption became most clear in the policies proposed for agroindustrial and technological development. In particular, a higher order of peasant organization could make the participation of peasant producers in agroindustrial development viable, especially in the production of the basic foods in the CBR distributed by the public sector.

Some of the most pronounced inequalities in the structure of the food industry were the result of technological disparities between small family enterprises and large transnational agroindustrial complexes. SAM studies showed, however, that, based on a reorganization of the scientific-technological community, the nation could have the capacity to develop its own technology within each of the phases of the food chain and could—and can—respond to the need to develop coherent but heterogeneous technology within the agroindustrial chain. In this way, the foundation could be laid for an effective policy for dealing with transnational corporations.[2]

2. See "La estructura scientifico-tecnólogica y el Sistema Alimentario Mexicano" and "Las empresas transnacionales y el Sistema Alimentario Mexicano" (mimeographs, Mexico City, SAM, 1981).

SAM's strategic propositions saw the integration of the food sector as an important lever for self-reliant development. The Mexican food chain could, in the short and medium run, grow more rapidly than the rest of the economy and thus could distribute income to its wage earners and small self-employed producers. Paradoxically, these food producers were also the consumers with the lowest food intakes.

The self-sustained development of the food sector implied a redefinition of Mexico's commercial, technological, and financial insertion into the international sphere, and a reduction of its food vulnerability. This redefinition took into account a serious analysis of the evolution of the world food system, in particular the large food-producing regions such as the United States, the EEC, India, China, and, most important, Latin America. Based on these reflections, Mexico shared the SAM approach with other countries. Aggressive proposals and actions were begun to achieve food security in Latin America, in particular in Central America, and to redefine relations with the larger regions, especially the EEC. These efforts were also directed toward the different relevant international institutions, such as the World Food Council, the Food and Agriculture Organization, and the Economic Commission for Latin America. With the same orientation, SAM initiated and participated in international projects on concrete aspects of food security, such as the establishment of the International Bank of Phytogenic Resources and the design of alternative food systems for various countries.

Some Reflections on the Introduction of SAM

The poor harvests of 1979 exacerbated the agricultural crisis that had been holding Mexico back for more than a decade. When this developed into a social problem, the lack of a defined agricultural policy became more evident. In January 1980 the team of advisers who had been working on an analysis of the agrofood problem was called on by the president to turn their work into a proposal for a strategy. The group's research and design of alternative policies were still in process, although the basic conceptions that shaped SAM were clearly defined. Although risking prematurity, the advisers seized the political moment and told the president that they could have the new policy ready within two months. To hesitate would have been to cede the initiative to others and to condemn SAM to the crowded shelves of unimplemented academic studies. The process of translating analysis into policy measures and operational programs forced the team to consolidate the work achieved thus far into a precise combination of guidelines, goals, and instruments that would take aim at the central problems of the moment and have an impact on the key points needed for the transition toward a new, alternative food system.

By starting from the proposition that the agricultural production goals were based on the consumption needs of the population, SAM's systems approach met with no resistance from the bureaucratic apparatus. By March, guidelines and goals were developed for national and regional levels by commodity, and for financial and input requirements. The first challenge to be met was to transform the food problem—in its manifestations as malnutrition, undernutrition, and inadequate production—into a political problem. This would have to be taken on by various social sectors, fundamental among them the Mexican state, as a first step. When SAM was publicly announced, an editorialist remarked that in contrast to El Cid, who won his last battle after his death, SAM won its first battle before its birth: it converted the food problem into a political problem, creating a consensus around the need to eradicate the nation's malnutrition.

SAM came about as a short- and medium-run strategy. Therefore, instead of being assigned its own resources, it sought to reorient the actions of the state apparatus and to add additional resources to strengthen programs that already pointed in the right direction. SAM sought to modify some of the programs of existing institutions substantially in order to sustain its propositions. Thus, in the medium and long run, it would carry out a profound institutional restructuring that would permit a more integrated state apparatus, opening spaces in the political arena for joint action with the various organizations of civil society.

The road from strategy to operational program was quickly traveled, guided by the clearly defined objectives of the first stage. But the program was permanently subject to the constraints set by the weight of the Mexican state apparatus. It was thought that the possibility of rationalizing the food-related institutional apparatus and its actions would be determined by the process of reorganization from the systems approach perspective, involving not only all or most of the activities of the food chain but also, above all, the actions of the organizations of producers and consumers.

The process of building this new institutional rationality, as a process that identified food policy actions all along the food chain, would give coherence to the programmatic opening. Through strategic projects, it would also increase the coherence of and control over budgets. The capacity to carry out this rationalization would restore the state's capacity as an actor in, and interlocutor of the historical subject of, the strategy. It required—and requires—a political decision that had to go hand in hand with the creation of instruments and institutionalized spaces through which producer and consumer organizations would be able to plan and reach agreement with the state on how, what, when, how much, and under what conditions to produce, as well as how to assure consumption.

An indispensable element was to create forums at the level of production units (balance and planning assemblies) and at the municipal, regional, or district levels, all of which would converge with institutional management and allow the participation of civil society. In this way, it would be the civil organizations themselves that would, through their presence, make possible the process of rationalizing the state apparatus, controlling and guiding its activity. The state could then make an integrated response to the equally integrated social forms of production, consumption, and social participation.

The short bureaucratic life of SAM prevented the transformation of the state apparatus involved in the food system, and the implementing apparatus—created in order to carry out programs very different from those proposed by SAM—often made SAM's priorities lose coherence. The various state institutions found, in the dynamic launched by the SAM strategy, a channel for projects that did not always correspond to the priorities and critical path set by the strategy or, in some extreme cases, that simply contradicted its basic logic.

This did not prevent the strategic concepts from making headway within the Mexican state, and they have been taken on increasingly by organizations in civil society. The social consensus around the need to confront the food problem systematically continues to echo in theoretical and operational discussions. It is for this reason that this examination of SAM, and a reconsideration of the role of the food chain in the Mexican economy, seems today more current than ever—because of the crisis, not only in Mexico, but in all of Latin America.

We believe that SAM, like El Cid, will win its final battle after death.

PART III

IMPLEMENTATION

CHAPTER 4

State-Owned Enterprises:
Food Policy Implementers

James E. Austin and Jonathan Fox

The planners of the Mexican Food System (SAM) looked at Mexico's food problem through the lens of an integrated food systems approach, as described in Chapter 3. Their diagnosis and policy recommendations were based on a theoretical foundation that saw the processes of access to agricultural inputs, food production, commercialization, processing, and distribution as part of a single system. They used this approach to determine where policy makers could and should have an impact on that system in order to change it. To implement these decisions, Mexico's food policy makers chose to use a wide range of powerful state-owned enterprises (SOEs) active at the different stages in that system. The rationale was that the SOEs were in place and could mobilize immediately to implement the strategy. SAM's producer incentives and consumer subsidies were policy decisions made at the macro level, and the SOEs could serve as the conduits. The effectiveness of the new food strategy would be significantly determined by the substance and form of the response of the SOEs as implementing agencies. Could they, and would they, respond as the SAM strategists planned?

In this chapter we describe and analyze the role of Mexico's SOEs as policy instruments in carrying out the 1980–1982 national strategy of self-sufficiency in basic grains. First food sector SOEs are described. Next we discuss the SOEs' possible responses to SAM and document and analyze the actual responses. In the final section we draw conclusions regarding the use of SOEs as food policy implementers.

SOEs in the Food Sector: Roles and Prevalence

The Mexican government has increasingly used SOEs in all of its economic sectors (Hill, 1981; Barenstein, 1982; SPP, 1982; Cohen and

61

Thirty, 1983). The number of federal- and state-level SOEs grew from 84 in 1970 to 966 in 1982, excluding banks, which were nationalized in that year (*Mercado de Valores,* 1982). Almost one-third of these SOEs are in the food sector.

Food Sector SOEs: Definition. The food sector can be usefully viewed as a system encompassing the broad range of productive activities and inputs needed to produce, transform, and transport food to the ultimate point of consumption. The literature on SOE's has generally divided them into narrower, more traditional sectoral categories, such as agriculture, manufacturing, or commerce. This segmented approach is inadequate for analyzing food policy implementation, which by its very nature requires actions across (and among) the segments. Using the traditional sectoral categories also tends to cloak the importance of SOEs to food policy. Estimates of the importance of SOEs in the economy frequently relate their value-added to the GDP. For example, Jones et al. (1982) state that SOEs make a small contribution to the agricultural sector compared to their role in other sectors; they cite figures of 1.8% for India and 0.2% for South Korea, versus about 15% in manufacturing in these countries. Similarly, SOEs in Mexico play a very minor role in actual on-farm production. This comparison is misleading, however, because it does not take into account the importance of the SOEs in providing manufactured inputs to agriculture, in commercializing and processing the agricultural outputs, in supplying key logistic and financial services, and in shaping market prices at each stage of the production and distribution process.

We define food sector SOEs, therefore, as those engaged in any of the following activities: production and supply of agricultural inputs, such as seeds, agrochemicals, equipment, technical assistance; agricultural production; purchasing, transport, storage, processing or distribution of agricultural raw materials or transformed food stuffs; and provision of financial resources, such as credit and insurance. Table 4.1 lists the SOEs that were involved in the implementation of the SAM program; it reveals that these enterprises were important actors in all stages of the food system except the direct production of basic grains (see the Glossary at end of book for English translation of SOE names).

SOEs are also heavily involved in other aspects of Mexican agriculture, such as the production, commercialization, and processing of coffee, sugar, fish, and fruit, but these foodstuffs fell outside SAM's primary focus on basic grains and, therefore, outside our analysis. We do not examine the important role played by the central government through the Ministry of Agriculture and Water Resources (SARH). SARH controls one key production input, water, through its construction and distribution of Mexico's irrigation resources. It also provides much research and technical assistance. General macro policies—exchange rates, wage levels, and price controls, for example, also affect food policy; these are addressed in Chapters 13 and 14.

Table 4.1. Grain market shares of SOEs as of 1979

Activity	SOE	Share
Inputs		
Credit	BANRURAL FIRA	SOE share of total value, 75.4%. BANRURAL share of total credit, 52%. FIRA share of private credit, 48.6%.[1] BANRURAL share of area harvested: 27.4%.[2] Maize share of BANRURAL area harvested: 43%.[3]
Agrochemicals	FERTIMEX	Share of fertilizer production, 100%. Share of international trade, 100%. Import share of national consumption, 20%.[4] Distribution through SOEs, 53%.[5] #1 national insecticide producer, #2 herbicide producer.
Seeds	PRONASE	Share of certified maize and bean seed production, 90%. Certified wheat, 43%. Certified sorghum, 10%.[6]
Tractors	SIDENA, FTA	SIDENA-Ford market share, 30%.[7]
Crop insurance	ANAGSA	Share of basic grain area insured, 49%.
Technical assistance	BANRURAL FIRA CONASUPO	Share of area sown with maize covered, 48%.[8]
Production Grains	PRONAGRA	Share of national rice production, 6%.[9]
Commercialization		
Procurement	CONASUPO	SOE share purchased of national maize production sold, 23.1%. SOE share of maize demand, 33.1%. SOE share of grain imports, 100%. Import share of national maize consumption, 9.8%.[10]
Warehousing	ANDSA, CONASUPO-BORUCONSA	CONASUPO share of national warehouse capacity, 40%.[11] ANDSA share of CONASUPO capacity, 68%.[12] BORUCONSA share of national CONASUPO maize purchases, 73%.[13] Share of BORUCONSA maize received with PACE transportation subsidy, 19.8% (1978–1979).[14]
Wholesaling	CONASUPO-IMPECSA	SOE coverage of national food retailers, 15.8%.[15]
Retailing	CONASUPO-DICONSA	SOE share of national retail market for basic consumer goods, 9%. Coverage: 7500 towns, 25 million consumers.[16] Rural share of number of DICONSA outlets, 71.7%.[17]

Table 4.1. (continued)

Activity	SOE	Share
Processing		
Grains, oilseeds	ICONSA	SOE share of national oilseed milling capacity, 10.5%. SOE share of vegetable fat production, 10%. SOE share of wheat milling, 7.4%.[18] ICONSA share of basic food product market (Alianza brand), 6%.[19]
Maize	MINSA	SOE share of maize flour production, 27%.[20]
Bread	TRICONSA	SOE share of Mexico City bread production, 40%.[21]
Milk	LICONSA	SOE share of national milk production, 17%. Share of imports, 100%. Import share of national consumption, 11.8%.[22]
Animal feed	ALBAMEX	SOE share of mixed feed market, 6% (4th largest producer).
Formulated foods	NUTRIMEX	No data

Source: [1]Patron Guerra and Fuentes Navarro, 1982.   [2]SARH, Memo. 1982b. [3]FEP, 1981.   [4]Based on FERTIMEX data.   [5]FERTIMEX, 1980.   [6]SARH, 1981. [7]*AMIA Boletín*, Jan. 1981.   [8]FEP, 1981.   [9]Cabrera Morales, 1982; FEP, 1982. [10]CONASUPO, May 1982.   [11]CONASUPO, 1980.   [12]SPP, 1981.   [13]CONASUPO, 1982; FEP, 1981.   [14]Rubio Canales, 1982.   [15]CONASUPO, *Sistema C*, Nov.–Dec. 1981. [16]CONASUPO, 1980.   [17]FEP, 1981.   [18]CONASUPO, 1982a.   [19]CONASUPO, *Sistema C*, May 1982.   [20]CONASUPO, 1982a.   [21]CONASUPO, *Sistema C*, March 1982. [22]Santoyo Meza and Urquiaga, 1982.

SOE–Central Government Organizational Ties. Food sector SOEs are little different from SOEs in other sectors in their wide variety of relations with the central government. Some appear highly autonomous, like FIRA; others seem closely linked to particular ministries, as is the case of PRO-NASE and BANRURAL to SARH. On closer examination, however, degree of autonomy may not be conditioned primarily by organizational structure; political alignments within the ruling coalition, complementary or conflicting policy orientations, and degree of SOE dependence on central-government financing may be more important factors.

FIRA, for example, may appear to operate highly autonomously compared to other SOEs in the food sector, but it should be recalled that it operates under the policy direction of another SOE, the Banco de México, which often shares the policy orientation of the finance ministry. In addition, FIRA has greater financial autonomy because of its international donors. Thus, while an SOE appears to operate independently of functionally related parts of the state apparatus, it may in fact be following a policy direction set elsewhere in the government.

SOEs may be managed highly efficiently, and their directors may be in political favor, but if their function is to deliver goods and services at subsidized prices, they must turn to the central government for funds to cover the resultant deficits. This necessarily increases the SOEs' dependence on the central government. Conversely, SOEs that generate large surpluses, such as the oil company PEMEX, may achieve substantial autonomy even without being run efficiently. The drive to increase the financial autonomy of the enterprise may be one reason why some of the more entrepreneurial SOEs in the food sector serve more affluent market segments in addition to those in need of subsidies (e.g., DICONSA, LICONSA, ALBAMEX). The issue of central government–SOE relations is dealt with below in the context of the discussion of SOE response to SAM.

Although President López Portillo began his administrative reform upon assuming office in 1976 (Bailey, 1980), SOEs were not reorganized into sectors under the new ministries until 1982. Most food sector SOEs then fell under the ministries of agriculture, (ALBAMEX, NUTRIMEX, PRONAGRA, PRONASE), finance (ANAGSA, BANRURAL), industrial development (FERTIMEX, SIDENA), and commerce (CONASUPO). (For a complete listing of SOEs by sector, see *Mercado de Valores,* Sept. 13 and 20, 1982.)

Food Sector SOEs: Economic Importance. Some of the available indicators of the economic importance of SOEs in the food sector include number, employment, trade, agroindustrial investment, and market shares.

1. *Number:* Approximately 30% of 1982's 966 SOEs were directly related to the food system.
2. *Employment:* SOEs in the food sector are large employers. CONASUPO and its affiliates employed approximately 24,000 people in 1979, making it one of the largest employers in Mexico (CONASUPO, 1980a). Some of SOEs are highly capital-intensive, however, because of the techniques of production chosen to manufacture, for example, fertilizer, animal feed, or bread.
3. *Trade:* CONASUPO controls all of Mexico's international grain trade, and other SOEs figure heavily in the commercialization and trade of key crops such as coffee and sugar.
4. *Investment in agroindustry:* Government ownership of the food-processing industry rose from 18.3% in 1970 to 21.7% in 1975. The SOE percentage of gross fixed investment in food processing, however, rose from 20.1% in 1970 to 88.6% in 1975. The SOE share of value-added, however, was only 6.7% in 1970 and 10.5% in 1975. This is partly because the food processing share within SOE value-added as a whole was 3.3% in 1970 and 3.4% in 1975 (SPP, 1982: 4.1.3.3.4; 4.2.1.6.8).
5. *Market shares:* To give a sense of the role of SOEs in the food system

before SAM, Table 4.1 presents some indicators of their relative importance within their particular activities. It is clear that SOEs are significant actors throughout the food system.

SOEs and SAM: The Implementation Experience

As pointed out in Chapter 3, the SAM strategy rejected the notion that it
was in the national interest to continue its growing dependence on imported (largely U.S.) foodstuffs. Instead of following the dictates of the law of
"comparative advantage" of exporting oil and importing grains, Mexico
would sow oil revenues in the countryside. The stated goal was not only to
revitalize food production, but to modernize and redistribute income, particularly to small farmers. This was to increase national autonomy while
reinvigorating a state-peasant alliance that dated back to the revolution.
In addition, increased access to subsidized food was to meet nutritional
needs and to defuse the tensions deriving from rising but unmet expectations.

The implementation task raised the question, Would the SOEs have the
capability and motivation to respond to the new strategy? As stated previously, we contend that the effectiveness of the new strategy was significantly determined by the substance and form of the response of the
SOEs. Furthermore, full implementation of the strategy would require
changes in the structures and administrative procedures of the SOEs
whose actions were not previously in accordance with the SAM strategy
(see Chandler, 1962; Ickis, 1978).

Three categories of factors could facilitate or inhibit policy implementation: temporal context, bureaucratic incentives, and economic resources. At a conceptual level these three categories imply a general institutional behavior model for SOEs. Their response will be shaped by the
social and political situation out of which the new strategy arises, the set of
political or other incentives that motivate the SOEs to support or resist implementation, and the level of economic resources available, which affects
the capacity for implementation.

Temporal Context

Facilitating Factors. SAM emerged in the aftermath of, and partially in
response to, the shortfalls in the 1979 grain crop. This may have created a
"crisis-response context" that in turn created a greater sense of urgency
and propensity to act on part of the SOEs.

Inhibiting Factors. This same "crisis-response context," which increased the need to act, also compressed the time frame for implementation. This sharply constrained the possibility of comprehensive implementation for three reasons: (1) It reduced the time available to the

SAM strategists to consult with, and elicit the cooperation of, the SOEs; (2) SAM was announced with less than three years remaining in the administration, limiting the ability of those SOEs that did try to implement SAM to make structural changes in their organizations; (3) The decision-making process and its short time frame prevented the active participation of peasant producers, SAM's state primary target group, in the strategic formulation or the implementation design of the policy.

Bureaucratic Incentives

Facilitating Factors. Although the SOEs function as autonomous legal entities, the political structure and patronage system make them highly responsive to the political priorities emanating from the presidency. Thus, by clearly manifesting that SAM was a top political priority of the president, the policy makers were able to create a strong incentive for SOE managers to implement the strategy vigorously. The SAM planners hoped that this would compensate for the time constraints. Furthermore, those SOEs that already had adopted, on their own, strategies similar to SAM's were probably able to accelerate their growth more than they would have otherwise.

Inhibiting Factors. The emergence of the SAM group within the office of the presidency gave rise to a potential competitor within the arena of bureaucratic politics. Some entities may have felt threatened by the existence of the SAM group and therefore tried to reduce its power, perhaps by resisting or diverting implementation of the strategy. The fact that the SAM group was a policy-formulating entity rather than an operating entity made it dependent on the existing ministries and SOEs for implementation. This gave leverage to the latter and thus raised the potential for bureaucratic competition impeding implementation. Outright opposition to SAM, however, was not expected to be politically viable because it would mean openly rejecting the president's decision. This was a particularly sensitive situation because of the upcoming candidate-selection process for the 1982 change in administration. This constraint, plus the inhibiting factors mentioned above in the context of timing, created an incentive for "window-dressing" behavior. The possibility existed that the SOEs would designate many of their activities as SAM, while in reality only relabelling existing activities rather than making any substantive changes to implement the strategy.

Availability of Economic Resources

Facilitating Factors. The SAM strategy called for heavy use of subsidies for both producers and consumers. This required a large outlay of resources. The willingness of the president to channel revenue to the implementers of SAM to cover these needs eased implementation by reducing SOE

resistance. The SAM strategists were able to co-opt many of the SOEs by
(1) ensuring that their regular operations would not suffer from the subsidized prices by ensuring reimbursement of deficits, (2) allowing them to
expand their operations and hence their sphere of influence, and (3) enabling them to add the SAM target group without necessarily abandoning
their traditional clients and constituencies.

Inhibiting Factors. As economic resources became scarcer, the capacity and willingness of the SOEs to implement the strategy decreased. Thus
the SOE performance for SAM was hindered when the 1982 austerity
measures greatly reduced the resources available.

SOE Responses

To examine the possible SOE responses to SAM, we analyze the actions
of the SOEs during the 1980–1982 period. The analysis is based on
published and unpublished documents, as well as direct, structured interviews with SOE managers, SAM strategists, and industry experts.

Our analysis first documents the extent of the responses to SAM of each
of the SOEs operating in the input, processing, and commercialization
stages. In general, we find that the empirical evidence reveals a strong,
positive, although varied response by the SOEs.

Inputs

PRONASE: Seeds. SAM's package of production inputs included 75%
subsidies on the price of improved and selected grain and bean varieties,
leading to a large jump in demand and a strong response by the SOE supplier. PRONASE output of certified seed shot up to 183,000 tons in 1980,
106% over 1979. Production continued to grow to 235,000 tons in 1981, and
to 215,000 in 1982, an increase of 141% over 1979 (FEP, 1984:521).

In order to evaluate the PRONASE response to SAM, it is necessary to
disaggregate the growth in seed production by crop. Certified maize seed
production grew 559% between 1979 and 1980, from 8000 tons of certified
seed to 53,000, fell slightly to 44,000 tons in 1981 and then to 17,000 tons in
1982. Bean seed production was very low before SAM, only 5300 tons in
1979. By 1981 PRONASE was producing 40,000 tons, an increase of 658%.
Rice seed production rose to 23,000 tons in 1981, 86% over 1979. Wheat,
because of its earlier importance, grew only 180%, from 45,000 to 126,000
tons between 1979 and 1982. Some crops not considered basic to SAM also
experienced significant increases in PRONASE output, particularly barley, soybeans, and sorghum (FEP, 1981:521).

Most of this growth took place during three crop cycles. Special implementation actions to maintain quality control, as well as rapid expansion of PRONASE's network of reception, processing, and storage infra-

structure, were required. Reception capacity, for example, was reported to increase from 4000 tons in 1979 to 91,195 tons in 1982 (PRONASE, 1982).

The *criollo* maize program tried to improve the productivity of local varieties; it was applied by an estimated one million peasants on 2.5 million hectares in 1982. The share of nonreproducing hybrids in relation to freely pollinating certified seed rose from 2% in 1979 to 45% in 1981 (PRONASE, 1982). This suggests an increased integration into the market for purchased inputs on the part of a substantial fraction of maize producers. The area sown with improved varieties grew about two and a half times between 1977 and 1981. PRONASE estimated that use of improved seed results in productivity increases of at least 15% (assuming the seed is appropriate to the zone and that quality control holds up). Some observers, however, have expressed concern about the possible loss of *criollo* genetic stock, as well as small producer autonomy, as a result of this process.

By 1981 the area sown with PRONASE seeds had risen to an estimated 3.0 million hectares, 150% over 1977 (PRONASE, 1982:114). PRONASE's growth is accounted for by the favorable convergence of SAM's production and input subsidies with PRONASE's increased investment in its reception, processing, and storage network. PRONASE's close identification with SARH may have accelerated its growth; PRONASE could be seen as having been one of SARH's principal implementing vehicles for SAM. Increased use of high-yield varieties was a central part of SARH's own policy agenda. As an indication of the priority SARH accorded PRONASE during SAM, the incoming 1982 administration made the former minister of agriculture the new director of PRONASE. In conclusion, PRONASE responded quantitatively in a major way, but with imperfections that may have serious long-run implications. For a more detailed analysis of the seed industry, see Chapter 6.

FERTIMEX: Agrochemicals. FERTIMEX's ability to respond quickly to SAM's increase in demand was limited, because FERTIMEX plants were already working at close to capacity, and new plant investments require both large capital outlays and long construction periods. One way FERTIMEX dealt with the announcement of SAM was to increase the truck fleet to speed delivery and to make procedural changes necessary to reduce inventory supplies from 30 days of demand to 15 days. As one high FERTIMEX official put it, "we learned of the existence of SAM precisely on the 18th day of March, 1980." FERTIMEX also quickly increased its import levels.

SAM's production input policies for basic grains cut FERTIMEX prices to about 20% below prevailing prices. An additional 10% transport subsidy was also given. This 10% extra was available to all rain-fed farmers and to irrigated farmers with 20 hectares or less. This second discount was the only significant SAM subsidy with access explicitly limited by size of producer, but 20 irrigated hectares is relatively large. The discounts created a huge economic incentive to use more fertilizer. The challenge to FERTIMEX was to meet the increased demand.

The increases in imports during the SAM period were relatively costly. As one official recalled, "we had to enter the international market in desperation," that is, without bargaining power. However, a series of large investments in increased capacity begun in 1977 soon began to come onstream. In 1982 FERTIMEX had an annual installed capacity of 4.4 million tons, and the plants under construction in that year were to add 1.7 million more in the short term.

Although FERTIMEX is an autonomous state enterprise, the central government compensated it for the losses incurred by the pricing policies and the need to import. Fertilizer prices were estimated to have been 26% below production costs in late 1981. The government also financed FERTIMEX's capital spending and debt service charges. Fertilizer production increased 9.4% from 1979 to 1980, from 2.5 to 2.8 million tons. Volume continued to increase in 1981 to 3.4 million tons and in 1982 to 3.6 million tons, increases of 22.7 and 6%, respectively. The deficit in national production grew to 23% in 1980 and then fell to 16 and 18% in 1981 and 1982. Imports totaled 78,000 tons in 1982 (FERTIMEX, 1982:66, 95; 1982b). Sales volume, which includes imports as well, grew 22.3% in 1980 and another 14.8 and 11.9% in 1981 and 1982 (FEP, 1984:302).

SAM's impact on FERTIMEX is shown by the increase in sales of NPK, a fertilizer often used for rain-fed grain crops. Sales of NPK rose 8.2% in 1980, 13.7% in 1981, and 14% in 1982. By 1982 sales were 33% greater than had been projected for that year. The effect on the total area treated with fertilizer was more uneven, rising 9.5% in 1980, only 0.3% in 1981, and then 22.5% in 1982. This rate of change may be partly accounted for by the 1981 increase in the average amount of fertilizer applied per unit area, which probably went up due to favorable rainfall, and partly by 1982's inflation-induced drop in the real price (FERTIMEX, 1982b).

SAM coincided with the implementation of unrelated major changes in the system of national fertilizer distribution. In an attempt to increase efficiency, FERTIMEX shifted from reliance on commission agents to more direct and institutional sales. Between 1979 and 1980, sales through commission agents fell 42.8%, while sales to BANRURAL rose 36.8%, from 22.8 to 28.2% of total sales. A network of state-level enterprises was also created, and their share rose 47.8%, to 24% of the 1980 total (FERTIMEX, 1980). Limits on markups made provision of decentralized storage and trucking facilities less profitable for private entrepreneurs, which may have increased BANRURAL's importance as a supplier of fertilizer for small producers.

In conclusion, FERTIMEX came up against the technical constraint of bulky investments, was not consulted at the operational level, and was pressured by time. FERTIMEX responded by changing procedures and by resorting to costly, last-minute imports.

One high official noted that "perhaps our participation would have been a bit more efficient if they had invited us in advance.... We were practically ignored, in spite of the fact that SAM, no doubt, wouldn't have gotten anywhere without fertilizers." According to a top SAM planner, the director of FERTIMEX had been consulted in advance, but word apparently did not filter down to the operational level in the form of preparations.

FERTIMEX did respond. Supply was not a constraint on implementation, although it was costly. Finally, there were procedural and some structural changes in distribution, but these may not have been sufficient to reach the smaller farms effectively.

SIDENA, FTA: Tractors. Total public and private tractor sales (excluding SIDENA) increased 12% from 14,000 to 15,700 between 1979 and 1980, and another 15.4% in 1981 to 18,000 units. In 1982, as SAM budget cuts tightened credit, reducing demand to 13,400 units, sales fell 25.9% (*AMIA Boletín* 205, Jan. 1983).

SIDENA, with the only small tractor on the market, experienced increased demand, but not as much as producers of medium and large tractors. In 1980, for example, production of small tractors (under 60 HP) increased 2.1%, while output of medium-sized tractors grew 25% and of large tractors 18% over 1979 (*Mercado de Valores,* 21 Dec. 1981).

SOE participation in tractors increased during SAM when the Agricultural Tractor Factory (FTA) was formed in 1981 as a joint venture between the state development bank NAFINSA (60%) and Ford Motor Company (40%). Ford entered this project as part of the Mexicanization process for transnational companies. Previously, SIDENA had manufactured for Ford on a subcontracting basis. Ford held a major share of the market. FTA's first factory was projected to produce 9000 units in 1983 (*Mexican-American Review,* Oct. 1982).

There was a definite increase in area using mechanized farming during SAM, particularly in the irrigated areas. The irrigated area that was totally mechanized in crop year 1980–1981 reached 2.9 million hectares, 37% over 1978–1979. The irrigated area partially mechanized increased 21% over 1978–1979 to 1.3 million hectares, while the small amount of non-mechanized area under irrigation increased 37% to 0.2 million hectares (FEP, I, 1982:41). This expansion was related more to SAM's indiscriminate increase in credit than to a particular state role in tractor production.

In conclusion, farm machinery SOE changes directly caused by SAM were modest, while the private sector response to the market stimuli was substantial (see Chapter 10).

BANRURAL: Credit. BANRURAL was one of the first SOEs consulted by SAM, receiving what one SAM planner called "special attention." The planners wanted to see how much flexibility there was for a quick response to make credit available in time for the spring 1980 planting decisions. BANRURAL quickly carried out a pilot project to see how the packet of inputs and incentives worked. The response was encouraging, and the agency moved ahead.

For BANRURAL, the March 1980 announcement of SAM meant the immediate reduction of interest rates for any producer of maize or beans to 12%. Below-market rates were available for other crops as well. Given that there was no readjustment clause for inflation, which was running at over 20%, these interest rates were negative in real terms.

BANRURAL credits in 1981 were 62% over 1980 and 119% over 1979 (FEP, IV, 1981:203). Inflation was 28% in 1980 and 20% in 1979, so the increase in credit available was substantial in real terms.

The area receiving BANRURAL credits increased significantly during SAM. Area financed in 1980 was 49% more than in 1979, reaching 4.8 million hectares. In 1981 the area financed increased another 31.5%, to 6.3 million hectares, and in 1982 it hit a high of 7.2 million hectares. Area financed expanded more rapidly than the also-increasing area planted, leading to a higher BANRURAL share of area sown (FEP, 1984:528).

Within the crop sector, the attention to maize and bean producers increased. Area sown with maize and beans in 1980 was 68% greater than in 1979, rising from 1.8 million to 3.0 million hectares. This rose another 38% in 1981 to 4.2 million hectares, and a high of 4.4 million hectares was financed in 1982 (see Chapter 5).

BANRURAL also served a substantially larger number of producers during SAM. It financed 17% more producers in 1980 than in 1979, rising from 1.24 million to 1.45 million. In 1981 the number rose 33% to 1.65 million and in 1982 hit 1.7 million. Of these, 93% were *ejidatarios* (producers who work land that they have a legal right to exploit but cannot sell; see Chapter 2) (SARH, 1982b:468).

BANRURAL's pattern of credit allocation among types of loans also changed somewhat. The share of short-term crop loans, which had fallen to 70% in 1980, rose to 73% in 1981 and 79% in 1982. Agricultural investment loans, on the other hand, fell from 24% in 1980 to 19% in 1982. Long-term credit is dedicated to increasing capitalization, which is potentially the principal source of sustained increased productivity. Little goes to peasant producers, however. Whether long-term credit promotes basic grains is a function of what type of production is in fact capitalized. This involves examining credit use by type of loan.

The crop-livestock breakdown shows a pattern similar to the changing short-term–long-term shares. Within short-term credit, the crop share fell to 61.7% in 1980, rising to 64.5% in 1981. The livestock share, on the other hand, had risen from 16.3% in 1977 to 26.7% in 1980, falling to 23.4% in 1981. Within long-term credit, the crop share fell from 73.8% in 1977 to

58.2% in 1980, rising to only 60.3% in 1981. In real terms, long-term live-stock credit grew from 1977 to 1981 at an annual rate of 35%, that for the agroindustry at 56.1%, and for crops at only 16.6% (Reyes, 1982:6, 7). Given that small producers tend to receive only short-term crop loans, a substantial proportion of lending apparently went to producers who had alternative sources of formal credit, and whose solvency reduced their need for subsidized rates of interest. During SAM, however, BANRURAL increasingly favored short-term crop loans, reversing the earlier trend.

In addition to providing subsidized credit, BANRURAL also administers nineteen trust funds. Two of them, the Shared-Risk Fund (FIRCO) and the Rural Promotion Fund (FIPROR), were created in 1980 and 1981 with the purpose of assisting in the implementation of SAM. The FIRCO was, in the words of a SAM strategist, "100% SAM." It attempted to induce traditional producers on rain-fed land to adopt new technological production methods by insuring their investment cost against crop failure. The insurance took into account the average value of the harvest, that is, the farmer's revenue risk, not just the cash value of the inputs applied. The goal was to assume the risks involved in adopting the new technologies in areas particularly vulnerable to weather variations. FIRCO participants received seeds and fertilizer inputs at discounts even greater than the regular SAM subsidies.

FIRCO played an important role in SAM's public relations, as an example of López Portillo's intended renovation of the historical state-peasant alliance. FIRCO coverage, however, was fairly limited, reaching a high point of 78,000 hectares in 1981. This was only 1.37% of the area covered by conventional crop insurance through ANAGSA, and it dropped almost 50% in 1982 to only 35,000 hectares (ANAGSA, 1982:43).

According to a former top ANAGSA official, BANRURAL had difficulty administering the shared-risk program and often encouraged producers to sign up with ANAGSA rather than set up another account with FIRCO. SAM created FIRCO through BANRURAL rather than through ANAGSA because planners saw ANAGSA as too inflexible. It appears, however, that SAM was unable to encourage BANRURAL to push FIRCO, leading at least one SAM planner to conclude that the shared-risk crop insurance program should have been carried out by ANAGSA rather than an entirely new agency.

The changes in credit allocation described above indicate that there was a major shift in the orientation of BANRURAL's activities when SAM became a national priority. Until that time, credit expansion primarily benefited luxury and industrial crops and livestock instead of basic grains. In 1980 the direction of expansion shifted toward rain-fed and grain-producing beneficiaries. The change was in the pattern of growth, however; it was not a redistribution away from the previously privileged sectors, which continued to receive real increases. While the data show increasing attention to maize and beans, Reyes's (1982) information shows that, in financial terms, livestock still was more important relative to crops at the end of López Portillo's term than at the beginning. Future analyses of the pattern of post-1982 cutbacks will show the degree to which the SAM-induced changes were structural, as opposed to simply being a result of temporary access to large amounts of resources.

In terms of administrative procedures, BANRURAL made some changes to fit SAM's political and economic goal of increasing attention to the rain-fed grain producers. The goals were to increase the efficiency of resource use, to reduce bureaucratic limits to credit access, and to improve the timing and amount of credit provided. Implementation measures included (1) decentralizing authority to shift resources between different crops, activities, and areas, in accordance with SAM priorities, to the regional level, (2) systematically publicizing credit-application information in advance, (3) automatically renewing credit for proven clients, and (4) doing research at the regional level to improve long-term credit use. BANRURAL also announced an attempt to increase the degree to which peasant needs were taken into account. Peasant credit requirements, expressed through local branches, were to be passed up to the local Irrigated or Rain-fed District Committee, which was to increase coordination with other state agricultural agencies, as well as with peasant organizations. The branch operational plan was then passed up to be integrated into the regional and national credit allocation plans.

By the late 1970s, BANRURAL had acquired the reputation of being a massive bureaucracy, corrupt and insensitive to peasant needs. Its national network of 634 branches was one of the principal institutions through which the government related to the peasant population on a regular basis. BANRURAL's 1980–1982 broadening of access to crop loans must be viewed, not only in the context of SAM's economic goals, but also in terms of the attempted renewal of the state-peasant alliance.

BANRURAL responded to SAM by flooding the countryside with credit. The amounts of land and number of producers financed reached all-time highs. While the establishment of the SAM strategy led to important increases in agricultural credit, there were still major problems in policy coordination, for both formulation and implementation, between the different government agriculture agencies. According to one BANRURAL analyst, production did not reach its optimal level because of "deficiencies regarding the complementarity of use of credit, insurance, and inputs" (for a further analysis of BANRURAL lending, see Chapter 5).

FIRA: Credit. FIRA, like BANRURAL, offered interest rates of 12% to any maize and bean producer. Other rates were slightly differentiated according to types of producer and activity. Middle- and high-income producers paid between 26% and 29% for loans for crops other than maize and beans. These remained the rates in 1982, after 1981's 30% inflation rate and 1982's 90–100% rate. The nominal scaling, then, was very gradual in real terms and negative interest rates prevailed.

Both FIRA's number of beneficiaries and amount of credit disbursed increased markedly during SAM. Between 1979 and 1980 the number of FIRA credit recipients shot up 79%, from 237,000 to 424,000. The share of producers considered to be of low income increased to 66.6%, but the number of higher-income producers still increased 44%. In 1981 the total number of recipients rose to 511,000, 116% over 1979, while the number of higher-income recipients fell for the first time. In 1981 FIRA created a new category of credit recipient, "other types of producers," which was essentially an admission that its hitherto middle-income category actually included a substantial number of quite high income producers. These high-income producers received almost 15% of 1981 credits (FIRA, 1982:50). During SAM, FIRA greatly increased both the amount and proportion of attention to smaller producers while still increasing attention to medium and large producers.

SAM affected FIRA's activity in terms of types of operations as well. Total FIRA credits increased 83% between the 1978–1979 and 1980–1981 crop years, with the proportion of credit going to annual crops growing from 42 to 50%. The share benefiting basic grains increased at the highest rate, 156.5%, from 21 to 30% of FIRA credit. Oilseeds and beef also increased their shares to 15 and 17%, respectively (Patron Guerra and Fuentes Navarro, 1982:6).

Implementation of SAM also involved nonfinancial operational changes. The medium-term maize and bean program was designed to increase technical as well as financial assistance. The program involved 67,000 producers in 1982, 86% of whom were considered to be of low income. They received 45% of the resources. The program as a whole covered 854,000 hectares of maize and 275,000 hectares of beans.

FIRA also introduced administrative changes to increase support services for low income rain-fed producers. New mechanisms for commercial credit were introduced to promote the new Agricultural Development Law (Ley de Fomento Agropecuario). Because of the rapid expansion of credit, however, only an estimated 25% of 1981 credits were actually supervised by FIRA. The rest were reportedly supervised either by BANRURAL or not at all.

In order to evaluate its maize and bean program, FIRA drew a representative sample from 97,000 clients during 1980–1981. All were considered to be of low or middle income. FIRA defined low income as an annual net income of under 1000 times the daily minimum wage of the region, while medium income was considered to be annual net earnings of between 1000 and 3000 times the local minimum wage. It should be noted that some observers consider this cutoff to be relatively high, since few rural wage workers are employed year-round and most do not receive the minimum wage when they do work. The calculation of net income is another possible complication. Since it is in the producer's interest to undervalue the estimate of net income in order to receive discounted credit, the actual implementation of the income cutoffs is very sensitive to local relations between producers and branch bank officials.

The poll results indicate that 40% of the sample had not received credit during the previous year, and 75% of them were of low income. Of the total, 35% increased their area cultivated, and 64% of this group were of low income. In addition, 90% of the area incorporated into production belonged to producers with holdings of more than 10 hectares. In terms of productivity, 46% of the borrowers increased their crop yield, 61% of whom were of low income. The correspondence between long-term credit and yield was indicated by the finding that 92% of the long-term borrowers noted yield increases, which primarily came from irrigation and mechanization.

The FIRA study provides important data regarding access to inputs and commercialization. Of those producers who sold their production, only 46% had access to the guaranteed price or its equivalent. The great majority of these (70%) were not low-income producers. This group showed a large increase in prices received and share of harvest marketed in comparison to the previous year. In terms of inputs, a direct relation was found between access, degree of organization of the borrowers, and income. The producers without access tended to be unorganized and of low income.

FIRA, like BANRURAL, was able to use the large increase in financial resources to expand its loans to the new target group, as well as to its old clientele. As the FIRA poll and BANRURAL data indicate, the public credit institutions poured credit into the countryside; this enabled the politics of credit allocation to be a positive-sum game.

ANAGSA: Insurance. As part of SAM, ANAGSA reduced its premiums from 20 to 3% for rain-fed maize, beans, rice, and wheat, as well as for irrigated maize and beans on plots under 20 hectares. With the Ministry of Finance covering the cost, area covered and amount of coverage both shot up. All publicly financed crop loans must carry ANAGSA insurance; an expansion in credit thus means an expansion in insurance.

In 1980 ANAGSA insured 4.6 million hectares, 54.1% over 1979. Insured area increased another 48.4% in 1981 to 6.9 million hectares. The coverage in 1982 reached a record 7.6 million hectares, a 10.6% increase (FEP, 1983:454).

SAM's focus on basic grains changed ANAGSA's mix of crop coverage. In 1979, 48% of ANAGSA's insured area was in maize, beans, rice, and wheat, rising to 69.6% in 1981. This meant a 240% increase in area, from 1.4 million hectares in maize, beans, rice, and wheat in 1979 to 4.8 million hectares in 1981 (FEP, 1983:454). The share of insured maize area jumped from an average of 40% during the late 1970s to 69% in 1980 (FEP, I, 1981:322).

ANAGSA's expansion of crop coverage was accompanied by a rapid increase in livestock and head-of-household life insurance coverage as well. While the expansion of livestock insurance was not a SAM goal, expansion of access to life insurance was consistent with SAM's more political goals. A top ANAGSA official held that all categories of insurance went up because of the general increase in promotion.

In order to cope with the massive expansion in its coverage, ANAGSA carried out major changes in its structure and procedures, including regional decentralization and increases in personnel and supervision in the field. In part because of the newly increased coverage, ANAGSA needed to ascertain each stage of investment that might be claimed as a loss. An ANAGSA official held that supervision of credit use by both bank and insurance officials reduces the opportunities for corruption through false claims.

SAM created the political conditions that allowed ANAGSA to succeed with its decade-long effort to get the legislature to adopt a new agricultural insurance law, which went into effect in early 1981. Its provisions included the principle of "hectare lost, hectare paid," instead of the previous all-or-nothing loss principle. Coverage was also extended to the producer's entire outlay. Before, coverage began only if 75% of the seed germinated, and then it only insured 70% of the crop cost in irrigated zones and 50–60% in dry land. For the first time, BANRURAL loans could be paid back in full for crops that were lost. This dramatically increased its recovery rate. According to an ANAGSA source, the new law increased BANRURAL's recovery rates from 65% to over 90%. ANAGSA earned a financial surplus in 1980 and 1981 for the first time in ten years, but the implementation of the new law involved shifting BANRURAL's losses to ANAGSA's accounts, which were then covered by the central government.

Agricultural Production

PRONAGRA: Farming. While PRONAGRA's growth increased rapidly because of SAM's push for production of basic grains, it did not take on national significance. Area farmed in 1982 reached 127,000 hectares, a 236% increase over 1979. Production grew even more; 281,000 tons in 1982 was a 1300% increase over 1979, consisting overwhelmingly of rice, in addition to some maize and sorghum (Cabrera Morales, 1982).

Commercialization

CONASUPO: Crop Purchasing. One of the most important single measures in the SAM program was the increase in guaranteed prices for basic food crops. The support prices were set by a consensus of a commission of the Agricultural Cabinet, which was made up of top representatives from the ministries of Agriculture, Finance, and Internal Commerce, as well as CONASUPO, BANRURAL, and, informally between 1980 and 1982, the SAM leadership. The commission had generally discussed proposals for price increases coming from SARH, which reportedly often met resistance from the Ministry of Internal Commerce, since it would be responsible for having to increase tortilla subsidies. BANRURAL reportedly tended to support SARH, and CONASUPO fell in between. According to one member of the commission, SAM "arrived with a lot of political force," shifting the center of gravity of discussion greatly in favor of increased prices to stimulate production.

In 1980 nominal prices were raised 28% for maize, 55% for beans, 18% for wheat, and 24% for sorghum. Since inflation was 28% in 1980, this meant more of a halt in the decline of the prices than increases in constant terms. Adjusted for inflation, beans increased 20%, but maize only 1%, while wheat dropped 7% and sorghum 2%. In 1981, however, nominal prices were raised again, 47% for maize, 33% for beans, 31% for wheat, and 36% for sorghum. Increases in constant-price terms were 15% for maize, 4% for beans, 1% for wheat, and 6% for sorghum. By comparison, with the 1965–1969 price average as an index, the 1980 index of prices in real terms for maize was still 81, rising to 94 in 1981. It should be noted that while sorghum support prices tended to keep up, SAM strategists recognized that that policy did not slow sorghum's competition with maize. Increased maize production by commercially oriented producers depends to some degree on a favorable price relationship with the highly substitutable sorghum.

Mexico's inflation rate in 1982 was unexpectedly high, reaching 90–100%. Nevertheless, the support price for maize was raised only 35% at a time when inflation was expected to hit 50 to 60% that year, which meant a drop to an all-time low. Other nominal support-price increases followed the same pattern, leading to the lowest constant prices offered in decades. (SARH/DGEA, "La Determinación de los Precios de Garantía para los Productos del Campo," November 1982).

The earlier SAM price increases led to substantial increases in CONASUPO's share of national crop markets. In 1981 CONASUPO bought 19.7% of that year's bumper maize crop, 2.9 million tons. This was more than three times the amount purchased in 1980, which was only 7% of a much smaller crop. CONASUPO's share of beans was 13.6% in 1980 and 35.8% in 1981, wheat held to 40% in 1981, and the share of sorghum production shot up to 37.8% (CONASUPO, *Sistema C,* May 1982).

In 1980, the concentration of maize purchase in five states increased to 81.2%, as Chiapas increased in importance and Tamaulipas and México decreased. In 1981, however, the top five states' share decreased to 68.9%, while the total volume purchased tripled (CONASUPO, *Sistema C,* July 1982).

On the demand side, CONASUPO's share was substantially higher because of increased imports. Its share of the national maize supply was 26.2% in 1980 and 25.4% in 1981. In response to the poor harvest of the 1979–1980 crop, however, maize imports in 1980 tripled to 3.2 million tons, and almost 2.5 million tons more were imported in 1981. Because of the record crop of 1981, this led to the build-up of record maize reserves, 1.95 million tons (CONASUPO, *Sistema C,* May 1982). The total food import bill did fall, however, from over U.S. $3 billion in 1980 to $2 billion in 1981 and an estimated $1 billion in 1982 (*Business Week,* 20 Dec. 1982).

The extent of the import crisis of 1979–1980 was not only a major cause of the adoption of the SAM strategy, it also led to a change in the way Mexico imported grain. The magnitude of the imports gave Mexico's trade decisions increased influence in the world grain market. In 1980, CONASUPO created the Foreign Trade Coordinating Body, which studied and coordinated the use of futures markets in international grain trading. CONASUPO purchased approximately one-third of its grain imports on the futures market in 1981. After the unit thereby saved several million dollars, it was converted from an office to a full department.

The Foreign Trade Coordinating Body was formed in coordination with, but not as a response to, SAM. Its director had previously worked in the office of the presidency, the institutional location of the SAM planners. CONASUPO's formation of a unit to enter the grain futures market systematically was not inconsistent with the concurrent strategy aimed at self-sufficiency. Given the vagaries of the weather in Mexico, even a consistently applied self-sufficiency strategy could not achieve *total* freedom from the need to import, and the new unit better equipped CONASUPO for the management of import flows at any volume (Austin and Hoadley, 1987).

ANDSA (CONASUPO): Storage. The 1979–1980 import crunch strained ANDSA's capacity, as it did all of Mexico's storage and transportation infrastructure. There was an investment in additional capacity during SAM, increasing 3.4% between 1979 and 1981. In 1982 ANDSA's capacity grew another 10.1% to 4.4 million tons (Merino Castrejon, 1982:15).

BORUCONSA (CONASUPO): Storage. BORUCONSA's importance increased substantially because of SAM. Between 1979 and 1981, the number of rural warehouses increased 10.3%, from 1528 to 1686, while total capacity grew 31.3%, from 1.49 million to 1.96 million tons, aside from the reception centers. The actual tonnage of crops in storage rose 79.4% between 1979 and 1981, expanding the utilization of capacity from 29.6 to 40.4%. This increase reflects the combined effects of the massive 1981 production increase and CONASUPO's increased intervention in the market.

Through the Rural Commercialization Assistance Program (PACE), BORUCONSA subsidized producers' maize transportation costs in an attempt to broaden effective access to the official price. During 1979 and 1980, the PACE program underwent a change of orientation that led to an increase in its share of BORUCONSA purchases to 42.8% for the 1979–1980 crop year. The procedural changes included the extension of reimbursements to include initial processing services and access to the transportation subsidy for small private farmers in addition to *ejidal* producers, exclusively for producers of rain-fed maize and beans. There was no limit on the size of producer that could take advantage of the program, but since 1982, PACE covered only the first 50 tons of maize delivered. Between 1980 and 1981, the amount of maize purchased through the PACE program jumped more than 400%, from 198,000 to 807,000 tons. PACE then accounted for 27.7% of total national CONASUPO maize purchases. Maize purchases receiving PACE services increased another 110% in the 1981–1982 cycle to 1.7 million tons, about 80% of BORUCONSA's total purchases and almost as much as BORUCONSA's total pre-SAM high in 1978. The numbers of producers involved increased from 11,708 in 1980 to 339,000 in the 1981–1982 cycle, which largely accounts for the massive increase in amount of maize serviced (Rubio Canales, 1982).

During SAM, PACE grew from a small program extended to a limited number of states to one of national scope with the goal of providing a full range of commercialization services to rain-fed producers. PACE's growth was reinforced by the change in national strategy, which in turn facilitated PACE's procedural changes and increased orientation towards SAM's target group.

IMPECSA (CONASUPO): Wholesaling. IMPECSA's participation in wholesale marketing to private reatilers increased substantially during SAM. The number of establishments attended rose from 40,000 to 130,000 between 1979 and 1982, raising IMPECSA's share of total retailers served from 15.8% to an estimated 52.8%. The most rapid growth period was between 1979 and 1980, when sales quadrupled, but sales tripled from 1980 to 1982. IMPECSA's expansion was an important part of SAM's increased distribution of the subsidized basic market basket of food (CONASUPO, *Sistema C,* 1(3), Nov-Dec. 1981).

DICONSA (CONASUPO): Retailing. Before the institution of SAM, DICONSA, like IMPECSA, had begun (in conjunction with CO-PLAMAR, a government undertaking aimed at assisting rural low-income groups) a large expansion of its rural distribution network. It increased its rate of growth significantly in conjunction with SAM. To begin with, SAM defined its target population for consumption subsidies in terms very similar to DICONSA's. The DICONSA-COPLAMAR target group of those in extreme poverty was 20.8 million, while SAM considered 19 million Mexicans to be severely malnourished. In terms of geographic distribution of this population, 96% of SAM's designated "critical popula-tion" was similarly designated by DICONSA-COPLAMAR.

SAM targeted all low-income rural communities for distribution cover-age except those of less than 500 inhabitants or not accessible year-round. The rural expansion of DICONSA's network indicated increasing cover-age of the estimated 13 million malnourished in the countryside. The total number of DICONSA outlets grew from 6660 in 1979 (71.7% rural) to 8369 in 1980 (76.4% rural). By the end of 1982, DICONSA had established 11,201 stores, 9049 of which were rural (81%). Rural sales increased from 17.7% of total sales in 1979 to 18.3% in 1980 and 21% in 1982 (FEP, 1983: 178). DICONSA estimated that its share of the rural food market in-creased from 11% in 1980 to 17% in 1982. The rural DICONSA price of the basic market basket of food was estimated at 30 to 35% below the rural open-market price (DICONSA, 1982).

As part of its expanded rural network, DICONSA changed its pro-cedures to achieve greater community participation in its operations. The increased local input into solving problems and determining needs was to improve efficiency and control over the flow of goods. Peasant com-munities assembled and elected councils, which were to make operational decisions together with DICONSA. The CONASUPO-COPLAMAR pro-gram considered these Community Supply Councils to be a form of cores-ponsibility over the task of ensuring the delivery of subsidized food to the target groups. According to a DICONSA planner, the program was par-ticularly successful in those areas where communities organized region-wide and took the initiative to deal with problems. Some elements of DICONSA's large administrative and operational apparatus, however, did not share the planners' orientation toward a program of such social character. In addition, necessary coordination with the ministries of Agriculture and Finance was difficult. As the DICONSA planner des-cribed the limitations to SAM in general, "this was the big problem—even the President could not break the inertia of the bureaucracy."

DICONSA responded massively to the change in national strategy, in particular where the SAM approach converged with its own. Since DICONSA did not require special funding through SAM, however, SAM's contribution to DICONSA's expansion was through political rather than economic resources. SAM and DICONSA shared a common view of the proper role of the state in the countryside, and of the need for a comprehensive approach to the food system, from production through consumption.

Processing

The processing activities were handled by four other CONASUPO sub-sidaries (ICONSA, MINSA, TRICONSA, and LICONSA) and three ad-ditional SOEs.

ICONSA: Food Processing. ICONSA produced significant shares of most of the processed products in the SAM basic market basket of food. Sales in 1980 were 51% over 1979, and in 1981 they were 26% over 1980. Volume of production increased 27% in 1980 and 13% in 1981, reaching 831,000 tons. Substantial investment increased ICONSA's share of national food-processing capacity during SAM. By 1981 its share of oilseed-milling capacity rose to 16.5%, oil refining to 21%, and vegetable fat to 14%. ICONSA's share of wheat flour production grew to 8.9% of the national market, and maize flour to 8.7%. By 1982 ICONSA estimated that its products reached 6 million largely low-income consumers, through IMPECSA and DICONSA channels (CONASUPO, *Sistema C,* May-June 1982).

MINSA: Maize Flour. Between 1980 and 1982, MINSA production grew at 5%, from 285,000 to 300,000 tons. This increased its share of the maize flour market from 27.9 to 28.5%. The share of MINSA production distributed through IMPECSA and DICONSA increased slightly to 23.5% of production in 1982 (CONASUPO, 1982a:46).

TRICONSA: Bread. Total bread production increased 14% between 1979 and 1982. After a 6% drop in TRICONSA production in 1980, which cut the market share from 7.7 to 7%, by 1981 production had increased 15%, bringing the market share to 7.8% (CONASUPO, 1982b: annex 22).

LICONSA: Milk. Milk consumption continued its rapid growth into the SAM period, during which LICONSA's share of the market increased substantially. While consumption of rehydrated and pasteurized milk increased at over 4% annually during SAM, LICONSA's share of sales rose to 17% in 1980 and 23% in 1981. In 1980 approximately 1.0 million liters per day were reconstituted from powdered milk, enriched, and sold to selected low-income families through 370 special stores in Mexico City at 29% of the price of commercial milk (CONASUPO, 1980). LICONSA production of concentrated milk increased 174% from 1979 to 1981, more than doubling its share of the market from 21% to 44% (CONASUPO, 1982b: annex 22). This rapid growth increased the deficit in national production. In 1980, for example, imports increased 148% over 1979 (Santoyo and Urquiaga, 1982).

Since milk was one of the basic products targeted for self-sufficiency, the SAM strategy considered LICONSA "an efficient mechanism . . . to stimulate primary production" (CONASUPO, *Sistema C,* Jan.-Feb. 1982). LICONSA promoted fresh-milk production within the context of SAM's encouragement of vertically integrated food systems; it expanded its network of refrigerated collection centers to forty and provided support prices and inputs to stimulate small- and medium-sized dairy producers. LICONSA's role within national production grew; the amount of fresh milk collected increased 140% in 1981 and another 80% in 1982 (CONASUPO, 1982b: annex 26).

ALBAMEX, FERMEX: Animal Feed. After a slight drop in 1981, ALBAMEX's production of balanced feed reached 420,000 tons in 1982, 20% over 1980. The increase in grain production during SAM led to an increase in ALBAMEX's capacity utilization. ALBAMEX's growth was constrained, however, by its inability to break into the most important market segment for feed—poultry. In order to compete with the transnationals' completely vertically integrated line of poultry stock and inputs, ALBAMEX planned to branch out into the provision of poultry stock through its new subsidiary, NUTRIMEX.

FERMEX was created in 1981 as a joint venture between ALBAMEX and Japanese capital (40%). It was designed to lessen Mexico's dependence on foreign technology in the livestock sector by producing two amino acids, primarily lycine, for the feed industry.

NUTRIMEX: Fortified Foods. NUTRIMEX considered itself "100% SAM"; it was created in direct response to SAM. Its goal was to improve the nutritional content of the Mexican diet. It began food-enrichment projects to target specific zones and commodities. Yucatán was the first state to receive sugar fortified with vitamins A and C and niacin. Since SOEs process most Mexican sugar, access was not a problem. The technology to produce fortified sugar, which is no different in color or flavor, was available and not overly complex. By 1982 NUTRIMEX had the production capacity to fortify 20% of national maize flour production with tryptophan, an amino acid that maize lacks. The fortified flour was sold through the DICONSA network. Unlike sugar processing, maize processing is highly decentralized and dominated by the private sector, making distribution of fortified maize flour more difficult. NUTRIMEX's other main food product was a rehydratedable powdered baby food made from soy protein concentrate, imported nonfat dry milk, and sugar.

NUTRIMEX also adopted a strategy to develop breeder stock in poultry and swine, but this was stalled by the 1982 budget cuts. The goal had been to supply, eventually through small producers, fresh milk and 10% of the national demand for meat, with an emphasis on the "low end" of the market. SAM planners hoped that if SOEs could offer a vertically integrated line of livestock inputs, then dependence on transnational corporations could be lessened.

NUTRIMEX distributed its food products through two channels. Low-income and rural consumers were reached through small stores supplied or operated by DICONSA and IMPECSA. NUTRIMEX also sold through private supermarket chains. Although intended to benefit low-income people, it tried to avoid the perception of its products as inferior goods. Nutrimex intended to show the viability of manufacturing and marketing nutritionally fortified foods, in the hope that if its products were competitive, fortification would be adopted by the private sector.

Patterns of SOE Responses

In this section we categorize SOE responses during the SAM period and discuss the determinants of these responses (see Table 4.2). These SOE actions were not always entirely due to SAM: sometimes they were responses to other pressures. Responses to SAM were both qualitative and quantitative. Qualitative change is here defined as the creation of a totally new SOE activity, a major change in the clientele served by the existing activity, or a major change in the way economic activity is carried out (i.e., administrative procedures). Quantitative change is defined as an increase in output, sales, market share, or other indicators of economic activity during the period 1980–1982.

A comparison of SOE actions before and after the adoption of the SAM policy shows that most food sector SOEs increased their level of activity, some quite substantially, but without changing their basic orientation. Others did change the nature of their activities quite fundamentally during the López Portillo administration, but several had changed their orientation well before SAM was adopted (e.g., IMPECSA, DICONSA). Others (e.g., BANRURAL, BORUCONSA-PACE) increased their scope of activity dramatically without neglecting their traditional clientele, but because this expansion increased attention substantially to new groups, it has to be considered both qualitative and quantitative.

The designation of qualitative change does not imply structural or permanent change. The degree to which structural change occurred is much more difficult to assess and would probably be made clearer with an analysis of the nature of the post-1982 cutbacks in both amount and nature of SOE activity. NUTRIMEX, for example, was unable to follow through with its planned expansion in direct livestock production because of budget cuts. BANRURAL, under pressure to lend only to "credit-worthy" producers, cut back access to credit substantially in 1983. The de-

Table 4.2. Variation in SOE response to SAM

Quantitative change only	Quantitative and qualitative change
FERTIMEX	PRONASE
SIDENA	BANRURAL
PRONAGRA	FIRA
ANDSA	ANAGSA
IMPECSA	FTA
ICONSA	BORUCONSA
MINSA	DICONSA
TRICONSA	LICONSA
ALBAMEX	FERMEX
	NUTRIMEX

gree of structural change could be measured by future examination into which of the qualitative SAM changes remained intact after 1982.

Our analysis of the SOE responses are based on a combination of recorded data on 1980–1982 SOE actions and structured interviews with direct participants in SAM's formulation and implementation. Among these participants are high-level managers of three-fourths of the principal SOEs active in the food sector. We evaluate the factors facilitating and inhibiting implementation in terms of the three categories used earlier: temporal context, bureaucratic and political incentives, and availability of economic resources. There are no precise boundaries between these categories; each feeds back into the others, as we see below.

Temporal Context. The 1979 grain shortfall, as important as it may have been at the level of policy making, did not create a "crisis-response context" that in turn spurred the SOEs' propensity to act. In the eyes of most SOE managers, 1979 was simply another bad crop year, natural to rain-fed agriculture.

On the inhibiting side, the timing of the decision to adopt the SAM strategy caused three kinds of problems. First, there were technical constraints that limited SOEs' ability to act. Fertilizer production capacity, for example, cannot be rapidly increased on short notice. Second, the short time frame limited the SAM strategists' ability to consult fully at the operational level of the SOEs. Where there was consultation before the March 1980 announcement, SOE response was facilitated. But there were missed opportunities and inefficiencies where consultation did not take place, as in the placement of the shared-risk crop insurance program in BANRURAL rather than ANAGSA, or in the financial cost of not planning ahead with FERTIMEX. Third, the adoption of the SAM strategy with only two and one-half years remaining in the administration limited the possibility of SOE structural changes. Furthermore, where fundamental changes were necessary to achieve SAM goals, the short time frame also reduced the incentive to make them, since most top personnel change with the change in administration.

Bureaucratic and Political Incentives. Regarding bureaucratic incentives, there was unanimity that the presidential priority SAM received was crucial. The presidential announcement of SAM in a strongly nationalist context helped to overcome some of the problems caused by the lack of consultation at the operational level. There was, as a result, a consensus around the desirability of national food self-sufficiency (although there was strong debate after 1982 about its feasibility and cost). This meant that the food sector SOEs all proclaimed their implementation of SAM. SAM's political life was prolonged well into the 1982 lame-duck period by the success of the 1981 harvest, but it ran into serious problems when the 1982 crop shortfall became apparent and a new round of massive grain imports was required. Since so much political and economic capital was invested in pursuit of aggregate production increases, the program as a whole was judged in the context of unmet 1982 quantitative goals.

Because the presidential initiative did not extend to changes in the operations of particular SOEs whose orientations were not necessarily consistent with SAM, the SAM label was often put on activities associated with pre-SAM food policy. This brings us to the point at which the issue of the importance of the presidential blessing overlaps with the second major bureaucratic factor affecting SOE response: previous SOE policy orientation. SARH, for example, was concerned not only because SAM could be seen as a bureaucratic competitor, but also because SARH had its own definite policy agenda, one that differed in many important ways from SAM's. SARH's approach tended to emphasize production efficiency and commercial producers, while SAM planners tended to be more interested in improving income distribution. As a result of the way SAM was translated from the drawing board to the countryside, some SOE administrators and food system analysts considered SAM to have been, in effect, co-opted by SARH.

The SOEs that responded most strongly to SAM were those whose own orientation coincided with, or at least complemented, the SAM strategy. The growth of DICONSA, IMPECSA, and BORUCONSA-PACE reflected their increased political importance resulting from policy orientations that coincided closely with the SAM strategy. The most important SOE that responded strongly with a complementary but distinct orientation was ANAGSA. The increased emphasis on agriculture after 1980 permitted ANAGSA to get the legislature to approve the only law not initiated by the presidency. That law changed ANAGSA's procedures, increased insurance coverage, and shifted much of the financial burden of nonpayment of crop loans from the producers and BANRURAL to ANAGSA.

The third major political factor inhibiting implementation of SAM was the lack of participation by the target group. Organized pariticipation by small producers would have created political incentives for SOEs to give them increased attention, as well as disincentives for SOEs that neglected peasant needs. Had new peasant credit recipients received increased BANRURAL attention because of a basic change in SOE procedures that integrated them into allocation decisions, the target group's vulnerability to the post-1982 cutbacks would have been less. Instead, it appears that they received increased attention only because of a temporary availability of extra resources from the central government.

Economic Resources. SOEs with policy orientations different from SAM's, particularly those that served primarily producers other than peasants who sow grain, were not forced to abandon their more privileged constituencies, either politically or economically. The policy was to proclaim "We are all SAM" rather than to encourage divisiveness between small and large landowners, or between peasants and ranchers. FIRA and

BANRURAL, for example, were able to respond to SAM's change in priorities by changing the distribution of their growing budgets, not by reducing or substantially changing their credit-allocation process. BANRURAL's percentage of crop loans for grain producers by 1982 was not very different from what it had been at the beginning of the administration, but by increasing the amount and importance given to agricultural credit, SAM was able to influence the credit's priorities.

Central government transfers to food sector SOEs permitted the SAM policy makers to choose indiscriminate rather than targeted subsidies, which in turn created an important incentive for SOE implementation of SAM. Subsidies with access limited on the basis of need, however, would have reduced SAM's political acceptability to the beneficaries of previous agricultural policy, that is, wealthier commercial producers. Targeted subsidies would also have required fundamental changes in the structures and procedures of SOEs whose pre-SAM orientation was designed to benefit other types of producers. Whatever the combination of short-term political and economic pressures to choose indiscriminate subsidies as a policy tool, the choice left the SAM strategy extremely vulnerable to the budget cuts that began in 1981 and 1982. The data surveyed above show that almost all SOE activity grew less in 1982 than in 1981. The drop in real prices offered for basic grains was perhaps most striking, hurting those least able to switch to more profitable crops, the small peasant producers. Those qualitative changes that were carried out, such as the broadening of credit and insurance access to peasants, except for those embodied in law, were particularly vulnerable to budget cuts. This was precisely because of the implementation strategy of adding on new groups instead of changing the structures and procedures through which SOEs allocated goods and services in the food sector.

Conclusions and Implications for Policy Implementation

From our research and findings one can draw some general conclusions and implications regarding the use of SOEs as implementers of Mexican food policy.

Responsiveness. The Mexican food sector SOEs demonstrated a clear capacity to respond significantly to major shifts in food policy. They were able to greatly augment their levels of activity in terms of production and delivery of goods and services and the number of clients served. The SOEs also showed that they could respond relatively quickly.

The SOEs should not be viewed, however, as instruments that respond automatically, willingly, or even efficiently. Their responsiveness is shaped by the positive and negative incentives for change that they face. Each SOE has its own institutional agenda, and its managers have individual priorities. Any new policy thrust impinges on these institutional and individual agendas and priorities, and the degree of congruence shapes the responsiveness. It should also be recognized that SOEs have a natural tendency to bend the policy in the implementation process toward their own inclinations, which may deviate from the original intentions of the policy makers. This implies that policy makers should attempt to assess the degree of congruence of the policy with the SOEs, shape a set of incentives to motivate adherence, and devise a control system that can monitor implementation.

Conclusions regarding the efficiency with which the SOEs can implement food policy cannot be derived from our research. Chapter 11, on production and costs of SAM, addresses the general issue. Anecdotal evidence points to examples of corruption and wastefulness in implementing SAM, although these may have been insignificant compared to the magnitude of the program. Nonetheless, they do point to the need for a strong audit and control system for the financial aspects of SOE operations.

Political Support. Within the Mexican political system a major shift in food policy strategy must have strong and visible presidential support if it is to elicit a positive response from the SOEs. Although SOEs operate to a significant degree as autonomous economic entities, they and their managers, who are political appointees rather than tenured civil servants, are quite responsive to presidential priorities. Food policy emanating, for example, from the secretary of agriculture, will simply not elicit the same degree of response from the SOEs as that coming from the presidency.

Economic Resources. The SOEs do have the capacity to transform increased economic resources into increased goods and services. Added financial resources allow SOEs to implement new policies less painfully. The SOEs can increase their services to the new priority group (small producers) without decreasing their services to a traditional clientele (larger farmers). This facilitates the SOE's task of managing its multiple constituencies and therefore its receptivity toward the new policy.

The negative side is the fiscal burden. When the subsidies that are channeled through SOEs are generalized rather than targeted, the government pays a heavy price. On grounds of social equity there is little justification for subsidizing higher-income producers or consumers. Politically it may be deemed necessary, but its costs can be severe. If the government decides to institute selective subsidies, then it should expect possible resistance from those SOEs that may be forced to reduce benefits to traditional clients.

Timing. The timing of policy shifts affects the SOE's efficiency and motivation. The shorter the lead time between the policy decision and its implementation, the greater the likelihood of inefficiency. Similarly, the greater the divergence of the new food policy from the old, the more chance for problems in the adjustment process. New strategies generally require new structures or administrative procedures in order to be implemented effectively and efficiently. The SOEs require time to make such organizational changes.

In addition to the lead-time effects on efficiency, there is the impact on SOE behavior due to the timing of the policy change in relation to the political cycle. The later the policy is announced during the six-year presidential term, the more uncertain is the SOE commitment to the change. There is a tendency to become more risk-averse or cautious as the change in administration moves closer. In part this is because of the traditional lack of continuity of policies across *sexenios,* and therefore the reluctance to be overly identified with a policy that might not be acceptable to the incoming president. Thus, in terms of SOE response, the earlier in the political cycle one can launch the new strategy, the better.

Participation. To help ensure the effectiveness and efficiency of the implementation of policy, SOEs should be involved in the process of formulating policy. As key implementers, SOEs have a clearer understanding of the realities in the field, and these realities should be considered in policy formation to ensure that the proposed actions are feasible. Macro policy must be fused with micro realities, and SOEs can help in the planning of the fusion.

Not only should the top managers of SOEs be consulted, but the operation-level officials as well. Participation by SOE personnel increases their commitment to the policy and thereby makes adherence more probable as well as more efficient.

Effective policy implementation also requires joint understanding and commitment between SOEs and clients. If the policy involves a shift in attention from one clientele to another, however, the mobilization and participation of the new beneficiaries in program decision making and operations may be essential to implementing the policy change. This may involve friction and conflict in the struggle to develop new modes of interaction, but if the clientele are to change, so too must the SOEs. In the end, it is the degree of commitment and consensus among policy makers, SOEs, and clients that significantly determines the effectiveness of food policy implementation.

Bibliography

AMIA Boletin. Jan. 1983. Mexico City: Asociación Mexicana de le Industria Automotriz.

ANAGSA. 1982. "Estructura, organización y funciones de la aseguradora nacional agrícola y ganadera." Mexico City: COPIDER-Harvard Workshop.

Austin, James, and Kenneth Hoadley. 1987. "State Trading and the Futures Market." In Bruce F. Johnston, Cassio Luiselli, Roger Norton, Celso Cartas Contreras, eds., *U.S.-Mexico Relations: Agriculture and Rural Development.* Stanford: Stanford University Press.

Bailey, John. 1980. "The Presidency, Bureaucracy, and Administration Reform in Mexico: The Case of the Secretariat of Programming and Budget." *Inter-American Economic Review* 34(1).

Banco de México. 1981. *Informe anual 1981.* Mexico City: BM.

Barenstein, Jorge. 1982. *La gestión de empresas públicas en Mexico.* Mexico City: CIDE.

Barkin, David, and Blanca Suarez. 1982. *El fin de la autosuficiencia alimentaria.* Mexico City: Centro de Ecodesarollo/Nuevo Imagen.

Cabrera Morales, Jorge. 1982. *Que es PRONAGRA?* Mexico City: PRONAGRA.

Chandler, Alfred. 1962. *Strategy and Structure.* Cambridge: MIT Press.

Cohen, Rafael, and Kent Thirty. 1983. "Mexico: Crisis of Confidence." Harvard Business School Case 0-383-148.

CONASUPO. 1980. *Informe anual.* Mexico City: CONASUPO.

_____. 1980-1982. *Sistema C.* Mexico City: CONASUPO.

_____. 1982a. *La intervención del estado en el abasto y la Regulación del mercado de productos básicos.* Mexico City: Coordinación de Información y Publicidad de Conasupo.

_____. 1982b. "Producción nacional, consumo nacional aparente y Operaciones CONASUPO en compras y ventas, 1965-1981." Mimeograph. CONASUPO.

_____. 1982c. "Producción anual y comercialización de granos y semillas oleaginosas por CONASUPO, compras-ventas por entidad federativa, 1971-1981." Mimeograph. CONASUPO.

DICONSA. 1982. "Informe de DICONSA-COPLAMAR." COPIDER-Harvard Workshop. FEP. 1981-1984. *Informe de Gobierno, 1980.* Mexico City: Federal Executive Power.

FERTIMEX. 1980. *Memoria, 1980.* Mexico City: FERTIMEX.

_____.1982a. *Testimonio de una administración, 1976-1982.* Mexico City: FERTIMEX.

_____. 1982b. "Apoyo al SAM." COPIDER-Harvard Workshop.

FIRA. 1982a. *Informe anual—1981.* Mexico City: Banco de México.

_____. 1982b. *"Datos básicos."* COPIDER-Harvard Workshop.

Hill, Raymond. 1981. "State Enterprise and Income Distribution in Mexico." Economics Department Working Paper, Woodrow Wilson School, Princeton University.

Ickis, John. 1978. "Strategy and Structure in Rural Development." D.B.A. diss., Harvard Graduate School of Business Administration.

Jones, Leroy, et al., eds. 1982. *Public Enterprise in Less-Developed Countries.* New York: Cambridge University Press.

Luiselli, Cassio. 1980. "Agricultura y alimentación: Premisas para una nueva estrategia." In Nora Lustig, ed., *Panorama y Perspectivas de le economía mexicana.* Mexico City: El Colegio de Mexico.

Mercado de Valores. 1980-1982. Mexico City: NAFINSA.

Merino Castrejon, José. 1982. "Almacenamiento de productos básicos para la alimentación." COPIDER-Harvard Workshop.

Patron Guerra, Fernando, and Jesus Fuentes Navarro. 1982. "Acciones y coordinación FIRA en relación con el Sistema Alimentario Mexicano, 1980-82." COPIDER-Harvard Workshop.

PRONASE. 1982. *Origen, desarrollo y producción de productora nacional de semillas.* Mexico City: SARH/PRONASE.

Reyes, Antonio. 1982. "La estructura y dinámica crediticia oficial en el marco del Sistema Alimentario Mexicano." COPIDER-Harvard Workshop.

Rubio Canales, Abraham. 1982. "La participación del programa de apoyo a la comercialización rural (PACE) dentro de la estrategia SAM." COPIDER-Harvard Workshop.

Santoyo Meza, Enrique, and José Urquiaga. 1982. "Sistema LICONSA-CIDER." COPIDER-Harvard Workshop.

SARH. *Informe de labores.* 1981. Mexico City: Secretaría de Agricultura y Recursos Hidraulicos.

———. 1982a. "La determinación de los precios de garantía para los productos del campo." Mexico City: Secretaría de Agricultura y Recursos Hidraulicos.

———. 1982b. *Memoria 1977–1982.* Mexico City: Secretaria de Agricultura y Recursos Hidraulicos.

SPP. 1981. *El sector alimentario en Mexico.* Mexico City: Secretaria de Programación y Presupuesto.

———. 1982. *El papel del sector público en la economia mexicana.* Mexico City: Secretaría de Programación y Presupuesto.

CHAPTER 5

Channeling Credit to the Countryside

Raul Pessah

Agricultural and livestock credit has traditionally been scarce and expensive, and for several decades Mexico's financial system has been supporting primarily the most dynamic and profitable sectors. This emphasis changed under SAM, which saw credit as central to implementing its strategy of increasing poor peasants' production of basic grains. In order to assess how well SAM's new strategy was actually implemented by the government's agricultural banks, we need to look at Mexico's earlier agricultural credit allocation and specifically to compare the SAM period of 1980–1982 to the years immediately preceding it in President López Portillo's 1976–1982 term in office.

Historical Backdrop

The financial system was historically developed as the core of the postwar development model fostered in most countries that, like Mexico, were at the turning point between colonial past and a late-emerging capitalism. In this model, financial resources were devoted to achieving import substitution, accelerated industrial growth, and, consequently, the support of the infrastructure required to sustain urban development.

In this context, little credit flowed to farmers, since agriculture and livestock production require longer terms and offer an uncertain recovery. Commercial banks were more inclined to support industry and trade than primary activities. This left the bulk of peasant producers in the hands of local moneylenders.

The Mexican state, as manager of the economy, tried to fill this vacuum by creating and operating agricultural development banks. It financed this sector's activities by creating Banco Nacional de Crédito Agrícola in 1926, Banco Nacional de Crédito Ejidal in 1936, and Banco Nacional Agropecuario in 1965.

For three and a half decades, the growth of this sector's financing was slower than both the dynamics of the sector itself and the demographic growth. This no doubt contributed to the crisis in basic food production in the beginning of the 1970s. From 1970 to 1975, credits extended by the three official banks, at constant prices, grew jointly at an average annual rate of 15%. Although this reflects more dynamism than in previous years, in terms of production the results did not show similiar growth, because the official banks' lack of coordination prevented them from implementing an adequate financial policy. This, in turn, caused a series of irregularities: duplication of attention to clients, mainly those settled in the most productive areas; abandonment of most of the large rain-fed areas; and inadequate credit control and follow-up methods, which caused the diversion of resources to other ends. All this gave rise to a large overdue loan portfolio.

In spite of the determined participation of the Mexican state, overall amounts of financing channeled to the countryside in the past two decades accounted for only 13% of the total, whereas the population for which it was intended accounted for over 40% of the country's total. Only 30% of rural producers had access to institutional credit, whether public or private.

Within the sector, financing was swiftly channeled to profitable enclaves of irrigated agriculture and commercial livestock which already enjoyed public investment and developed infrastructure. Regionally imbalanced rural credit accentuated the gap between, on one hand, commercial and export-oriented enclaves and, on the other, peasant areas typically producing basic foodstuffs for on-farm consumption and for the supply of the domestic market.

With the model's exhaustion, the inequity and inequality between the urban and rural societies—and more dramatically within the sectors of the latter—were clearly exposed. This decay affected production levels of the agricultural and livestock sector, whose imports reached alarming levels during the 1970s and whose poor performance then even had an impact on production levels of the Mexican economy as a whole.

In examining Mexico's credit system immediately before and during SAM, we can see two periods of development: 1976–1979 and 1980–1982. During the first, the three sectoral banks merged and combined their functions and systems in order to control official rural financing in a uniform fashion. During the second period, there were efforts to reorient credit-granting policies toward the support of basic grain production and rain-fed producers with limited ability to repay.

Structural and Functional Changes

The first key question to examine in the context of the evolving nature of the state's agricultural finance system is, What structural and functional changes occurred in the system during SAM?

BANRURAL was constituted on July 7, 1975, by a presidential decree that provided for the merger of Banco Nacional de Crédito Agrícola, S. A., and Banco Nacional de Crédito Ejidal, S. A. de C.V., with Banco Nacional Agropecuario, which changed its name to Banco Nacional de Crédito Rural, S. A. On that date BANRURAL had twelve regional banks and two affiliates. The diversity of operational and accounting systems used by the three banks led to a complex merger process in which several new administrative and functional structures materialized. The accelerated growth of credit operations that began in 1980 when SAM was launched required the institution to expand at a time when the merger had not yet been fully consolidated. The resultant functional and structural changes for BANRURAL during SAM were considerable.

The bank's infrastructure increased, from 412 branch offices to 646 in a period of only five years, to include the 200 branches created within the framework of the Intersectoral Coordination Pact for the support of marginal areas (with CONASUPO, COPLAMAR, SARH, and SRA). This started operating during the spring-summer cycle of 1981.

The staff of 21,470 in 1977 increased to 24,580 in 1980 and 26,655 in 1982. This growth showed a macrocephalic tendency, since the central staff grew more rapidly than the regional offices, and the latter showed a higher figure for office employees than for field staff. In 1982, field inspectors, technicians, supervisors, and promotors of the institution who were in direct contact with producers totaled 6462 and accounted for 24% of the system, as compared to 20,192 office employees who held 76% of total jobs within BANRURAL as a whole. The personnel expansion was accompanied by an increasingly large and elaborate administrative structure. In 1976 there were 27 areas, from managerial to director general levels. By the beginning of 1980 the number of areas had gradually grown to 35, with the creation of four general subdepartments and ten managerial areas. However, in June 1980—with the new administration—the number of areas increased to 150 with the creation of two adjunct departments, six new general subdepartments, a general controllership, a general subcontrollership, two general coordinations, five area subdepartments, and managerial areas increased to 50.

With the purpose of operating functionally within the standards of BANRURAL, counterpart areas were created in the twelve regional banks. This new structure strengthened the functions of regulation, control, planning, budgeting, finance, and resource collection of the bank.

The Integrated Programming, Budgeting, Control, and Evaluation System (SIPPCE) was designed, and for evaluation purposes a subsystem was implemented to report monthly on advances in the operations of the twelve regional banks. Similarly, there were annual meetings for self-evaluation of BANRURAL, with the participation of the highest officials within the system.

A great step was taken in the area of electronic processing for the accounting systems and in the follow-up of operations with the design of the Integrated Accounting Systems (SICO), the Working Capital system (SICA), and the Medium-Term Credit System (SICRE).

The main actions and standards adopted by BANRURAL to respond efficiently to SAM's strategy were the following:

Establishment of a mechanism allowing regional banks to make transfers of resources between production lines and geographic areas, provided these were made according to priorities indicated by the SAM strategy.

Systematic launching of intensive campaigns to publicize the requirements to be met in credit applications, and the periods within which these were to be filed, in order to reduce the problems of numerous, simultaneous applications tied to the agricultural cycle.

Extension of branch office managers' powers to clear applications from clientele with a sound credit history with the bank.

Integration of technical cadres to expedite the granting of medium-term credits in each regional bank. These were to be in charge of formulating, supervising, and evaluating the projects made within the jurisdiction of their bank, with sufficient mobility to support branches.

Organization of accounting courses for creditors to help them keep an adequate control of their statements.

Within the trust area, changes were made to avoid duplication and to increase efficiency. Of particular importance was the creation of two new trust funds in 1981: Fideicomiso de Riesgo Compartido (FIRCO) and Fideicomiso de Promoción Rural (FIPROR). FIRCO carried out three programs: one of shared risk, whose purpose was to induce rain-fed producers to adopt technologies adapted to the agricultural and weather conditions of their area. It absorbed the cost of additional investments in the event that production and productivity were not achieved, and it guaranteed producers a production volume equivalent to the average obtained during the previous five agricultural cycles. With the programs of incentives for the production of maize, beans, wheat, and rice, SAM incentives reduced prices for seeds, fertilizers, and pesticides, and beneficiaries received cheaper credit and insurance for their crops. The purpose of the inputs program was to increase the use of technology in cultivating rain-fed maize and beans. It granted credit for technological packages, such as mechanization activities, improved seeds, fertilizers, and pesticides, which peasants used according to their requirements and the recommendations of the Ministry of Agriculture. FIPROR's main objective was to help make peasants in backward areas with little rainfall and no physical or service infrastructure eligible for credit.

With the transformation of BANRURAL into a multiple banking system in January 1982, its function expanded to include mortgage credits for rural *ejido* housing. This was achieved by keeping the structure of regional banks, which are empowered to do whatever operations are necessary for the system. Likewise, the potential of the institution to attract resources from the public through traditional demand deposits, savings accounts, and placement of securities was increased. Regional banks acquired great importance, considering the amounts of money obtained during the period of our analysis.

Dynamics and Structure of Credits Granted

From the foregoing description, it is clear that, during SAM, BAN-RURAL experienced major organizational changes in terms of size, structure, responsibilities, and activities. The next key question is, How did the credit flows change during SAM? An answer requires disaggregation of lending along several dimensions: (1) by subsector; did the credit flows to agriculture (a high SAM priority) increase relative to livestock (a low SAM priority)? (2) by credit term; were the short- and medium-term credits being allocated to SAM's priority areas? (3) by land type; were the dryland farmers, SAM's target group, getting a larger share of the financing relative to the irrigated farms? (4) by crop type; were the priority basic staples receiving more credit? (5) by credit size; were the loan sizes adequate to acquire SAM's technical package? (6) by regional credit distribution; was credit allocation under SAM able to alter the traditional regional concentration? Each of these questions is addressed in the following sections in a comparison of the credit allocation during the 1980–1982 SAM period with that during the 1976–1979 period.

Subsector

Total credit resources that BANRURAL channeled to agricultural and livestock activities from 1977 to 1982 (Table 5.1) shows an upward trend until 1981 and negative growth in 1982. If we take 1976 as the agricultural year of reference, credits granted increased, until 1981, at an average annual rate of 9.1% (Table 5.2). However, growth was uneven; the pace fluctuated at rates of 0.1% in 1978 and 12.2% in 1979, with an average annual growth from 1977 to 1979 of 6.0%. In SAM's first year, 1980, the highest real-term growth rate of the presidential term was recorded, hitting 15.8%. It increased at 8.8% in 1981, and in 1982, in the face of austerity in SAM's last year, it showed a negative rate of 16.5%.

An analysis by subsector also shows unequal growth. Credit to agriculture shows significant positive growth rates in 1980 (9.1%) and in 1981 (16.3%), which were consistent with the SAM strategy, but there was a drop in 1982 of 14.1%. On the other hand, livestock credit grew at an accelerated

Table 5.1. Credit applied by activity, 1976–1982 (constant 1976 prices, millions of pesos)

	1976	1977	1978	1979	1980	1981	1982[a]
Agriculture	12,398.9	13,438.6	12,734.8	12,800.5	13,972.0	16,250.0	13,950.0
Livestock	2,816.3	2,966.6	3,049.8	4,794.9	6,124.4	6,065.0	4,410.3
Industry	1,153.3	1,347.0	1,700.7	1,714.7	1,764.4	2,102.9	2,192.5
Other[b]	1,875.2	1,071.0	1,361.3	1,828.1	2,638.6	2,240.5	1,698.0
	18,243.7	18,823.2	18,846.7	21,138.2	24,499.4	26,658.4	22,250.8

[a]Figures reported as of April 30, 1983.
[b]Other: poultry, apiculture, etc.

Table 5.2. Evolution and average annual growth rates of credit by activity, 1977–1982 (percentage)

	1977	1978	1979	1977–1979	1980	1981	1982	1980–1982	1977–1981
Agriculture	8.4	−5.2	0.5	−2.4	9.1	16.3	−14.1	−0.1	4.8
Livestock	0.0	2.8	57.2	27.0	27.6	−0.9	−27.3	−15.1	19.5
Industry	16.8	26.3	0.8	12.8	2.7	19.2	4.3	11.5	11.8
Other	−42.8	27.1	34.3	31.0	44.3	−15.1	−47.9	−19.8	20.0
Composite	3.1	0.1	12.2	6.0	15.8	8.8	−16.5	−4.7	9.1

rate until 1980, at an annual rate of 57.2% in 1979 and 27.6% in 1980, and it recorded decreases of 0.9% in 1981 and 27.3% in 1982. For its part, the agroindustrial subsector increased consistently until 1982.

Agriculture, which the 1980 SAM strategy saw as the major activity to support with credit, had a smaller share in 1980–1982 than it had in 1977–1979. Agricultural credit absorbed 71.4% of credits granted by BAN-RURAL in 1977, and it is that same year when its greatest share was recorded. In 1978 agriculture received 67.6% of credits granted, and for the following two years its share decreased to 60.6% and 57%, respectively. In 1981 and 1982, although its share improved at 61.0% and 62.7%, reflecting the SAM priority, it did not reach the higher levels of previous years.

Credit to cattle ranching enjoyed 15.7% of the total in 1977 and increased its share gradually to 16.2% in 1978, 22.7% in 1979, 25% in 1980, and 22.8% in 1981. Although cattle's share of livestock credit in 1982 decreased to 19.8%, its relative importance was still higher than 1977. Agroindustrial credit showed fluctuations between 7.2% and 9.9% in its share; the latter is the level reached during the last year of the series. Thus, although increased credit flowed to agriculture during SAM, reversing the downward trend in its share relative to livestock, the share was still less than what it had been in 1978.

Credit Term

In the analysis above we consider both working capital and medium-term credits. However, these have different development dynamics—because of different recovery terms (clearly higher in the case of medium-term credit) and different purposes (working capital vs. capitalization of the productive unit). Hence we must analyze their evolution separately.

The analysis reveals that, consistent with the SAM strategy, working-capital loans did increase, especially for basic grains. Medium-term loans decreased relative to short-trm loans, but the livestock sector increased its share of medium-term credits, a result not entirely consistent with the SAM orientation. The structure of credits granted during the whole presidential term shows a predominance of working-capital credits, which had a tendency to decrease their relative share in the first years of the period from 81.4% in 1977 to 70.3% in 1980. However, in the following two years their relative importance recovered and increased to 73.3% in 1981 and 75.9% in 1982. Medium-term credits showed the opposite tendency. Their relative importance increased during the first three years, from 15.2 to 23.5%, and then gradually decreased to 20.0% in 1982 (Tables 5.3 and 5.4 show the evolution of working-capital credits; Tables 5.5 and 5.6 present the data for medium-term credits).

In order to avoid the distortion caused by budgetary cuts of the atypical year of 1982, evolution can be analyzed to 1981 to better reflect trends due to SAM strategies. Thus, during the period 1977–1981, working-capital credits grew at an average annual rate of 2.8% and medium-term credits, 20.0% (see Tables 5.4 and 5.6). However, when analyzing the pre-SAM period (1977–1979) we see that working-capital credits decreased, whereas during SAM they increased 7.1% in 1980 and 18.8% in 1981. Medium-term credits had grown 32.0% during the 1977–1979 period, but this growth rate then declined in 1980 and 1981. In 1982, because of general budget cuts, both working-capital and medium-term credits declined.

Resources channeled to agricultural working-capital credits, which because of their amounts are the most important loans of their kind, decreased at a rate of 6.3% from 1977 to 1979 (Table 5.4). Given the strategic importance attached to them by SAM, they were able to reach positive rates in 1980 (7.1%) and 1981 (18.8%), but they fell again in 1982 (−13.7%). The average annual growth rate between 1977 and 1981 was 2.8%, thus giving agricultural working-capital credit the smaller rate of the period, after livestock working-capital credit, which grew by 15.6%, "other working-capital credits," which recorded an annual average rate of 30.0%, and industrial working-capital credits, which grew by 7.1%.

From 1977 to 1979, credits destined for maize and beans (Group I) showed a slight decrease in their share of agricultural working-capital credit, since in the first year of the period they acconted for 21.2% and in the last, 20.7%, while crops included in Group II increased from 29.0% to 34.4%, and Group III decreased from 49.8% to 44.9% (Table 5.3).

From 1980, the share of amounts channeled to the SAM priority crops (Group I) increased to 30.5%, crops of Group II accounted for 30.7%, and the share of Group III decreased to 38.8%. During the last two years there was an increase in the share of agricultural working-captial credits destined for the SAM crops in Group I, accounting for 41.1% in 1981 and 40.7% for 1982.

The main tendency of medium-term credit was an increased support of livestock activities (Table 5.5). Its share was 20.8% in 1977, increasing to 31.2% in 1979, 32.9% in 1980, and 30.4% in 1981. Although it decreased to 26.8% in 1982, its relative share was still at a level higher than during the 1977–1979 period. It appears, then, that the livestock subsector was the one that benefited most from medium-term credit. On the other hand, the

Table 5.3. Working-Capital credit by crop, 1976–1982 (constant 1976 prices, millions of pesos)

Crop	1976	1977	1978	1979	1980	1981	1982[a]
Group I							
Maize	1,840.0	1,934.2	1,906.1	1,581.7	2,492.5	3,789.8	3,671.9
Beans	753.4	461.6	478.0	471.0	751.9	1,405.2	699.6
	2,593.4	2,395.8	2,384.1	2,052.7	3,244.4	5,195.0	4,371.5
Group II							
Sesame	132.9	126.6	130.2	255.1	228.4	92.5	66.4
Rice	351.4	324.5	221.1	354.3	394.7	505.4	506.6
Safflower	217.1	300.7	424.0	430.6	241.2	237.4	114.5
Sorghum	1,260.9	1,123.1	1,035.5	1,015.4	1,210.6	1,387.4	1,215.9
Soybean	195.0	227.9	217.3	526.0	209.8	441.7	557.8
Wheat	1,301.5	1,178.6	1,190.4	838.8	982.3	1,085.8	1,208.7
	3,458.8	3,281.4	3,218.5	3,420.2	3,267.0	3,750.2	3,648.0
Group III							
Cotton	2,069.1	3,131.4	2,595.2	2,406.8	2,052.8	1,761.4	1,057.5
Other crops	2,181.6	2,291.7	2,182.2	2,059.7	2,077.7	1,908.0	1,732.0
	4,370.1	5,644.5	4,806.3	4,466.5	4,130.5	3,669.4	2,789.9
Other items	119.4	221.4	28.9	—	—	—	—
Total	10,442.3	11,321.7	10,408.9	9,939.4	10,641.9	12,644.6	10,809.4

[a]Figures as of April 30, 1983.

Table 5.4. Evolution and average annual growth rates of working-capital credit by crop, 1977–1982 (percentage)

Crop	1977	1978	1979	1977–1979	1980	1981	1982	1980–1982	1977–1981
Group I									
Maize	5.1	-1.4	-17.0	-9.6	57.5	52.1	-3.1	21.0	18.3
Beans	-38.6	3.5	-1.5	1.0	59.7	86.8	-50.2	-3.5	32.0
Composite	-7.6	-0.5	-13.9	-7.4	58.0	60.1	-15.8	16.1	21.3
Group II									
Sesame	-4.5	2.4	96.2	42.0	-10.6	-59.2	-28.2	-46.0	-7.5
Rice	-7.4	-32.0	60.2	4.5	11.6	27.8	0.2	13.3	11.7
Safflower	38.7	40.9	1.7	19.7	-44.1	-1.7	-59.9	-31.0	-5.8
Sorghum	-10.9	-7.7	-2.0	-4.9	19.3	14.5	-12.3	0.2	5.4
Soybean	16.9	-4.8	142.4	52.0	-60.1	110.5	26.2	63.0	18.0
Wheat	-9.4	0.9	-29.5	-15.6	17.0	10.6	11.3	11.0	-2.0
Composite	-5.1	-2.0	6.3	2.1	-4.5	14.8	-2.7	5.7	3.4
Group III									
Cotton	51.3	-17.1	-7.2	-12.3	-14.7	-14.2	-40.0	-28.0	-13.4
Other crops	5.0	-4.8	-5.6	-5.2	0.8	-9.2	-8.2	-8.7	-4.5
Composite	29.2	-14.8	-7.1	-11.0	-7.5	-11.2	-23.9	-17.8	-10.2
Other items	85.7	-86.9	—	—	—	—	—	—	—
Composite	8.4	-8.1	-4.5	-6.3	7.1	18.8	-13.7	0.8	2.8

Table 5.5. Medium-term credit applied by activity, 1976–1982 (constant 1976 prices, millions of pesos)

	1976	1977	1978	1979	1980	1981	1982
Agriculture	1,976.6	2,117.0	2,326.0	2,861.1	3,330.1	3,635.1	2,747.8
Livestock	730.8	594.6	791.5	1,550.0	1,889.3	1,830.6	1,193.5
Industry	177.2	79.3	263.9	280.9	349.1	436.7	454.4
Other	59.6	76.4	90.3	283.6	173.0	129.4	57.8
	2,944.2	2,870.3	3,471.7	4,975.6	5,741.5	6,031.8	4,453.5

Table 5.6. Evolution and average annual growth rates of medium-term credit applied by activity, 19
1982

	1977	1978	1979	1977–1979	1980	1981	1982	1980–1982	1977 1981
Agriculture	7.1	9.9	23.0	16.2	16.4	9.3	−24.4	−9.1	14.5
Livestock	18.6	33.1	95.8	61.0	21.9	−2.8	−35.0	−20.0	32.0
Industry	55.2	332.8	6.4	88.0	24.3	25.9	4.0	14.4	54.0
Other	−28.2	18.2	314.1	92.0	−39.0	−25.2	−43.7	−160.0	14.1
Composite	−2.6	21.0	43.3	32.0	15.4	5.0	−26.2	−11.9	20.0

Table 5.7. Share of area financed by BANRURAL of national harvested area, 1976–1982 (thousands of hectares, percentage in parentheses)

	1976		1977		1978	
	National	BAN-RURAL	National	BAN-RURAL	National	BAN-RURAL
Group I						
Maize	6783.2	1333.8 (19.7)	7374.3	1374.6 (18.6)	7183.9	1337.5 (18.6)
Beans	1315.8	574.5 (43.7)	1613.4	414.9 (25.7)	1580.2	403.2 (25.5)
	8099.0	1908.3 (23.6)	8987.7	1789.5 (19.9)	8764.1	1740.7 (19.9)
Group II						
Sesame	198.0	80.0 (40.4)	204.6	76.1 (37.2)	243.8	83.2 (34.1)
Rice	159.4	81.0 (50.8)	173.5	71.2 (41.0)	120.7	52.7 (43.7)
Safflower	184.9	104.6 (66.6)	399.7	163.9 (41.0)	429.1	181.0 (42.2)
Soybean	172.4	76.2 (49.2)	314.2	77.4 (24.6)	216.4	87.9 (40.6)
Sorghum	1251.1	571.8 (45.7)	1367.8	537.0 (39.3)	1396.6	543.2 (38.9)
Wheat	894.1	362.6 (40.6)	708.4	346.7 (48.9)	758.8	324.0 (42.7)
	2859.9	1276.2 (44.6)	3168.2	1272.3 (40.2)	3165.4	1272.0 (40.2)
Group III						
Cotton	235.0	140.6 (59.8)	393.3	210.4 (53.5)	347.0	187.4 (54.0)
Other	3548.8	628.5 (17.7)	3280.8	694.6 (21.2)	4165.5	576.4 (13.8)
	3783.8	769.1 (20.3)	3674.1	905.0 (24.6)	4512.5	763.8 (16.9)
Total	14742.7	3953.6 (26.8)	15830.0	3966.8 (25.1)	16442.0	3776.5 (23.0)

agricultrual medium-term credit share was drastically cut from 73.8% in 1977 to 57.5% in 1979. Its recovery to absorb 61.7% of medium-term loans in 1982 still left it far from the level it had reached at the beginning of the presidential term. Agroindustrial medium-term credits increased considerably, from 2.8% of the total medium-term credit in 1977 to 10.2% in 1982. The evolution of medium-term credit at constant prices (Table 5.6) showed an average annual growth rate between 1977 and 1981 of 14.5% for the agricultural subsector, 32.0% for livestock, and 54.0% for agro-industry.

In general terms, the second three-year period (1980–1982) showed a decrease in all subsectors except industry, which recorded a positive rate of 14.4%. It is worth stressing that both agricultural and livestock medium-term credits recorded their highest increases in 1979. The support to the livestock subsector with medium-term credit began to grow in 1978, at 33.1%, reached 95.8% in 1979, and in 1980 continued to grow by 21.9%. In the case of agriculture, the same trend appeared at a less rapid rate (9.9% in 1978, 23.0% in 1979, and 16.4% in 1980).

Land Type

BANRURAL greatly expanded the area financed during SAM, and, as sought by the SAM strategy, the share to rain-fed land rose significantly.

Table 5.7. (continued)

	1979		1980		1981		1982
National	BAN-RURAL	National	BAN-RURAL	National	BAN-RURAL	National	BAN-RURAL
5916.0	1366.7 (23.1)	6955.2	2290.1 (32.7)	8150.2	3098.9 (30.0)	8742.6	3280.7 (37.5)
988.3	441.6 (44.7)	1763.3	753.7 (42.7)	2150.2	1096.7 (51.0)	2222.2	1153.4 (51.9)
6904.3	1808.3 (26.2)	8718.5	3043.8 (34.5)	10300.4	4195.6 (40.7)	10694.8	4434.1 (41.1)
321.2	138.3 (43.1)	282.3	140.5 (49.8)	150.5	68.9 (45.8)	205.3	66.4 (32.3)
150.4	71.5 (47.5)	132.0	95.4 (72.3)	179.6	127.7 (71.1)	196.9	156.5 (79.4)
494.2	191.0 (38.0)	392.2	202.2 (51.6)	390.5	239.3 (61.3)	207.0	144.9 (72.4)
427.7	149.0 (34.8)	153.7	69.0 (44.9)	377.8	127.0 (33.6)	403.5	167.2 (41.4)
1215.9	611.9 (50.3)	1578.6	805.7 (51.0)	1767.3	902.7 (51.1)	1939.0	907.4 (46.8)
599.9	259.6 (43.3)	738.5	334.2 (45.3)	861.1	446.8 (51.9)	1076.8	565.6 (52.5)
3209.3	1421.3 (44.3)	3277.3	1647.0 (50.3)	3726.8	1912.4 (51.3)	4028.5	2008.0 (49.8)
376.8	203.9 (54.1)	237.7	186.0 (78.3)	355.0	173.9 (49.0)	280.0	105.3 (37.6)
4240.1	605.8 (14.3)	5590.7	713.2 (12.8)	4193.8	679.3 (16.2)	4776.7	694.2 (14.5)
4616.9	809.7 (17.5)	5828.4	899.2 (15.4)	4548.8	853.2 (18.8)	5056.7	799.5 (15.8)
14730.5	4039.3 (27.4)	17824.2	5590.0 (31.4)	18576.0	6961.2 (37.5)	20050.0	7241.6 (36.1)

Table 5.8. Area financed, by land type, 1977–1982 (thousands of hectares, percentage in parentheses)

	1977	1978	1979	1980	1981	1982
Irrigated	1,213.7 (30.6)	1,263.1 (33.4)	1,300.3 (32.2)	1,284.3 (23.0)	1,450.6 (20.8)	1,601.6 (22.1)
Rain-fed	2,753.1 (69.4)	2,513.4 (66.6)	2,739.0 (67.8)	4,305.7 (77.0)	5,510.6 (79.2)	5,647.0 (77.9)
	3,966.8	3,776.5	4,039.3	5.590.0	6,961.2	7,243.6

mained practically the same (Table 5.7), growing from 3,967,000 to 4,039,000 hectares. Nevertheless, the balance between irrigated land and rain-fed areas changed. Financed irrigated area increased from 1,213,700 to 1,300,000 hectares, or 33%, while financed rain-fed area decreased slightly during the same period (see Table 5.8). In contrast, during the SAM years financed area grew substantially, reaching 5,590,000 hectares in 1980, an increase of 38% over 1979, 6,961,000 hectares in 1981, 25% over the previous year, and 7,200,000 hectares in 1982. The share of irrigated area financed decreased from 33% in 1979 to 23% in 1980, and to 20.8% in 1981. Its relative share increased to 22.1% in 1982. For this same year, 25,459 billion pesos were destined to irrigated areas. This amount respresented 47% of the agricultural working-capital credit applied for the year, while rain-fed areas—which accounted for 78% of the accredited area—received 53% of the amount applied.

Crop Type

Looking at areas financed by crops (Table 5.9), we see that in 1977 and 1979 maize and bean (Group I) area financed remained at 45%; the share of accredited area for the second-priority crops (Group II) was between 32 and 35%, and the remainder of crops (Group III) between 23 and 20%. During the SAM years, there was an accelerated growth of the area covered by BANRURAL. The relative importance of Group I crops increased from 1.8 million hectares of maize and beans financed in 1979 to 3.0 million in 1980 and 4.2 million in 1981. In 1982, 4.4 million hectares of Group I crops were financed. Thus the share of these crops within the structure of financed area increased year after year during the second half of the presidential term to 54.4, 60.3, and 61.2%. In the case of Group II there was also an increase in financed area in absolute terms. Its relative importance, however, decreased from 29.5% in 1980 to 27.5% in 1982, with the remainder of the crops showing a similar tendency, with figures from 16.1 to 11.1%. Clearly, BANRURAL's response was consistent with SAM's crop priorities.

Table 5.9. Area financed by crop, 1976–1982 (thousands of hectares, percentage in parentheses)

	1976	1977	1978	1979	1980	1981	1982[a]
Group I							
Maize	1,333.8 (33.7)	1,374.6 (34.7)	1,337.5 (35.4)	1,366.7 (33.8)	2,290.1 (41.0)	3,098.9 (44.5)	3,280.7 (45.3)
Beans	574.5 (14.5)	414.9 (10.5)	403.2 (10.7)	441.6 (10.9)	753.7 (13.5)	1,096.7 (15.8)	1,153.4 (15.9)
	1,908.3 (48.3)	1,789.5 (45.1)	1,740.7 (46.1)	1,808.3 (44.8)	3,043.8 (54.4)	4,195.6 (60.3)	4,434.1 (61.2)
Group II							
Sesame	80.0 (2.0)	76.1 (1.9)	83.2 (2.2)	138.3 (3.4)	140.5 (2.5)	68.9 (1.0)	66.4 (0.9)
Rice	81.0 (2.0)	71.2 (1.8)	52.7 (1.4)	71.5 (1.8)	95.4 (1.7)	127.7 (1.8)	156.5 (2.1)
Safflower	104.6 (2.6)	163.9 (4.1)	181.0 (4.8)	191.0 (4.7)	202.2 (3.6)	239.3 (3.4)	144.9 (2.0)
Sorghum	571.8 (14.5)	537.0 (13.5)	543.2 (14.4)	611.9 (15.1)	805.7 (14.4)	902.7 (13.0)	907.4 (12.5)
Soybean	76.2 (1.9)	77.4 (2.0)	87.9 (2.3)	149.0 (3.7)	69.0 (1.2)	127.0 (1.8)	167.2 (2.3)
Wheat	362.6 (9.2)	346.7 (8.7)	324.0 (8.6)	259.6 (6.4)	334.2 (6.0)	446.8 (6.4)	565.6 (7.8)
	1,276.2 (32.2)	1,272.3 (32.1)	1,272.0 (33.7)	1,421.3 (35.2)	1,647.0 (29.5)	1,912.4 (27.5)	2,008.0 (27.7)
Group III							
Cotton	140.6 (3.6)	210.4 (5.3)	187.4 (5.0)	203.9 (5.0)	186.0 (3.3)	173.9 (2.5)	105.3 (1.5)
Alfalfa	23.8 (0.6)	29.6 (0.7)	28.0 (0.7)	28.0 (0.7)	29.8 (0.5)	36.1 (0.5)	39.3 (0.5)
Potatoes	7.0 (0.2)	5.1 (0.1)	6.9 (0.2)	8.8 (0.2)	8.2 (0.1)	8.9 (0.1)	10.6 (0.2)
Other	597.7 (15.1)	659.9 (16.6)	541.5 (14.3)	569.0 (14.1)	675.2 (12.1)	634.3 (9.1)	644.3 (8.9)
	769.1 (19.5)	905.0 (22.8)	763.8 (20.2)	809.7 (20.0)	899.2 (16.1)	853.2 (12.3)	799.5 (11.1)
Total	3,953.6	3,966.8	3,776.5	4,039.3	5,590.0	6,961.2	7,241.6

[a]Figures as of April 30, 1983.

Table 5.10. Credit quotas by crop, 1976–1982 (constant 1976 prices, pesos/hectare)

	1976	1977	1978	1979	1976–1979	1980	1981	1982	1980–1982
Maize	1,379	1,407	1,421	1,157	1,341	1,088	1,223	1,119	1,143
Beans	1,311	1,112	1,185	1,066	1,168	977	1,281	606	961
Wheat	3,589	1,664	3,674	3,231	3,039	2,939	2,430	2,175	2,515
Rice	4,338	1,870	4,195	4,956	3,840	2,990	3,958	3,237	3,395
Sorghum	2,205	2,091	1,902	1,659	1,964	1,502	1,537	1,292	1,444
Sesame	1,661	1,663	1,564	1,844	1,683	597	1,343	1,000	980
Safflower	2,075	1,834	2,342	2,254	2,126	1,193	992	790	992
Soybean	2,559	2,943	2,459	3,530	2,798	3,040	3,478	3,336	3,285
Cotton	14,716	14,881	13,848	11,802	13,862	11,037	10,129	10,043	10,403
Average	3,763				3,536				2,791

Credit Size

Although BANRURAL was financing more rain-fed land planted in SAM target crops, the standard unit size of its loans was shrinking. In effect, more farmers were getting less. This may have reduced their ability to adopt the full technological packages contemplated in the SAM strategy.

The size of credit quotas were differentiated by crops, regions, and technological level. They were defined at the beginning of the agricultural cycle and took into account variations in production costs, though they did not cover the costs of preparing the land, sowing it, and so on. Since the operating plans of the bank were subject to an authorized budget, the variations recorded in production costs during the agricultural cycle were not taken into account. A retrospective analysis of credit quotas applied by unit can be based on the relation between amounts of credit granted and accredited area by line or crop. Thus an average of quotas by crop and agricultural years is built. In our analysis we see that the average applied to nine crops (maize, beans, wheat, rice, sorghum, sesame, safflower, soybean, and cotton) decreased in real terms using constant 1976 prices (see Table 5.10). At constant prices, the average quota of the years 1976–1979 was 3536 pesos per accredited hectare, while in 1980–1982 this quota fell to 2791 pesos. In real terms, each producer received 21.1% less for every hectare cultivated during the latter period. Of the nine products, only soybeans recorded a slight increase in average quota.

When considering credit quotas of priority crops, we find 1407 pesos per hectare in 1977 for maize, increasing to 1421 pesos in 1978 and then decreasing to 1157 pesos in 1979 and 1088 pesos in 1980. In 1982 maize producers financed by BANRURAL received 78% of the 1978 quota. The same situation is seen in the case of wheat, whose average quota decreased from 3039 to 2515 pesos from the first to the second three-year period, and rice, from 3840 to 3395 pesos per hectare. The increase in financed area that took place during these years is, therefore, accompanied by a real de-

Table 5.11. Evolution and average annual growth rates of credit quotas by crop, 1977–1982

	1977	1978	1979	1980	1981	1982
Maize	2.0	1.0	−18.6	−6.0	12.4	−8.5
Beans	−15.2	6.6	−10.0	−6.5	28.5	−52.7
Wheat	−53.7	220.8	−12.1	−9.0	−17.3	−10.5
Rice	−56.9	224.3	18.1	−39.7	32.4	−18.2
Sorghum	−5.2	−9.0	−12.8	−9.5	2.3	16.0
Sesame	0.1	−6.0	17.9	−68.0	225.0	−25.5
Safflower	−11.6	27.7	−3.8	−47.1	−16.8	−20.4
Soybean	15.0	−16.4	43.5	−13.9	14.4	−4.1
Cotton	0.0	−6.9	−14.8	−6.5	−8.2	−0.8

crease in intensity of agricultural working-capital per unit of area and per creditor.

In annual evolution (Table 5.11), maize credit quotas decreased by 18.6% in 1979 and 8.5% in 1982, the last two years of each three-year period, while beans recorded decreases of 10.0 and 52.7% during the same years. It should be noted that this latter percentage is the highest of all crops. In the case of wheat and rice, a significant increase of credit quotas was recorded in 1978, 220.8 and 224.3%, respectively. These same products showed the largest decreases in 1977: 53.7 and 56.9%.

The other crops had an annual evolution of credit quotas that leveled off their relative importance within priority groups, with the exception of cotton, which showed slight decreases at a moderate pace (14.8% in 1979 and 0.8% in 1982).

Regional Credit Distribution

Far from correcting the historical regional asymmetry, development credit has been a major factor in shaping it. In 1982, 45.7% of credit operations concentrated in six states: Baja California Norte, Chihuahua, Jalisco, Sinaloa, Sonora, and Tamaulipas (Table 5.12). Historically, these same states have been the beneficiaries of public investments that fostered the development of agricultural and livestock entrepreneurial enclaves. For this reason, a major part of commercial bank credit was allocated to these entities; development banks played a complementary role, because their function was to reach the areas where commercial credit did not flow.

Another criterion of concern for credit distribution is the ratio of credit to available working area in each region. In this sense we see that, in these six states, the share of credit is in every case greater than the share of arable land registered by the sector's census:

Sonora received 11.7% of the credit resources of BANRURAL, 6.2 billion pesos (current prices), in spite of the fact that the 1970 census indicates that the state has 3.5% of the arable land in the country. Of the

Table 5.12. Working-capital credit distribution by state, 1982 (percentage in parentheses)

State	Arable area[a] (thousand hectares)	Financed area[b] (thousand hectares)	Amount (million pesos)
Aguascalientes	137.3 (0.6)	79.1 (1.1)	422.5 (0.8)
Baja California Norte	323.7 (1.4)	161.9 (2.2)	2,613.9 (4.9)
Baja California Sur	64.7 (0.3)	30.7 (0.4)	558.7 (1.0)
Campeche	262.8 (1.1)	127.6 (1.7)	1,049.0 (2.0)
Chiapas	1,801.4 (7.8)	346.1 (4.9)	2,777.2 (5.2)
Chihuahua	1,067.6 (4.6)	345.7 (4.9)	2,528.2 (4.8)
Coahuila	470.0 (2.0)	101.3 (1.4)	2,086.6 (3.9)
Colima	179.0 (0.8)	39.8 (0.5)	325.4 (0.7)
Distrito Federal	25.5 (0.1)	1.3 (0.0)	20.9 (0.0)
Durango	717.3 (3.1)	290.5 (4.0)	1,219.4 (2.3)
Guanajuato	1,095.3 (4.7)	242.2 (3.4)	1,942.6 (3.7)
Guerrero	885.6 (3.8)	206.1 (3.0)	1,180.9 (2.2)
Hidalgo	587.1 (2.5)	123.2 (1.7)	367.3 (0.7)
Jalisco	1,442.5 (6.2)	457.5 (6.3)	3,421.8 (6.5)
México	643.0 (2.8)	211.7 (3.0)	1,503.0 (2.8)
Michoacán	1,058.7 (4.6)	426.1 (6.0)	3,112.7 (5.9)
Morelos	124.6 (0.5)	59.3 (0.8)	543.6 (1.0)
Nayarit	427.3 (1.8)	124.9 (1.7)	1,233.2 (2.3)
Nuevo León	322.7 (1.4)	164.0 (2.3)	615.4 (1.2)
Oaxaca	1,015.2 (4.4)	188.8 (2.6)	926.3 (1.7)
Puebla	893.9 (3.9)	259.3 (3.6)	1,268.4 (2.4)
Querétaro	209.8 (0.9)	52.8 (0.7)	239.8 (0.5)
Quintana Roo	16.1 (0.1)	131.4 (1.8)	578.0 (1.1)
San Luis Potosí	711.4 (3.1)	225.6 (3.1)	769.1 (1.4)
Sinaloa	1,024.6 (4.4)	533.3 (7.4)	5,534.9 (10.5)
Sonora	812.7 (3.5)	328.5 (4.5)	6,182.4 (11.7)
Tabasco	1,021.7 (4.4)	34.9 (0.5)	312.1 (0.6)
Tamaulipas	1,073.8 (4.6)	754.9 (10.5)	3,846.5 (7.3)
Tlaxcala	235.6 (1.0)	88.5 (1.2)	317.5 (0.6)
Veracruz	2,868.8 (12.4)	428.6 (6.0)	3,072.8 (5.8)
Yucatán	604.9 (2.6)	133.0 (1.8)	487.4 (0.9)
Zacatecas	978.6 (4.2)	499.4 (7.0)	1,923.7 (3.6)
	23,103.4	7,198.0	52,981.2

[a]Source: *Resumen General del V Censo Agrícola, Ganadero y Ejidal,* 1970.
[b]Agricultural year 1982–83.

7.2 million hectares financed by the bank in 1982, Sonora got 328,000 (4.5%). This reveals the highly technical and capital-intensive nature of the production process; this is why the percentage of credit granted is 2.5 times the percentage of the area.

Sinaloa, with 4.4% of the nation's arable land, obtained 10.5% of the credit granted during the agricultural year.

With 3.4 billion pesos, Jalisco received 6.5% of BANRURAL's credit for an area of 457,500 hectares, a little over 30% of the arable land in the state.

With 4.6% of the nation's arable land, Tamaulipas absorbed 7.3% of the credit.

Chihuahua holds 4.6% of arable land and received 4.8% of the credit.

Baja California Norte received 4.9% of the credit, with BANRURAL financing almost 50% of the arable land of the state which accounts for 1.4% of the nation's total. With the exception of wheat, all crops significantly supported in this state were not considered among the priorities set by SAM; 40% of the area financed by BANRURAL was destined to "other crops," followed by cotton and alfalfa.

To complete this privileged group, three states have to be added to the list: Chiapas, Michoacán, and Coahuila, with 5.2%, 5.9%, and 3.9% of agricultural working-capital credit.

Thus, in the agricultural period of 1982, nine states received over 60% of the credit support for working capital given by BANRURAL. At the other extreme, the nine states that received the least support, 7.1% of the credit granted, were Aguascalientes (0.8%), Baja California Sur (1.0%), Colima (0.7%), Hidalgo (0.7%), Morelos (1.0%), Nuevo León (1.2%), Querétaro (0.5%), Tabasco (0.6%), and Tlaxcala (0.6%). The middle group, with the remaining thirteen states, absorbed 32% of the agricultural working-capital credit. Table 5.13 summarizes this unequal distribution.

Table 5.13. Area and credit shares

State	Arable area (%)	Credit (%)	Correlation coefficient[a]
High credit concentration Baja California Norte, Chihuaha, Jalisco, Sinaloa, Sonora, Tamaulipas, Coahuila, Michoacá, Chiapas	39.1	60.7	1.55
Medium credit concentration Campeche, Durango, Guerrero, Nayarit, Puebla, Veracruz, Zacatecas, Quintana Roo, Oaxaca, San Luis Potosí, Yucatán, Guanajuato, México	48.5	32.2	0.66
Low credit concentration Aguascalientes, Baja California Sur, Colima, Hidalgo, Morelos, Nuevo León, Querétaro, Tabasco, Tlaxcala	12.4	7.1	0.57

[a]This ratio correlates the arable area with credit granted by group. Even if it has no real quantitative significance, it does suggest that the medium-concentration group is closer to the low- than to the high-concentration group. It also reveals that the credit concentration per unit of area is much denser in the high-credit-concentration group.

Conclusions and Perspectives

The stagnation seen during BANRURAL's first years (1976–1978) was overcome in 1979 through the substantial increase in operations in real terms, which largely benefited the livestock sector. The adoption of the SAM strategy in 1980 caused a shift from livestock toward agriculture. In 1981 real growth in agricultural credit was greater than for livestock. Nevertheless, this major push was insufficient to reverse the trend seen during the period as a whole. At the period's end, growth rates for the agricultural subsector were lower; they lagged behind livestock and agroindustry, just as they had in the years before SAM was implemented. However, the downward curve of medium-term livestock credit that started in 1981 was steeper than that of agriculture. During that same year, medium-term agricultural credit showed a positive increase that, in 1982, decreased at a slower pace than that for livestock. If this continues during the coming years, agricultural credit will be the sectoral leader in financing.

Under SAM, credit recorded an accelerated growth from 1980. Greater growth was seen in the areas of top-priority crops, maize and beans, which in 1982 amounted to 61.2% of the whole financed area and accounted for 40.7% of agricultural working-capital credit. However, the decrease in real terms in the values of credit quotas suggests that expansion was not necessarily reflected in the introduction of modern technological patterns nor, therefore, in higher productivity of the financed areas. Such a phenomenon is particularly clear in rain-fed areas, which accounted for over three-fourths of the area financed in 1982 but which absorbed only half the amount destined for agricultural working capital during the same year. The tendency to concentrate credit regionally in states characterized by commercial production was lessened, but in no way reversed, during the years of SAM's implemention.

The macrocephalic growth of the institutional administrative structure was paradoxically accentuated in 1980–1982, negating the objectives of deconcentration and decentralization proposed as policy guidelines.

During these first six years of existence, the BANRURAL system grew and consolidated its structures. Its experience allowed it to become the leading organization in financing the nation's agricultural and livestock sectors, with greater penetration in the countryside through its large network of branch offices. Thus it appeared as the leading agency of the series of instruments available to the Mexican state to foster the integrated rural development and food objectives set by the new administration of President Miguel de la Madrid.

CHAPTER 6

SAM and Seeds

David Barkin

The announcement of the Mexican Food System (SAM) in 1980 might have been viewed as the dawn of a new era for an improved-seeds industry in Mexico. Unlike other sectors, this industry has never been expected to support a policy for national food self-sufficiency. In fact, the improved-seeds industry is a recent creation, and is itself a part of the modernizing process that precipitated Mexico's production crisis. The government's response to popular political demands to increase basic food production was to create SAM. SAM, in turn, generated new demands that produced unanticipated pressures on the national capacity to supply seeds.

A Historical Overview

In order to understand the situation, one must review the birth, growth, and consolidation of the Mexican improved-seeds industry. Without going through the long and fully documented history of the Green Revolution in Mexico, it is worth stressing that there was early official interest in improving the quality of seeds for the production of maize, "Mexico's wage crop." The index of purchased seeds rose from 29 in 1940 to 136 in 1965.[1]

The first stage in the history of the improved-seeds industry may be characterized as one of conflict and growth. Since the turn of the century, research efforts to improve the quality of seeds and the decision to under-

The research for this chapter was part of a more extensive study sponsored by the Centro de Ecodesarrollo with the full collaboration of Blanca Suarez; see David Barkin and Blanca Suarez, *El fin del principio: Las semillas y la seguridad alimentaria* (Mexico City: Centro de Ecodesarrollo and Ediciones Océano, S.A., 1985).

1. See Reed Hertford, "Sources of Change in Mexican Agricultural Production" (Ph.D. diss., University of Chicago, 1970), p. 20.

take their production reflected a determination inherited from Mexico's revolution: to focus on the peasant's fundamental crop—their own food. In 1934 the first commercial efforts in Mexico to produce improved seeds were undertaken by the Office of Experimental Fields of the Ministry of Agriculture. During the following years, researchers sought to improve the quality and variety of open-pollinated, indigenous (*criollo*) seeds, which could be reproduced in peasant fields.[2]

These specialists showed great sensitivity to the social impact of technical change. They rejected the route of developing hybrid seeds, contending that they were inappropriate for a country like Mexico, which was not prepared to reproduce and distribute seeds year after year in a quantity commensurate with national needs. They argued that even if their productivity were greater (and there was not adequate evidence of this), the land area on which they could be used would be so much smaller that total production increases would not justify—in social terms—the investment required to develop and produce hybrids. They channeled their energy and resources into improving the system of cultivation, based on better techniques with the use of *criollo* seeds. During the period in which these agronomists influenced the definition of research and agriculture extension policies, they succeeded in distributing seeds and improving cultivation techniques in one-fourth of the total area planted with maize and beans.

During the following years, both the interest and the commitment to increase production capacity for improved seeds created contradictory trends within the public sector, the only one in which the problem could be posed. The conflicts became evident in the prolonged debate during the second stage, in the 1940s and 1950s. While some agronomists advised that the country concentrate its efforts on the production of hybrid seeds, others proposed improving the qualities of *criollo* seeds, whose use was common among the peasants. Institutionally, the debate became one between the Agricultural Research Institute (Instituto de Investigaciones Agrícolas—IIA), successor to the first official efforts of the 1930s and advocate of *criollo* seed, and the Office of Special Studies (OEE), created as a result of an international agreement made in 1943 between Mexico's government and the Rockefeller Foundation.

The debate reflected an incipient battle between two development styles, a discussion that now—with the benefit of years of reflection—may be characterized as the academic manifestation of a deep social conflict. Today these differences are reflected in the divergence of views on the socioeconomic effects of the Green Revolution in rural societies. The arguments may be summarized as follows: (1) Advocates of the new technology assert that with the application of the package—which in-

2. *Criollo* seeds are those that evolved through the interaction of natural processes and human selection to produce varieties particularly adapted to the singular ecological and economic conditions in each peasant agricultural region.

cludes hybrid seeds (whenever possible), fertilizers, pesticides, and preferably, irrigation—soil productivity could be increased and thus could contribute to solving food problems. (2) Opponents admit the above but add that (a) the expansion of public expenditures on this package would divert experts and resources from the traditional producers to improve commercial agriculture in wealthier parts of the country, (b) the cost of the technological package is so high that a farmer or peasant wanting to adopt it would need access to agricultural credit (or have abundant working capital), and (c) the cost of the package over long periods—including years of good and bad harvests—would be so large that there would be no security that the net monetary yields, after deducting modern input costs and financial charges, would be greater for the producers.

In the third stage, beginning in 1961, the state's preference and the bias of the national scientific community became evident (coinciding with the preference of the international business community): they favored hybrid seeds. The merger of the OEE and IIA into the new National Institute of Agriculture Research (INIA) was conceived in the framework of the emerging Green Revolution; namely, to attack the problem of low agriculture productivity with new seeds and technological systems that would reduce the problems of food production on a worldwide scale. The official National Seed Producer (*Pronase*) was created to regulate the seed market as well as to produce new varieties on a commercial scale. It absorbed both the assets and the history of the National Maize Commission, created in the 1940s with a clear commitment to *criollo* maize varieties. There is no need to review here the results of this introduction of new seeds and technologies, which did increase productivity in the countryside and ensured a considerable increase in the production of foodstuffs wherever new technology was introduced. It is worth recalling, however, that in reality these efforts accentuated the social inequalities prevailing in the sector, further widening the gap between modern agriculture and traditional practices, and giving the state a tool to favor commercial agriculture at the expense of the peasant economy. The state assumed that the modern sector would be able to respond to the agriculture needs of the country.[3]

A fourth stage of the seed industry's development—the rise of the private sector—began with the stagnation of agricultural production. Although this stagnation started in 1965, the production crisis appeared only ten years later. The modernization of agriculture and new development policies created incentives for farmers to grow commercial products destined for export markets or to initiate agroindustries producing fruits,

3. There is abundant literature on the subject. See, for example, Andrew Pearse, *Seeds of Plenty, Seeds of Want: Social and Economic Implications of the Green Revolution* (Geneva: United National Research Institute for Social Development, 1980), and references quoted therein. Louis Goodman, at The Smithsonian Institution in Washington, D.C., is currently working on a major study of the Mexican seed industry.

vegetables, and animal feed. These crops displaced basic foodstuffs on irrigated and quality rain-fed lands and flourished at the expense of peasant and backyard production. The private seed industry concentrated its efforts on these new crops. New varieties of seeds developed by U.S. corporations were imported, and the production of certified seeds began slowly in the country. At that time, the private Mexican industry was so immature that vegetable-garden seeds produced in the country were exported wholesale, only to be reimported in packages for retail sale, a practice that is still common despite the industry's maturity. As for sorghum, its spectacular growth as a feed surpassed the industry's ability to produce seed, so imports have continued to be an important source of income for U.S. companies.

The seed industry started expanding significantly in the mid-1960s. At first, its activity was limited to importing and distributing seeds for the domestic market. Seed production started with the process of adapting varieties to local conditions and with modest investments in factories to treat, select, and package seeds. The domestic industry had a captive market since, as irrigated areas expanded and opportunities to cultivate them were given to commercial farmers, there was an automatic demand from agronomists for the new seeds. They did not have to be persuaded of their advantages, because during their training they learned that the increased yields (under good conditions) more than compensated for their higher cost.

At that time, the main imported species were vegetables, fruits, cotton, and alfalfa, crops characterized by their high rates of growth in the incipient commercial agriculture. During this initial period, the most outstanding participants were Asgrow, Northrup King, and DeKalb, companies that, with their presence and strength, determined the growth pattern of the dawning seed industry (see Table 6.1).

In contrast to the transnational corporations that settled in the country, specializing in lines that were particularly profitable for commercial agriculture, the state concentrated on producing new varieties of wheat and certain hybrid maize seeds. It was especially active in publicizing the new wheat seeds that resulted from the program developed jointly with the International Center for the Improvement of Maize and Wheat (CIM-MYT). This was successor to the OEE and an integral part of the world network of international agricultural research centers. Because of the biases in research on *criollo* seeds, inadequate resources were devoted to the continued improvement of maize. This discrimination was compounded by other policies that maintained low prices for maize (in order, among other things, to limit the rise of the workers' wages), thereby limiting the commercial perspectives of state-produced varieties of maize. Hybrid maize seeds were channeled to certain areas within modern sectors that cultivated the basic grain for particularly profitable industrial uses.

Table 6.1. Seed-producing transnational corporations in Mexico

	Date started	Main seed lines
Anderson Clayton (USA)	1965	cotton
Asgrow Mexicana (Upjohn-USA)	1961	sorghum, maize, vegetables
Ciba-Geigy (Funk, C-G-Swiss)	1974	sorghum, maize
Diamond Chemical (USA)	NA	safflower
Horizon (USA)	NA	sorghum
La Hacienda (Pioneer-USA)	1965	sorghum, maize
McFadden (USA)	NA	cotton
Northrup King (Sandoz-Swiss)	1961	sorghum, maize, vegetables
Pfizer (USA)	1964	NA
Semillas Agrícolas (Growers-USA)	1979	sorghum, maize
Semillas del Noroeste (Diamond Shamrock-USA)	1972	NA
Semillas Ferry Morse (Purex-USA)	1972	beans, vegetables
Semillas Híbridas (DeKalb-USA)	1961	sorghum, maize
Semillas Nacionales (Pacific Oilseeds-USA)	NA	sorghum, safflower
Semillas Security (USA)	NA	sorghum
Semillas Wac (USA)	NA	sorghum, maize

Source: Centro de Ecodesarrollo and data from Dirección General de Inversiones Extranjeras (Mexico City: SEPAFIN).

The state's role in the case of wheat was outstanding. Wheat yields increased—in less than a decade—from 700 kilograms per hectare to close to 4 tons.[4] The new state seed enterprise, PRONASE, began reproducing and distributing these seeds, but it subsequently turned over full responsibility for producing and distributing wheat seeds to a new link in the institutional chain of the Mexican agriculture sector, the producers' associations of the northwest. These groups were given both the responsibility and the rights over the seed, nationally and internationally.

The private producers' associations formed to comply with government regulations as well as to defend their economic interests and to undertake certain common activities, such as support for research and seed production, the purchase and sale of machinery, and marketing. When they first began producing wheat seeds, the northwestern farmer associations expanded their scope of action to become the largest worldwide exporters of these seeds, displacing several transnational corporations from the national market (e.g., Cargill). In recent years, these associations have diversified by increasing and decentralizing the production of open-pollinated seeds. In our field research, we succeeded in identifying at least 33 associations, all concentrated in the northern part of the country, in-

---

4. For further details on yields and expansion of grain crops, see David Barkin and Blanca Suarez, *El fin de la autosuficiencia alimentaria* (Mexico City: Centro de Ecodesarrollo and Nuevo Imagen, 1982).

volved in production and, at times, in the export of wheat, vegetables, and, lately, oilseeds. At present they supply a substantial part of Mexico's wheat seed, and not only for their members; their vegetable seed production is also considerable but is usually limited to their members.

The private sector did not limit itself to importing seeds. In spite of the obligation to secure a research permit (which is not easily issued) and of the time-consuming process for the approval of new varieties before their sale—imposed by law in 1961—transnational corporations settled in Mexico, in some cases joining national capital to produce seeds based on international technology, with foreign and national genetic material. New, smaller, national corporations such as Mexagro and Semillas Master also began to import seeds, while some produced small quantities of seeds in the country. In this way, the private sector of the improved-seeds industry came into being with a limited number of companies (three with national capital and three transnationals), and increased to twenty-five in the 1970s; of these, more than half are transnational corporations whose actions largely determine the behavior of the whole industry (see Table 6.2).

Because of the nature of the industry, the leadership of foreign private companies is not surprising. Historically, improved seeds for commercial agriculture—and especially hybrids—have come from the United States. This trade has always been anarchistic, as distributors and the farmers themselves purchase their seeds from U.S. sources and import them directly for their own use and resale. In fact, the presence of transnational

Table 6.2. Main domestic seed-producing corporations[a]

	Main seed line
Compañia Beneficiadora de Semillas	wheat, maize, vegetables, pastures
Impulsora Agrícola (1959)	barley
Mexagro Internacional (1965)	rice, soybean, safflower, beans, maize, vegetables
Monterrey y Compañia	soybean, safflower, sunflower
Semillas de Culiacan	safflower
Semillas Delicias	soybean
Semillas y Fertilizantes de Sinaloa (1960)	oilseeds
Semillas del Golfo (1975)	sorghum, soybean, maize
Semillas Master (1965)	sorghum, maize, vegetables
Semillas del Pacífico	cotton, safflower
Semillas Seleccionadas de Hermosillo	cotton, wheat
Semillas Sol (1979)	sunflower
Semillas El Yaqui	soybean

[a]Includes only seed-producing companies; excludes companies whose main activity is seed distribution or import; many of these offer open-pollinated seeds, selected from local grain merchants, for local sale.

Source: Direct research of Centro de Ecodesarrollo.

corporations is a result of efforts to improve the control of trade and competition in this new market. It is worth stressing that their operations in this country were not new; because of the country's great genetic variety, they had searched for Mexican varieties and progenitors of several species. They also took advantage of climatic variations to accelerate the process of experimentation with hybrids and the development of new varieties. The enactment of the 1961 legislation, the accelerated modernization of Mexican agriculture, and the increasing demand for high-yield seeds created great opportunities for companies that started to increase production of certified seeds from basic and registered varieties brought in from abroad and complemented their limited national production with imports. Their intervention succeeded in restructuring the market, since it generalized the use of trade names, strengthened a national network of distributors, and implemented many forms of marketing of agriculture inputs common in their countries of origin. Private corporation and national partners of foreign companies grew under the protection and encouragement that transnational corporations gave to the use of improved seeds and other elements of the Green Revolution's technological package.

In order to consolidate their position and to gain some degree of bargaining power with the federal government as it tried to regulate the industry's operations, AMSAC (Mexican Seed Association, A.C.) was founded in 1968. According to industry representatives, it was necessary to express clearly their wish to continue research, to produce new varieties, to market them, and to expand their share in the national market. Their lack of authorization to carry out private research was a serious obstacle—one that gave INIA a great advantage. The latter's lack of objectivity and slow pace in evaluating the different varieties produced by the private sector caused a permanent tension between the parties. AMSAC worked as an entrepreneurial association to press the authorities to facilitate and extend the scope of its associates' work. In spite of the atmosphere of mistrust and chauvinism that prevailed in the state bureaucracy regarding an industry dominated by foreign interests, AMSAC was quite effective in securing for some of its more important members (Asgrow, Ciba-Geigy, Horizon, La Hacienda, Northrup King, Ferry-Morse, DeKalb, Master, Impulsora Agrícola) official clearance to carry on their own research. It also succeeded in facilitating (though not accelerating) the process of evaluating new varieties.

In spite of obstacles, the private industry continued to expand toward areas of dynamic growth, that is, those dominated by modern farmers, commercial export crops, luxury products, or those destined for agroindustrial use (mainly fruits, vegetables, and sorghum). Its mechanisms to expand the markets and to penetrate the agriculture sector were very effective. They were mainly oriented toward strengthening a national network of independent distributors of agriculture inputs, including seeds, supported by national advertising and demonstration lots. In the face of bureaucratic restrictions, companies adopted different strategies: ignoring the formal obstacles to the development of their own research, pressing for expeditious approval of new varieties, or simply selling them illegally.

At a different level, PRONASE produced maize varieties and other species that resulted from INIA's research. It also had to orient its production toward farmers' demands, as clients of the official specialized banks were eager to improve their crops and to get subsidized credits. Initially, the agency explicitly defined its job in terms of basic crops, but with the passage of time and the consolidation of the private industry, it expanded its role to reflect its competitors' activities, investing in the production of a wide variety of species, especially vegetables. But these competitive activities were never effective in regulating the market, since PRONASE's operations did not guarantee adequate quality control. This dispersion diverted resources from its basic activity and weakened the whole structure. As the boom in sorghum persisted and grew, PRONASE also tried to compete in this market, by multiplying the hybrid seeds created by INIA. These efforts were also ineffective, because of both the lack of appropriate control over production in the countryside and the bureaucratic problems

Table 6.3. Certified seed production by PRONASE, 1970–1984

	1970	1971	1972	1973	1974	1975	1976
Basics							
Rice	1,520	1,808	1,210	2,799	3,686	11,726	5,495
Beans	830	1,667	4,807	1,447	5,675	13,894	9,925
Maize	8,011	5,645	4,555	4,662	7,789	14,839	17,069
Wheat	20,169	25,266	62,319	57,894	77,381	66,284	60,179
	30,530	34,486	72,891	66,802	94,531	106,743	92,668
Oilseeds							
Cotton	—	1,835	6,364	3,346	5,666	958	632
Safflower	917	1,282	1,148	704	2,963	4,290	249
Soy	163	5,658	4,916	13,011	14,381	9,449	6,089
	1,080	8,775	12,428	17,061	23,010	14,697	6,970
Industrial							
Oats	82	384	851	2,383	1,386	861	1,329
Barley	4	—	—	23	10	412	1,323
Sorghum	—	37	—	125	753	397	547
	86	421	851	2,531	2,149	1,670	3,199
Fruits and vegetables	31	9	27	85	25	556	127
Total	31,809	43,844	86,813	87,065	121,394	124,697	103,088

[a]Preliminary figures.
Source: PRONASE, Technical Division; and Second Governmental Report, 1984.

related to payments, handling of inventory, quality control, and dis-
private enterprise, the institutional obstacles were so formidable that the
commercial sector was not threatened.

PRONASE is a singular organization. Although it holds a privileged
position in the industry, with unlimited access to the research results of internationally recognized research institutions such as INIA and relatively
abundant financing to pursue its goals, its history is one of inefficiency,
incompetence at its highest levels, lack of an integrated policy to guide its
own activities, and the absence of a clear view of how the industry (or the
agricultural sector on which it is dependent and to which it is accountable) should evolve. In addition, this inability to fully utilize its potential
as a regulatory agency is compounded by the limitations imposed by the
nature of its relationship to INIA and BANRURAL, the state agricultural bank.

Although PRONASE has access to seed varieties "liberated" by INIA,
these are not always the most appropriate ones. Furthermore, its customers' demands often do not coincide with social needs. For example, it
has found little support for stimulating the use of *criollo* lines of maize and
the development of criteria for their selection and treatment. In its

Table 6.3. (continued)

1977	1978	1979	1980	1981	1982	1983	1984[a]
4,582	3,582	12,374	15,515	23,085	17,899	16,910	20,213
995	2,222	5,300	24,333	40,204	31,502	16,178	24,071
9,879	8,667	8,034	52,926	43,999	17,116	20,337	36,445
39,529	45,309	45,305	65,965	93,210	126,355	62,865	124,663
54,985	59,780	71,013	158,739	200,498	192,872	117,290	205,392
1,250	1,709	2,835	2,592	778	326	2	3,050
851	3,458	1,830	2,234	2,096	3,890	1,582	4,541
6,474	13,555	7,850	15,920	27,031	11,932	13,363	35,000
8,575	18,722	12,515	20,751	29,905	16,148	14,947	42,591
—	1,528	1,184	789	1,207	3,884	2,418	7,608
188	143	387	612	134	805	773	1,017
170	9,974	2,142	923	1,453	1,020	2,940	13,000
358	11,645	3,713	2,324	2,897	5,709	6,131	21,625
42	59	215	422	0	ND	78	370
64,246	91,908	89,316	183,278	235,246	215,467	138,763	272,090

Table 6.4. Relative shares of the certified-seed market, 1970–1977 (percentage)

	1970		1971		1972		1973	
	PRONASE	Private	PRONASE	Private	PRONASE	Private	PRONASE	Private
Basics								
Rice	38	62	42	58	52	48	91	9
Beans	100	—	93	7	100	—	99	1
Maize	87	13	81	19	54	46	76	24
Wheat	72	28	44	56	72	28	77	23
Composite	73	27	48	52	72	28	78	22
Oilseeds								
Cotton	—	100	9	91	30	70	16	84
Safflower	19	81	18	82	50	50	22	78
Soy	10	90	54	46	49	51	60	40
Composite	5	95	23	77	37	63	38	62
Industrial								
Oats	100	—	100	—	100	—	100	—
Barley	—	100	—	100	—	100	1	99
Sorghum	—	100	—	100	—	100	1	99
Composite	1	99	3	97	10	90	16	84
Fruits and vegetables	8	92	7	93	9	91	7	93
Total %	41	59	35	65	60	40	58	42

Source: David Barkin and Blanca Suarez, *El fin del principio* (Mexico City: Océano, 1985).

dealings with BANRURAL, PRONASE faces a structural obligation to respond to the bank's demands, which are not systematically structured to allow for adequate planning of seed production. The bank's contracts with PRONASE commit it to purchase "up to a certain amount," freeing the bank from any risk in planning and inventory management. These relations have clearly limited the operations of PRONASE, subjecting it to the demands of an inflexible financial organization and restricting the genetic resources it can draw upon to those available from a research program that is beyond its control. However, it is important to stress that, in spite of all these limitations, the greatest problems are inside the agency itself—problems that stem from a structure without clear operating guidelines and a lack of capable directors trained in management and technical strategies.

By the end of the 1970s, the seed industry was highly polarized. The government sector was producing a wide variety of products that had very little acceptance among commercial farmers. The private firms responded to the needs of commercial agriculture and contributed greatly to the productive transformations that generated the food crisis the country was

Table 6.4. (continued)

1974		1975		1976		1977	
PRONASE	Private	PRONASE	Private	PRONASE	Private	PRONASE	Private
75	25	52	48	62	38	100	—
96	4	76	24	92	8	84	16
60	40	75	25	62	38	72	28
45	55	40	60	40	60	49	51
48	52	47	53	47	53	55	45
40	60	12	88	7	93	6	94
48	52	37	63	6	94	12	88
69	31	42	58	20	80	11	89
56	44	34	66	16	84	10	90
99	1	60	40	100	—	—	100
—	100	4	96	10	98	—	100
5	95	4	96	2	98	—	100
11	89	8	92	6	94	1	99
3	97	47	53	24	76	16	84
46	54	42	58	35	65	27	73

about to experience.[5] As in many other areas of state intervention, competition between private and official producers generated mistrust. The official apparatus created problems of product quality and failures in production and distribution systems that seriously affected the capacity of private companies to correct their own internal problems. In addition, there was a great fear that, at any given time, the state would decide to enter the market on a massive scale, displacing private producers by means of subsidies and large imports of seeds produced by their competitors or even by affiliates in the United States. Tables 6.3 and 6.4 provide some quantitative characterization of PRONASE's production growth and that of the private sector in the 1970s.

5. There is a large literature on structural changes in Mexican agriculture. Mention could be made of the valuable *Documentos de trabajo para el desarrollo agroindustrial*, published in 1981 and 1982 by the Coordinación General de Desarrollo Agroindustrial of SARH, as a primary source of information. A summary of these transformations is found in David Barkin, "El uso de la tierra agrícola en Mexico," *Problemas del Desarrollo* 12(47/48) (August 1981–January, 1982): 59–85.

Until 1980, coexistence between the parties was precarious. It was clear that PRONASE was no real threat to the members of AMSAC: its production was too limited, too diversified, and poorly accepted. However, the private companies thought that it wielded great power, since it could break into the market at any time and could also manipulate the institutional framework to its advantage. In spite of this potential threat, the evidence indicates that the industry was not unduly constrained; private producers maintained rapid rates of expansion. Many, including some national firms, had gained prestige for their seeds and for their ability to ensure quality. The modernization of agriculture and the expansion of commercial production (in spite of the contraction and crisis in the cereal sector) augured well for the seed-producing companies. There was solid basis for optimism, and many planned to expand their processing plants and warehouses. Yet in 1980 AMSAC expressed its fears and doubts: the profound crisis in agriculture would force the state to take stronger measures, and the prevailing public mistrust of private, and especially foreign, economic power was a source of constant concern.

The industry was in conflict. When the Center for Ecodevelopment (Centro de Ecodesarrollo—CECODES) convened a meeting with the different parties to discuss the status of the industry and its dynamics in 1980, the need to redefine the pattern of coordination between the private and public sectors was evident.[6] It was also evident that a national policy to revitalize this basic industry was needed as part of the program to modernize agriculture and transform it in accordance with national needs. CECODES held this seminar to diagnose the disorganization of the industry and to develop a framework for collecting the information required for planning. Decisions on these issues were important because of the industry's key position in the operation of all available strategies for responding to the sectoral crisis. Although the seminar did not produce concrete results, it defined a series of issues that are still being debated.

Changes Induced by SAM in the Seed Industry

The announcement of SAM on March 18, 1980, caused a significant shift in the industry. On one hand, the increasing demand for basic foodstuff seeds induced a sharp jump in domestic production as well as in imports. On the other hand, it produced important changes in the types and varieties of seeds demanded as a result of the large subsidy that the government granted to maize. The social nature of this shift toward basic crops is also of great significance for our analysis.

---

6. See Ivan Restrepo, David Barkin, and Blanca Suarez, *Primer seminario sobre semillas mejoradas en Mexico* (Mexico City: Centro de Ecodesarrollo, 1980).

The Role of Seeds in SAM

SAM attempted, among its other goals, to increase production and productivity of basic grains, mainly maize and beans, but also wheat, rice, and sorghum. To this end, production subsidies of seeds and fertilizers complemented attractive support prices, credits, agricultural insurance, and so forth. Technical assistance in each agricultural area was also available.

As can be seen in official documents, the results of the 1981 harvest were spectacular (see Chapter 11).[7] Yet a more thorough analysis of this program's social impact poses questions about the efficiency of the process undertaken by a bureaucracy that lacked the effective mechanisms to utilize the existing mass peasant organizations and mobilize their political support to pursue its policy goals. Perhaps the most outstanding facet of the program was the substantial productive response of commercial agriculture to SAM's incentives. The limited disaggregated data available on regional responses of production point toward a possible social concentration of benefits among wealthier producers. Although the peasant population increased its production and utilized subsidies, as a social group it may not have received most of the benefits, since these were also directed toward the larger producers, that is, the commercial farmers.

One vital aspect of the program was the increase in unit yields per hectare. Seeds are an important input to increase productivity. The program assumed that better seeds and other inputs would have an immediate effect. At the same time, other projects were implemented to reduce losses due to pests and handling during transportation, storage, and processing. It was correctly foreseen that the seed demand would increase considerably in response to the 75% subsidy in the price of maize and bean seeds and a 30% discount in fertilizers. In the case of credit and agricultural insurance premiums, there were also heavy discounts. It is not surprising that both PRONASE and the private sector increased their supply and production levels to meet the increasing demand for seeds.

It is worth recalling that the use of improved seeds had already become general in other crops in Mexico. This was the case for wheat, sorghum, some oilseeds, and vegetables, whose markets continuously expanded and, with the stimulation from SAM, grew further. It is estimated that over 90% of the area cultivated with sorghum was using hybrid seeds, while 100% of the wheat areas was enjoying the improvements introduced as a consequence of the research of the Green Revolution. However, a substantial share of the wheat areas was cultivated with the producers' own seeds, because they are open-pollinated species that generally retain their fertility and productivity from one cycle to the next.

7. See SAM-SARH, *Informe de resultados del sector agropecuario y forestal, 1981* (Mexico City: SAM, 1982).

Nevertheless, in traditional crops where agricultural and economic policies have discriminated against peasant producers, as in the case of maize and beans, the use of modern inputs was quite limited. The portion of the crop using improved seeds (25%) in the 1930s as a result of the efforts of the Cardenas technicians has not been equaled in recent years. Their use has been limited to a small georgraphic area where the regular purchase of hybrid seeds was customary and where certified varieties are better adapted to local conditions. In spite of research in maize to create high-productivity (hybrid) varieties resistant to different pests and tolerant of some extreme weather conditions, very little advance has been achieved in their adaptation to the varied agroecologic mosaic characteristics of peasant agriculture, thus limiting their usefulness.

For this reason, it is not surprising that these problems were not overcome from one agricultural cycle to the next. Far from supplying seeds adapted to this varied mosaic, a considerable share of the seed in the market came from a small number of hybrid varieties. The private sector claimed that in 1981 it produced 11,500 tons of twenty-five maize varieties for use in irrigated and rain-fed areas, 10% above the 1976 level,[8] while PRONASE supplied twenty-seven hybrids, eight open-pollinated varieties, and five synthetic ones, accounting for 44,000 tons[9] (see Tables 6.3 and 6.4 for a detailed account of seed production during this period). The twofold increase in public-sector production in a single year was possible only because not all the varieties were hybrid. However, even this record production did not meet the total needs, which according to official estimates, reached almost 175,000 tons. (SARH's estimate and that of PRONASE are based on 50 kilograms of seed per hectare as a requirement for optimal cultivation.)

Other factors also contributed to the use of improved maize seeds. Credit institutions preferred to provide crop insurance for hybrids; since hybrids often require irrigation, fertilizers, and pesticides, they appear to offer less risk. Thus the extension of insurance to new producers' groups provoked a reduction in the use of *criollo* maize, which was replaced with hybrids. Most of the assistance programs reinforced these changes, since, in reality, local technicians showed a clear preference for hybrids in areas where they worked. Thus production and demand for improved maize seeds increased. An executive of an important private firm stated that "in reality, this agricultural year [1981] could not have been better for seed-producing companies, regardless of who and where they are, because increases were recorded almost in the whole country.... Even in areas like Tapachula [Chiapas], the demand for maize hybrids reached unexpected levels.... Thus," he concluded, "it is not that SAM was a big business, but

8. See Eduardo Correa, "Las semillas mejoradas, decisivas en el futuro del SAM, " in *Agro-sintesis* 12(1) (1981).
9. Information found in PRONASE's *Catálogo de variedades.*

rather a good governmental decision to increase production. Of course, those who had something to offer benefited."[10]

With the implementation of this program, the industry fostered a high-priority strategy at all levels: the promotion of hybrids as an integral part of a technological package that induced, and contributed to, the expansion of commercial forms of production in all parts of the Mexican countryside. Simultaneously, it reinforced the advance of profitable production at the expense of mass-consumption products, since the program also extended credit, higher guaranteed prices, and subsidies to these nonpriority products.

The Structural Impact of SAM on the Seed-Producing Industry

SAM policy had an important structural impact on the industry, which underwent an abrupt change as a result of the basic production incentives. There was an increase in demand for hybrid seeds that, supposedly, would be permanent. This increase is a tribute to the success with which peasant producers, even in remote areas, were informed of subsidies and of the link between productivity and improved seeds. Although they were not the ones contributing most to the quantitative response to incentives, they are the largest group of maize farmers in Mexico. Moreover, the medium-term impact of this promotional effort will be even greater, since the bad harvest of 1979 destroyed seed supplies, the 1981 subsidies induced peasants to sell their *criollo* seed as grain, and the 1982 frosts wreaked considerable damage on the crops. These factors will increase the need to purchase commercial seeds in the coming years.

Obviously, the increase in seed demand and the types and number of customers were not limited to peasant groups. Among larger farmers, however, the boom of maize cultivation ended with the program of incentives for other basic products. They showed no hesitation in changing their production once again to eminently commerical crops, the most profitable ones. In this same manner, it is to be expected that as other producers are incorporated into the hybrid seed "package" of modern agricultural inputs, they will be more readily inclined to change to other more profitable products, such as sorghum, and thus become permanent customers of improved-seed suppliers.

Just as SAM produced an unexpected opportunity for maize seed producers, now that the program has terminated we may expect a return to the accelerated pace of change toward the production of seeds for more profitable crops—those for exports, luxury consumption, and livestock. This implies a return to the production of seeds for vegetables and sorghum, a tendency that was confirmed in 1982 and 1983 with the new

10. From an interview with Octavio Vejar, sales manager of Semillas Hibridas, in *Agrosintesis* 12(11)(1981): 97–101.

PRONASE campaign to enter into the sorghum market through the purchase of basic seedlines from U.S. researchers for adaptation and reproduction in Mexico. For reasons discussed elsewhere, this should accelerate the trend toward the reorientation of the Mexican countryside toward livestock.[11]

Finally, a significant structural change in the industry may be foreseen. With the trend toward an increase in the use of improved seeds, especially hybrids, and the greater dependence on the world market for basic foodstuffs, one may anticipate the greater strength of the private sector. Worldwide trends toward the concentration of seed industries, as well as the pace of technological change, have so accelerated that national structures will not be able to remain separate from the increasing range of products likely to be required by national agricultures. National institutions might retain a certain degree of control over wheat and maize, but these will account for a decreasing share of the whole industry. Under these circumstances, PRONASE will be forced to react to the transformations in world markets, thereby inevitably remaining behind. PRONASE is thus losing its capacity to regulate the market, and the Mexican seed system is facing greater difficulties in guiding the development of a national industry.

Some Implications

SAM's varied implications for agriculture are explored in other chapters of this volume. Here we concentrate on those related to peasants and national genetic resources. SAM publicized and massively promoted the use of a modern technological package that was indiscriminately financed through subsidies with inadequate regard to its productivity. Among other factors, it sparked the move to increase production and productivity in areas devoted to basic crops.

Despite significant increases in the volume of food produced, there were problems. When peasants are induced to use industrial inputs and costly technologies, they are led to depend, cycle after cycle, on the seed and agrochemical markets. However, in view of their post-SAM limitations in obtaining adequate state subsidies and bank credits, it is difficult to guarantee their continuing access to these markets. On the other hand, it could be that by offering a significant subsidy for seed purchases, traditional patterns of seed use and conservation will be substantially changed. These tendencies are of interest not simply for their traditional nature, but also because they preserve genetic wealth and diversity. This is an issue of major concern because, as modernization of agriculture advances, there is a tendency to substitute or lose a good portion of the

11. See David Barkin, "El problema ganadero: Resultado de la modernización agropecuaria," in *Comercio y desarrollo* no. 10 (April–June 1979): 4–9.

genetic resources that have, in the past, stimulated agricultural development.

From the very remote past, peasants have preserved and selected their seeds for cultivation. Traditionally, they chose the seeds for the next agricultural cycle from each preceding harvest. With the passage of time, the conservation and reproduction of often unique seed varieties have continued, and together with natural selection, peasants have maintained the basic characteristics of the seeds without manipulating their genetic base.

In this context, the spread, promotion, and subsidy of the so-called miracle (hybrid) seed reduces the use of *criollo* varieties. Many valuable genetic traits may be lost, and the possibility of exploring the characteristics of natural evolution or induced mutation of this input destroyed. This new trend of commercial hybrid seed development, therefore, threatens food security and capacity by reducing genetic diversity. Furthermore, part of the genetic stock and technology to produce hybrids is controlled by a small group of large producers. The possibility of producing adequate supplies of food domestically is also conditioned by world market structures and the control of technology.

A related concern is the modernization process that increases national dependence on foreign supplies of food. For a long time, and with the outstanding exception of the short lifespan of SAM, Mexican agricultural and livestock policies have fostered the displacement of cereal production by other highly productive or export-oriented crops. Such a shift has stimulated import of large volumes of cereals for basic human and animal consumption. However, this could be a dangerous strategy, especially if it becomes generalized in the Third World, since these low-income countries will be exposed to the fluctuations of a volatile world market characterized by a high concentration of monopolistic power and political maneuverings.[12]

The problem of food security is of great importance in any discussion of seed production. The first difficulty is the contrast between the high adaptability and diversity of *criollo* varieties and the homogeneity of hybrids. The hybridization program aims to create this homogeneous product, but this course is inadvisable given the varied conditions that prevail in the Third World. The variety of, and frequent changes in, pests and weather conditions impose inordinate costs and technological challenges on a hybridization program. Furthermore, the assumption that hybrid seeds are more productive socially than *criollo* varieties is debatable. In the light of their high production costs, monopolistic control, and the package of complementary inputs they require, it is uncertain whether the costs of such a program are outweighed by the increased production of the hybrid-

12. See Eduardo Santos, "El mercado mundial de cereales: Las opciones para el Tercer Mundo," *Comercio Exterior* 33(6) (June 1983): 551–555.

seed system, especially in societies in which peasants constitute a substantial proportion of the population.

It is worth noting that in the past few years a program for the distribution of *criollo* maize was implemented through INIA and PRONASE, with the additional support of SAM. It was aimed at a short-term increase in maize production through the utilization of selected *criollo* varieties that were treated to reduce the incidence of diseases and animal damage. Thus, there have been attempts to implement a type of technology suitable to peasant resources and techniques, attempts that, at the same time, preserve and guarantee the conservation of germ plasm in traditional varieties. This is a more realistic but difficult alternative for agricultural innovation than the pursuit of hybridization. Microecological considerations are important, and they favor a wide range of varieties adapted to specific environmental and weather conditions. During the first years, at least, and in view of the lack of adequate planning, it seemed that at the operational level the program experienced substantial failures because of the lack of appropriate care of the *criollo* varieties distributed in the rain-fed areas.

The data on the results of the program in the different rain-fed areas are limited. Nevertheless, it is foreseeable that the results will be less favorable after the agricultural cycle of 1981 because of the disappearance of the largest subsidies for seeds and the great promotion of commercial hybrids that are displacing *criollo* varieties. With the severe drought of 1982, it may be that the homogenizing impact on the genetic stock will be aggravated for *criollo* seeds as the millenary mechanisms of selection, preservation, and differentiation are impeded by the transfer of material from one area to another through the official distribution network. Furthermore, the economic austerity policy of the new de la Madrid administration does not strengthen this type of agriculture.

With this panorama, the situation is quite alarming. Production increases and productivity are now top-priority issues, but it is also important to preserve, safeguard, and extend the genetic resources that represent the food potential for future generations. This requires the use of research and inputs that are more within the reach of small producers, ones that do not require the use of hybrids and the complete new technological package. In this manner, a certain degree of autonomy could be achieved, and the payment of foreign exchange for the use of technology would be limited by utilizing national genetic potential and human resources to develop alternative technologies. This would slow the rate of expansion of the new technology in the countryside, which has been displacing peasants from rural production. (This is not to say that technological improvements should not and could not be implemented to increase peasant productivity, but rather that the approach used so far has been largely oriented toward other goals.) This would require a policy of support and incentives much more specifically directed to long-term social needs.

Actors in the Seed-Producing Industry

The Official Sector

A deep conflict persists between the public sector and corporations within the industry. The current dominant group would make the industry a pillar of support for the transformation and modernization of agriculture, accentuating its inclination to agro-exports and livestock. In this program, as was suggested earlier, PRONASE would produce seeds for commercial products and would gradually reduce programs that presently amount to subsidies to the peasant sector. INIA would also work along these lines by reinforcing its work in vegetables, sorghum, and pastures.

The opposite view proposes greater emphasis on the peasant economy. It points toward the danger of dependence on foreign food and toward the absence of alternative uses for many of the human and material resources in peasant societies released by modernization. In Mexico, support for the peasant sector need not compete with the promotion of commercial agriculture, because they operate in different regions and generally are oriented toward different markets. This view advocates a decrease in the subsidies to export products, which, of themselves, are quite profitable. There are also ecological reasons for supporting this type of approach: the areas lagging behind in commercial development are large, and their decay due to the extension of modern techniques is imminent. Finally, the social costs of abandoning the peasant sector are great: this development model offers no alternatives of productive employment for the displaced in the economy. Inevitably, the modernization strategy will create social unrest and internal political conflict. But the group that argues this view is in the minority and lacks significant influence in the design of seed-production policies.

With the end of SAM, the conflict does not appear to be easily resolvable. The advocates of international exchange and specialization are in control. Thus, at the beginning of the present administration, the minister of agriculture reaffirmed a historical conviction that "perhaps it will be best for the country to produce only 70% or 80% of its maize and 90% of its bean requirements, and import the balance, in exchange for exporting other more profitable agricultural products. . . . If we give too great an importance [to the cultivation of maize], we could not diversify our agriculture and economy."[13]

Notwithstanding this declaration, a few months later the administration announced the National Food Program (PRONAL) (see Chapter 15). The new program aims to "ensure the attainment of the objectives of food sovereignty, as well as food and nutrition conditions that will allow for the

13. *Excelsior* (April 22, 1983): 10-A.

full development of the capacities and potentialities of every Mexican."[14] Even though the program does not offer concrete plans for the achievement of these goals, it is clear that, once again, the goal of food self-sufficiency is at the center of the stage. Thus it seems evident that, although powerful groups still insist on the imperative of following the principles of "comparative advantage" and international trade, the current economic crisis is forcing the government to consider alternative strategies that offer the fuller use of available national resources, even if their immediate financial cost is higher than the short-term cost of imports. No doubt the lack of foreign exchange and the idleness of lands and peasants has made it necessary to reevaluate the position of some prevailing groups regarding the external sector. The conflict, then, is still on the agenda.

The Private Sector

Members of AMSAC may enjoy another privileged period in their history. The boom of the demand for improved seeds is, very likely, only in its early stages. Real restrictions on research and productive activity have decreased, although there have been no changes in the legal problems.[15] At present, the private sector advocates the country's adherence to international patent agreements to protect its innovations, but others object that this will allow for the acquisition, as private property, of varieties that are today available to all. The international trend toward centralization of the industry in the hands of transnational corporations that participate in various agricultural activities, in the pharmaceutical industry, or in others has resulted in the incorporation of Mexican seed production into new worldwide circuits of production and marketing. For Mexico this implies, on one hand, access to new technologies, and, on the other, restrictions on the varieties that will be available in the market.

Direct Producers

At present, there is a tendency to support those that are most efficient and to limit resources that are channeled to traditional producers. This implies a further strengthening of improved-seeds production, because the best are identified with the use of a modern package and are given credits, assistance, and productive protection. To survive, peasants will

14. Poder Ejecutivo Federal, *Programa Nacional de Alimentación, 1983–1988* (Mexico City: PEF, 1983), p. ix.
15. Subsequent amendments to the current Mexican law on the production, certification, and trade of seeds do not significantly alter the organization or operation of the industry in the private or public sectors.

probably have to act and respond firmly to this situation. The forms and consequences of such a response are yet to be seen. But as far as seeds are concerned, there is the danger that the erosion of the stock of *criollo* seeds in the past few years will severely limit the alternatives of this social group in particular, and of the whole country, in a future where food self-suf- ficiency will again be on the agenda.

Conclusions

The Mexican seed industry has changed substantially in the past few years. With SAM, its already high rate of growth was further accelerated, imposing a change in crop patterns because of the importance accorded to maize seeds. The productive response to incentives was the virtual explo- sion of the official supply of improved maize seeds. Although there are still doubts regarding their quality and the impact on genetic reserves of the country, the experience clearly shows a productive potential that is usually underrated.

This review of the industry also highlights the conflict among the various industrial groups. The private sector follows, almost blindly, the dynamics imposed by the development of agriculture on a worldwide scale, neglecting the needs of traditional producers in favor of their wealthiest customers. The public sector is hesitant. Its mandate, to regu- late the seed market and to ensure adequate and timely supplies, could be met either through (1) direct competition with the private sector in pro- ducing its same lines or (2) an integrated production and research pro- gram in closer agreement with national food production and social needs. At present, PRONASE is attempting the first alternative, directly con- fronting the private sector, though with insufficient resources and without technical support or administrative and operational capacity to guarantee a quality product. Thus PRONASE imposes a certain instability on the in- dustry, by trying to compete and implicitly threatening a more rigid con- trol of the whole without really being able to exercise this control.

There is still a fundamental problem to be solved: the supply of basic foodstuffs for the domestic market. The satisfaction of the minimal food needs of the country still requires large amonts of foreign exchange, while parts of the peasant sector remain unproductive because of the lack of adequate supports. In the debate over economic policy, "comparative ad- vantages" of foreign trade are like mythological sirens—an attractive doc- trine, colorful but inevitably leading the country to declining degrees of self-sufficiency. The theory of "comparative advantage" offers no ade- quate response to the demands of economic policy, since the various mo- dalities of protectionism and monopolistic control of the world markets of primary products prevent its full operation. Given forecasts of future re- strictions in world food supplies, increasing demand, and deep national crisis in basic production, it is difficult to characterize the arguments of those advocating self-sufficiency as mere rhetoric or ideology.

The crumbling of this capacity for food production has profound implications. There is an implicit assumption in the official sector that peasants will continue to maintain a stock of *criollo* seeds and their genetic base, as they have done for centuries, in spite of destructive tendencies encouraged by state policies. As part of this same attitude, this policy accords to the new seed varieties—supplied by companies within the sector and by basic and applied research institutions—a central role in the solution of the problem of agricultural supply, regardless of their composition by crop. Blind trust in technology and in the market as tools to increase productivity and to maintain the balance between supply and demand, so characteristic of this period in economic policy, generates a disregard for those who foretell of ecological disaster from the pursuit of this course.

Seeds provide the genetic reserve for further production. A large number of traditional varieties are being displaced or have and will become extinct through the use of improved varieties. No matter how one examines the agricultural and livestock situation at the beginning of the decade, serious concerns arise regarding the will to recreate and maintain the country's capacity for food self-sufficiency. The scientific community has clearly shown that the monetary interests of private producers are not adequate to guarantee the preservation of genetic resources or to adapt supplies to social priorities.[16] Ironically, Mexico—a country with an institutional structure capable of guaranteeing the expression of collective interests, and one that has most insistently spoken of the urgency to maintain and expand the international collections of germ plasm—has not taken the necessary steps to create a consistent industrial policy in this sector.[17] Thus the seed industry clearly reflects the unmet challenge to create a coherent agricultural and industrial structure.

16. See International Union for the Conservation of Nature and Natural Resources, World Conservation Strategy (Geneva: IUCN, 1980).

17. Mexico took the initiative to present a resolution to the General Council of the FAO in 1981 to create an "International Germ Plasm Bank." This proposal was discussed again at the 1983 meetings and, although it received widespread support, opposition from transnational industries and some advanced countries prevented its adoption.

CHAPTER 7

Generating and Disseminating Technology

Gustavo Viniegra Gonzalez

Although by the end of 1982 SAM had succeeded in establishing a public and institutional consensus on the need for a food strategy centered on self-sufficiency, it had not succeeded in establishing a consensus on the best methods to achieve this goal. One of the most important issues, about which there seemed to be no agreement, was the generation and dissemination of technology in the countryside. Nevertheless, there was a consensus that technological changes were needed to increase productivity.

In a recent symposium, "The Food of the Future" (organized by the University Program on Food, PUAL, Mexico City, 1982), national and international experts seemed to differ on a primary point: the form of productive organization in the countryside that might be most successful in increasing productivity. One group of experts favored the transformation of Mexican agriculture toward a highly mechanized and extensive system in which the traditional *campesino* (peasant) organization would be replaced by a modern entrepreneurial organization, either private or public. Another group saw a need to reinforce, broaden, and rearrange traditional farming organization, based on a labor-intensive system, with moderate or low levels of mechanization appropriate to the kinds of community organization of each region.

In this chapter I do not provide a general analysis of these major currents of thought about rural technology but instead have the more

The author expresses his appreciation to the following collaborators: Gabriel Hernandez and Javier Orihuela, whose fieldwork in Morelos was partially financed by the National Institute for Adult Education (INEA); David Martinez and his colleagues of PREDEPAC, for their information on the Guanajuato community; Alberto Munguia and Gerardo Ramirez, for their observations in los Altos de Jalisco; and Luis Cabello, Daniel Pamanes, and Adolfo Norton, from the CIIDIR of Vicente Guerrero, for their information and collaboration on the Durango research.

limited aim of discussing a number of field experiences related to the tasks
of generating and disseminating technology for the production of grains,
and of different attempts, more and less fruitful, to promote the integration of crop and livestock production in some communities in northern
Mexico. The four case studies reveal different dimensions of strategy and
provide the basis for the following insights regarding the type of technology dissemination central to the SAM strategy:

Case 1. Rigidity of technological packages is a fundamental flaw in the
policy design; it is essential to recognize the heterogeneity among
small farmers, even within a single community.
Case 2. Macroeconomic policies can unintentionally discriminate
against small farmers if the industry structure and dynamics have
not been adequately analyzed; furthermore, it is critically important to link production technology to the processing and marketing situation.
Case 3. There are inadequate links between the bank and the research institutes and between the program designers and the farmers.
Case 4. It is feasible to integrate the research and dissemination of
technology.

These case studies reveal the texture of the real problems of generating
and disseminating technology right at the farmer's level. It is there, ultimately, that the fate of food policy implementation is decided.

Production of Basic Grains

Ever since Mexico began to import grains in 1973, it has recognized the
importance of increasing the production of cereals, particularly maize.
Unfortunately, in recent years the growth of areas devoted to the cultivation of these basic grains has stagnated. There are physical limitations on
the availability of irrigated cultivable lands or of those with good rainfall,
and there is significant competition for the use of these areas from other
crops, such as sorghum. In an attempt to overcome these limitations, the
National Institute of Agricultural Research (INIA, see Chapter 6) undertook the so-called Plan Puebla to increase the productivity of the maize-
growing zones through a technological package that involved (1) an increase in the level of chemical fertilization, (2) an increase in the density of
cultivated plants, (3) greater attention to weeding and pest control, and (4)
the planting of improved seeds in experimental fields. Plan Puebla served
as a major pilot project in the development of location-specific technology
and its dissemination. As it has been subjected to many other analyses, let
it suffice to say that SAM adopted this technological package as the main
method for increasing maize production, through credit measures, agri-

cultural insurance, and distribution of fertilizers. This was an attempt to increase productivity in spite of the existing limitation of the cultivated area.

One way to explore the application of this package is to study how it was applied in communities of different levels of development. In this section two illustrative communities are presented: (1) a community in the state of Morelos, characterized by its lack of water, which operated under the parceling system of *minifundios* (small plots of land) and was subject to numerous internal tensions and conflicts; (2) a community in the state of Guanajuato, with abundant resources (irrigation, machinery, etc.), which operated with compact, collective parcels but was subject to important limitations on the marketing of its products (Bartra, 1979).

The analysis of the first community shows the limitation of SAM's technological package. The analysis of the second community illustrates the institutional and economic limitations of the application of the agronomic package when it is not linked to the problems of marketing and industrialization of basic grains, even in very favorable cases in which the production goals are achieved and apparently optimal conditions for the organization of rural labor are found. My goal is to show the complexity of the social and productive fabric of rural communities and of the way in which the application of the technology is subject to the complexities of that fabric.

Case 1: Disintegration and Reintegration in a Rain-fed Community

Despite many studies on the state of Morelos, little has been said about the dynamic problems of the relations between the peasant strata and factions within a rural community, or of how these relations affect the acceptance or rejection of a certain technological change. The case presented here is derived from a three-year study of a community that was part of a project called "Community Education." The project attempted to mobilize the peasant community's own resources to establish adult education programs and to promote activities benefiting the community, specifically through agricultural production, health and nutrition, alternatives for marketing surplus production, and employment.

This community is representative of others in Morelos in transition from a traditional organization with indigenous roots to a peasant community absorbed into a mestizo society. At present, about two-thirds of the peasants neither possess land nor have access to *ejidal* lands, simply because of population growth.

This community has over 6000 inhabitants grouped in about 750 families. The land is divided mainly in three ways: small plots (*minifundios*) found mainly in the lowlands, *ejido* parcels, and properties of the old indigenous community, which are usually used as pasturelands or for common use in projects such as schools, cooperative farms, or recreational areas.

The *ejido* lands are usually not irrigated and are therefore dependent on uncertain rainfall, which, in turn, raises the risk of losses in the traditional crops such as corn and peanuts, and new crops such as grain sorghum. On the other hand, irrigated *minifundios* are usually devoted to commercial crops, such as vegetables (tomatoes, cucumbers, green tomatoes, squash) or sugarcane.

There are few apparent signs of differentiation by social class. For instance, there is no visible rural middle class; no professionals, merchants, or public officials have established themselves as a social stratum distinct from the peasantry; even school teachers have been strongly assimilated into rural life. However, there are signs of stratification among the peasants themselves. On one hand, there are poor peasants who usually are not entitled to direct use of the land and who sell their labor as occasional field workers. There is also a small group of rich peasants who own good lands, farming equipment and oxen, and even some tractors, and who devote efforts to commerce, usurious credit, and politics. One of the apparent functions of this group is to serve as intermediaries between the *ejido* producers and the credit and technical-assistance institutions, such as BANRURAL and the Ministry of Agriculture and Water Resources (SARH). They also serve as middlemen between the state administrative and political organization and the informal and traditional organizations of the peasants. Between these two groups are middle-level peasants, those who have access to farmland and can employ their families mainly in agriculture and employ the poor peasants, but who lack sufficient capital and organization to establish themselves independently.

In this social context, the adoption or rejection of a certain technology is usually the result of a series of commitments established between the different groups of peasants, who frequently have antagonistic interests. Also, these commitments are subject to the restrictions of regional pressures. In the case we are discussing, the following concrete interests may be mentioned:

1. The poor peasants wish to obtain cheap and available corn in return for their day's work.
2. The middle-level peasants wish to retain a significant part of the corn for household consumption and to sell their corn and sorghum surpluses at a good price.
3. The rich peasants hope to buy and collect the greatest possible amount of grain at the least price in order to resell it with a maximum profit, whether for community consumption or for regional trade.
4. The industrial groups, whether state or private, wish to obtain a secure

and cheap supply of grain to use as raw material in the manufacture of tortillas (maize) or balanced feed (sorghum) for livestock.
5. The political-administrative groups (regional bureaucracy) wish to maintain their control and privileged position as middlemen between the countryside and the city.

During the June-December cycle in 1981, the year of SAM's biggest push, the institutions (SARH, BANRURAL) tried to attain the goals established by SAM by implementing a technological package consisting of (1) improved seeds; (2) subsidized fertilizers; (3) insurance against losses from "acts of God" (droughts, etc.); (4) advance payment of wages for the cultivation of basic grains; and (5) technical assistance for improving crops (increase in the density of planting) and for fighting pests. Rain was abundant but irregular during this cycle and the peasants frequently complained about the late delivery of fertilizers. They also complained about the lack of technical assistance for fighting pests, mainly worms, that invaded the maize roots.

Several direct observations of this crop showed a marked difference between the yields obtained from the so-called improved variety distributed by SARH and those of the local *criolla* variety. Usually the improved variety was smaller and infested with mold (*pochcahuate*), while the *criolla* was less moldly and had a larger size and yield. Interviews with peasants suggested that most of them mistrusted the improved variety and were unwilling to plant it. But they were forced to take the technological package as a whole; otherwise they would have no access to agricultural insurance nor to an advance in wages, which was essential in order to free themselves, at least partly, from the need to resort to the local moneylender. Peasants usually need loans from the local moneylender, which they consider to be a great calamity. While the lender lends without written security and gives the money immediately and without further arrangements, he is part of the community, knows his customers well, and can bring pressure to bear in case of the customer's failure to pay. Moreover, he collects in the form of crops that are still green and, therefore, have a lower value than if he had collected in grain. This happens frequently, because the greatest need for money arises before the harvest, precisely because of the need to pay for wages and the family's subsistence. The moneylender thus receives a high rate of interest through the revaluation of the harvest, which can easily double or triple their value from the field state to the "grain" state, especially if the grain is sold during times of shortages.

Thus the middle-level peasant faces the choice between accepting a frequently inappropriate technological package with credit advantages, or implementing his own technology with great uncertainty about credit and the stability of the financial cycle. If one adds the existence of back debts to BANRURAL, which condition the renewal of the credits and the acceptance of the bank's instructions, one can see that the peasants have few options regarding the technological package transferred by the institutions to the countryside.

Nevertheless, it is interesting to see the emergence of new alternatives, which indicate on a modest scale the peasant organizations' great adaptive capacity. In one case, a group of twenty-eight families, assisted by external advisers, obtained a loan to drill a well, install an electric pump, and irrigate a limited area. One result of this organization, which curiously enough came through a program promoted by the Federal Electricity Commission, an agency outside the agricultural sector, has been a wider variety of crops, including vegetables. The families also have introduced soybeans as an alternative crop for household consumption. This group serves as an example of the possibility of achieving economic progress by local organization of the middle-level peasants. But its small size and isolation from the rest of the community prevent these results from becoming a relevant community alternative.

Another group of peasants, in this case in the rain-fed areas, organized themselves around another alternative: the establishment of a transport cooperative, that is, the collective purchase and operation of a small 3-ton truck. In order to do this, this group had to resort to financing from a private foundation, since BANRURAL or any other of the national banks rarely lent money for the purchase of motor transportation in the countryside. This purchase had a greater impact than the drilling of one isolated well, since the transportation of fertilizers, grain, people, or any other material is a primary need for the subsistence of the peasants. It is precisely through transportation that the middlemen exert their control on the crops or on the distribution of fertilizers. The impact achieved by this promotion was immediately felt by the *ejidatarios,* and in a recent election of the *ejido* board, the old board, which had been feared and resented by most of the peasants because of its corruption and collusion with the BANRURAL authorities and regional politics, was replaced by representatives of the transportation cooperative.

It would be premature to conclude that this election of the new *ejidal* board will produce radical change in the administration of the *ejido* and lead to a new stage of negotiation about the adoption of BANRURAL's technological packages—this time more appropriate to the local production conditions. Rather, these observations point out the complexity of the social fabric in which the adoption, improvement, generation, and dissemination of technology takes place.

Case 2: External Limits of Rural Development

As we have seen, considerable internal conflict in a rural community serves to limit its social and economic development. In this case, a small rural Guanajuato community (500 inhabitants) in which class differentiation among the peasants had not taken place was studied. Thanks to a fortunate coincidence of ecology, good institutional relations, and general harmony, the community had been able to take advantage of much institutional support, to the point of adopting an advanced collective production and marketing technology for cereals with high yields per harvest (5 tons per hectare for wheat and 7 tons per hectare for sorghum). Nevertheless, the community was subject to outside limiting factors such as low guaranteed prices, increases in the price of fuel and electricity, lack of credit for agroindustrial activities, and the existence of monopolies subsidized by the state.

This small community was formed as the direct result of an agrarian struggle. It was consituted as a settlement of ex-laborers and day workers of adjoining haciendas who obtained, 40 years earlier, the consitution of an *ejido* and a population center in a place where no community had existed before. Its small size, the existence of many interrelated families, and its revolutionary origin created an organization with few fundamental conflicts. This homogeneity facilitated the acceptance of collective organization in agricultural work, a form promoted by the federal government during the 1970–1976 administration but resisted by most of the agrarian communities throughout the country. Apparently, mass collectivization required a homogeneity of interests and forms of organization rarely present in a Mexican *ejido.* But in this case it was easy. Three collective groups were organized for mechanized planting and harvesting of cereals in compact areas, irrigated by wells with electric motors and worked with modern machinery obtained through a BANRURAL loan.

The high yield achieved by irrigation and mechanization, along with modern fertilization and cultivation techniques, led to yearly harvests of up to 12 tons of grain per hectare in two cycles. These results were accepted by most of the *ejidatarios,* despite the fact that some of them (38%) decided to remain outside the collective as individual *minifundistas.*

The advantages of collectivization were soon clear: sufficient credit, mechanized work that freed part of local labor without diminishing family earnings, high yields in comparison with the noncollectivized rainfed areas, and collective bargaining power when dealing with government institutions. However, there were some disadvantages, the main one being the long-term commitment to BANRURAL in respect to lines of production, since the existence of a debt prevented the free use of the land (e.g., to plant more profitable crops such as vegetables for export).

The commercial obligation to CONASUPO tied them to the conditions for the purchase and sale of grains established by the federal government. CONASUPO establishes prices that in many instances are lower than those offered by private merchants for wheat, sorghum, and maize. Moreover, CONASUPO had frequent delays in receiving products and in paying for crops. There were frequent problems in inspecting the grain and in establishing price penalties for defects such as humidity or insects.

To solve these problems, the peasants, with the advisory services of PREDEPAC (a private consulting service), decided to seek an agroindustrial alternative to increase the value of their harvests. They formed a cooperative for the production of concentrated balanced feed based on sorghum. They received advice from a nutrition specialist, obtained a formula suitable for local dairy cattle, and with their own resources began a pilot demonstration with 20 tons of locally produced sorghum.

The positive result of this experiment, measured by the formula's acceptance by local dairymen and by a profit of 20%, encouraged them to consider building a factory with the capacity to process most of the sorghum produced in the *ejido*. But an economic assessment revealed a fundamental problem: the project would not be profitable if the machinery was purchased under the current credit conditions (33% yearly interest) and if the sorghum was bought, without subsidy, at free-market price. This situation exists because large companies maintain their market position through a federal subsidy, which in practice facilitates the purchase of sorghum at a price lower than the market price and permits the sale of balanced animal feed at a price that would be ruinous for small- or medium-sized companies lacking this subsidy.

The community was able to solve this problem thanks to a grant from a European foundation for the purchase of the machinery. In this manner it achieved a favorable (pro forma) economic assessment, which allowed it to obtain the support of other foundations, such as the Ford Foundation, for the expertise necessary for agroindustrial development.

This case study displays well the closed web of subsidies, preferential interest rates, monopsonistic arrangements in the purchase of grains, contracts tied to the production of cereals, and inflated costs of machinery, fuels, electric power, and fertilizers that put the peasants in a disadvantageous position with institutions that, on one hand, favor increasing their primary productivity and reducing their risks, but, on the other hand, limit them to producing cheap raw materials. Beyond this, there are important external restrictions that considerably limit the prospects for integrated rural development with an agroindustrial technology in the hands of the peasants themselves. It is not a case of administrative corruption or inefficiency. It is a case of macroeconomic policy that in practice discriminates against the small company, whether privately owned or collective, and that definitely favors large capital, whether private or state. Thus the adoption and dissemination of technology cannot be seen as exclusively a problem of rural organization; it must also be considered as a problem of national and international interests, one in which the alternatives rarely favor a greater economic integration for all primary producers.

Integration of Agriculture and Livestock

An important part of the SAM strategy was to reduce the nutritional competition between humans and domestic animals. To this end, two main projects were presented: (1) the production of new tropical crops to substitute for sorghum, such as cassava, and (2) the use of agricultural and agroindustrial residues to replace cultivated forage.

The implementation of these two projects as part of SAM was on a much smaller scale than the support for basic grain production. Only pilot field experiments were carried out for the production of cassava, and few factories were established to mix corn stubble with cane molasses to distribute as livestock feed during the dry season.

In spite of the second project's limitations, it has great potential for the reduction of grain imports and for the use of cultivable lands appropriate for basic grains for cattle. For this reason, it is interesting to study the manner in which this technology was applied in the countryside; two cases were selected for study: (1) a plant to mix molasses with corn stubble, built by BANRURAL, and (2) a plant to process the stubble, built by the National Polytechnic Institute. The first case illustrates the shortage of scientific information, or of feasibility studies adapted to local needs, for the development of these projects at the service of rural communities. The second case was taken as an example of alternative technological development in which basic research was linked to the peasants' practical experience. These isolated examples also highlight more complex aspects of the relation between banking institutions and the peasants, of the latter's limited access to agroindustrial technology, and of the problems that appear during the formulation and assessment of preinvestment projects conducted by official banking institutions. They also point out new kinds of institutional relations that might be of interest in the development and dissemination of technology in the countryside.

Case 3: Vertical Promotion of an Agroindustrial Project

A factory to mix molasses with stubble was built in Los Altos de Jalisco in the western region of Mexico, where livestock raising has been rapidly developed for dairy production. The commercial use of this type of mixture is a common practice in the Mexican cattle business. The large milk-processing plants, like Nestle, frequently sell these products to the peasant in order to ensure their supply of milk. This trade is carried out by the middlemen, called *boteros,* who buy raw milk from the ranches, sell supplementary cattle feed on credit, and transport the milk for resale to the large milk processors of the region.

In this case, a union of *ejidos,* that is, a society of rural communities, arranged for the construction of a factory for the production of stubble mixed with molasses. Their intention was to eliminate the small producer's dependence on the milk-processing companies by using their own raw materials or cheaper ones.

The negotiations with BANRURAL led to a voluminous feasibility study that generated the technical guidelines for the construction of the molasses-mixing plant. The feasibility study was undertaken with credit that was later granted to the union of *ejidos,* but it was carried out without the participation of the credit recipients, who were not informed of its specific results. Analysis of the feasibility study revealed two important aspects: (1) The formulas for the molasses mixtures and stubble were not justified on the basis of scientific references or known studies. They simply suggested the use of 50% of each ingredient. (2) The marketing study was based only on the 1978 census data for the cattle inventory of the *ejido* union. The potential theoretical market was considered as real; that is, it was assumed that all the producers who were members of the union would constitute a captive market for the molasses-mixing plant. With these two estimates and an additional assumption of a daily consumption of 8 kilograms of mixture per head of cattle, the flow of raw materials and finished products, the size of the needed equipment and warehouse, and the factory's profitability were calculated. The large size of the project (30 million pesos in 1980) gave rise to a debt that was charged to the *ejido* union as the credit recipient. The peasants, for their part, lacked information on the study, the construction process of the factory, and the commercial operating conditions of the project.

An investigation, requested by the peasants, revealed that there was no log book of the construction of the installations and that, therefore, it was not possible to evaluate the justification of the building costs. The lack of information and control prevented the peasants from expressing an opinion on the basis of the project's assumptions, so their only options were to either accept or reject the credit at the end of the construction period. Rejection of the credit entailed the conversion of the factory into a warehouse for fertilizers or its operation as a state-owned industry. Acceptance of the credit implied the responsibility of paying a debt for equipment about whose technical and commercial operation they knew nothing.

The *ejido* union decided to accept the debt and tried to run the plant. The results were bad. The cattle found the mixtures of molasses and stubble unappetizing. The actual consumption of these mixtures was less than 8 kilograms a day. The price offered turned out to be much higher than that of other molasses-mixture feeds sold by other private or state industries. The net result was short-term insolvency of the *ejido* union and BANRURAL's intervention in the operation of the factory.

A local market study for dairy cattle feed revealed that molasses mixed with corn stubble and urea could reduce the price of concentrated feed by 30% compared to that based on grain flour and oilseeds, without changing the nutritional value of these mixtures very much. The foregoing observations seemed to indicate that it was feasible to operate a factory for molasses-mixed feed, with an important change in the conception of the process. It was necessary to include the warehouse and grain mill, the solids mixer, and the urea shaker in order to produce more complex and nutritionally better-balanced mixtures. Unfortunately, the official institutions showed no interest and provided no funds to restructure the project and convert it to a profitable enterprise. Therefore, a debt of 30 million pesos remained as part of the so-called overdue portfolio charged to the *ejido* union.

This example highlights two problems: (1) The formulators of the BAN-RURAL project lacked technical advice; there was no link between the research institutes and the banking officials of the region; (2) There was no evidence of a consultation mechanism between the project designers and the peasants. The first problem was linked to lack of scientific information in the specialized literature. A bibliographic study showed that, while much research had been done on the use of molasses and agricultural residues, research was still needed to determine the optimal composition for this type of diet. The lack of a consultation mechanism between the designers and the users of the project severely hindered the assessment of the project's actual commercial prospects. The debtors had no way of controlling the employees who were theoretically in their service, that is, BANRURAL's technicians; but they were liable for the errors committed by those technicians.

This authoritarianism in applying technology has been observed in several of the country's regions, and the results have been disastrous; most of the livestock projects under BANRURAL's control have been insolvent or barely profitable. In contrast, many successful projects have been managed by the peasants themselves.

Case 4: Technological Development in an Ejido *in Durango*

SAM officials showed great interest in solving the problem of the competition between livestock and agriculture for land use. This interest was mainfested in their unpublished internal documents, in statements to the newspapers, and in some of the president's addresses to the Congress of the Union, but SAM did not develop a broad program for livestock production with alternative feeds, nor did it produce sufficient projects to deal with the competition between agriculture and livestock. SAM did, however, foster one project in which there was an innovative process of integration between research and immediate practical development. This was the experiment of the Interdisciplinary Research Center for Integral Development of the Rural Community (CIIDIR) undertaken by the National Polytechnic Institute based in Vicente Guerrero, Durango, with an experimental station in San José de Tuitan, Durango.

The project has attempted to solve the problem of the lack of forage for cattle in a cattle-raising *ejido,* particularly during the dry months of December through May. Throughout most of northern Mexico, the cattle cycle depends on important regional intermediaries who buy lean cattle at a low price, when they weigh around 200 kilograms, then fatten them for 90 to 100 days, and resell them at a higher price to supply the large urban market. For calves weighing less than 200 kilograms there is the option of exportation to the United States for fattening. The difference in price between lean and fat cattle allows the capitalization of the fattening system in corrals and is the way the economic surplus of small cattle breeders is extracted by the livestock middlemen.

In the *ejido* cattle areas, irrigation is very rare, and corn and beans for household consumption are the main crops. Pastures are over-exploited, since the peasants are forced to accept an overload of animals in their summer pastures in order to earn a little more money through the sale of calves and culled adult animals.

The CIIDIR group decided to develop a method for intensive cattle fattening through the use of cactus fiber (*Opuntia* sp.) complemented with corn stubble, molasses, urea, and small amounts of cereal grain. The Industrial Microbiology Laboratory of the Autonomous Metropolitan University in Iztapalapa, Mexico City, developed the basis of this system. Their research provided the mixture formulas of crop leftovers, molasses, urea, and small amounts of concentrates best suited for good digestion of fibrous forage. To apply this research, a series of nutritional experiments were carried out in a corral located within the communal lands of the *ejido* and operated by *ejidatarios* paid by CIIDIR.

After nearly three years of experiment, an optimal ratio of digestible energy to raw fiber was confirmed. The most interesting fact was that the cost of a feed ration was brought down 30% by the inclusion of corn stubble, molasses, and urea.

This techonological development attracted the attention of the *ejidatarios,* since it took place in the town itself and its procedures were accessible to the peasants. Thus interest grew in forming a rural production company devoted to the intensive fattening of cattle. The association was organized about the end of 1981 and began to operate in 1982. Its first tests took place in the experimental corrals of CIIDIR. Their results were apparently positive, since cattle have since been fattened during the dry season without resorting to cultivated forages. It seems that the problem that worried SAM's technicians so greatly was solved locally.

This is a good model of integrated research and application. Technicians lived and worked daily with the peasants, allowing them to understand the limitations and interests of this class of producer. To avoid reinforcing local stratifications, the CIIDIR group assembled mixed groups of jointly owned cattle and cattle belonging to rich peasants. The project also attempted to assemble groups in which the rich peasants, being in a minority, were unable to control the group decisions.

Furthermore, methods to improve the use of pasture lands (feed troughs for forage supplements, reintroduction of forage, partial elimination of thorny and poisonous plants) were studied locally. These studies led to the identification of an alternative for *ejidal* cattle that consisted of the optimal use of pasturelands, particularly during the rainy season, as the cheapest method to feed cattle. Seasonal pasturing could now be supplemented by the use of agricultural stubble and agroindustrial by-products as the central basis for intensive feeding of cattle. Thus the peasants could have more flexibility in the use of their resources and more bargaining power with the cattle middlemen.

These experiments in Durango show the feasibility of integrating research and the dissemination of the technology, within the conditions of rural Mexico. Practical results can be obtained from such integration, and they should be easy to apply to local agricultural production systems. Another advantage of such integrated projects is the education of field researchers for sensitivity to social problems; in this way one of the most important factors limiting the technological development of the primary sector can be eliminated.

General Discussion of the Case Studies

SAM has passed into history. Its staff has disbanded, its image has deteriorated, and it has been replaced by the new National Food Program. However, the problems that concerned SAM still continue, and among them is the problem of the generation and dissemination of technology in the rural environment. One of the most striking facts to emerge from the experiences in the field is the lack of a reciprocal relationship between the peasants and the technological specialists. It appears that the technicians begin with the hypothesis that the only science and technology to increase production is that derived from the research carried out by centralized institutions, and that the only way to disseminate technology is through a model fostered by specialists trained in institutions of higher learning.

The recent experience of some international foundations and of national research institutes themselves has shown the need to rescue, assess, and promote many of the traditional practices of rural production. A well-known example is the use of alternative or associated crops of maize and beans. This practice was frowned upon for many years by national and foreign specialists. At present it has been recognized as a way of putting cultivated areas to better use. Another important example is the use of *criolla* seeds rather than improved seeds. This case has great importance because of the trend of replacing the traditional *criolla* maize seeds for hybrids isolated in experimental fields. Unfortunately, the observations of the Morelos community discussed earlier are not isolated; they seem to be all too common in the Mexican countryside. It appears that the great heterogeneity of the rain-fed areas, together with the shortages and irregularities in the distribution of fertilizers and other products, makes it increasingly important to reconsider the policy of replacing *criolla* seed with improved seeds; the former seem to have more adaptive flexibility to adverse conditions than the latter.

A lesson from these types of experiments seems to be that rigid institutional imposition can generate apparently positive responses in the adoption of technology, as in the Morelos case. This is due to the extratechnological causes, such as credit or agricultural insurance, or even to local political pressures that have nothing to do with the productive interests of the peasants (Warman, 1976).

The rapid and systematic destruction of traditional systems of production can produce disasters, as has already happened, even in the midst of industrialized countries. The growing evidence of possible ecological disaster because of the sudden and uncontrolled disturbance of millennia-old rural production systems is only one of the arguments used to point out the need for a greater reciprocity in the relations between peasants and technicians. Another problem stemming from the verticality of relations between technicians and peasants is the social, economic, and political inappropriateness of many production designs proposed by the rural development agencies—designs which may be technically correct, but which in practice are not adapted to local conditions. This problem has been broadly discussed by a number of researchers on rural social problems (Bartra, 1974; Warman, 1976; Szekely, 1978; Bartra, 1979; Esteva, 1980; Jauregui et al., 1980).

The example of the imposition of a design in the molasses-mixing plant in Jalisco is perhaps somewhat grotesque, but unfortunately it is quite common in the livestock and agroindustrial subsector. However, it might be fitting to think a little more about the example in Guanajuato in which the production objectives were easily attained. The rural organization permitted a rapid absorption of advanced agricultural technology. However, the interests of the producers were not totally met. And, even more important, the search for technical alternatives was limited by a hostile market apparently designed to discourage the technological advance of the peasant producers, who were left behind in very disadvantageous conditions, as simple producers of cheap raw materials. The case of Durango, on the other hand, shows that it is feasible to integrate research with the interests of the peasant producers. It also shows that there are some institutions with sufficient social sensitivity to take this course, although still in a timid and restricted manner. Unfortunately, we must admit that the Durango example is exceptional, and even within the National Polytechnic Institute itself there is some resistance to supporting the efforts of researchers linked with the peasants.

In spite of these difficulties, I should stress some innovations of the new federal administration (1982–1988). Among them is the new Shared-Risk Program of Social Interest Technology launched by the National Council of Science and Technology (CONACYT). This program allows, for the first time, an organized group of peasants to obtain inexpensive and direct financing in order to develop their own technology. The peasants finance only a small percentage of the total costs, which can be repaid in money, in kind, or in services. Coincidentally, the new director general of CONACYT was also governor of the state of Durango during the project discussed here, and he also supported the operation of the CIIDIR from the general directorate of the National Polytechnic Institute. Other high officials of CONACYT have also been exposed to similar experiences, like that of the National Research Institute on Biotic Resource. It could be argued that pilot experiments quite often have a greater effect on the training of officials than on the problems at which they are directed. This suggests that one way to make institutional changes that favor the interests of the peasants is to educate government officials to be more sensitive to these interests. But one cannot hide the fact that it is the great changes of economy and society, often derived from deep states of crisis such as the one we are now undergoing, that are the most effective promoters of a new society. And when that time comes, we need to have fertile, vigorous, and sufficient ideas for a new system of science and technology linked to the well-being of the producers.

Bibliography

Bartra, Armando. 1979. *La explotación del trabajo campesino por el capital.* Mexico City: Ed. Macehual.

Bartra, Roger. 1974. *Estructura agraria y clases sociales en México.* Mexico City: Ed. ERA.

Esteva, Gustavo. 1980. *La batalla del México rural.* Mexico City: Siglo XXI Editores.

Jauregui J., et al. 1980. *TABAMEX, un caso de integración vertical de la agricultura.* Mexico City: Ed. Nuevo Imagen.

Szekely, E. M. 1978. "La organización colectiva para la producción rural." *Comercio Exteriro* 27(12): 1471–1484.

Viniegra Gonzalez, Gustavo. 1982. "Más alimento, pero para quien?" Cuadernos de nutrición. Mexico City: LICONSA (CONASUPO).

Warman, Arturo. 1976. "Y venimos a contradecir: Los peasants de Morelos y el Estado Nacional." Mexico City: CISINAH, Ediciones La Casa Chata.

CHAPTER 8

The Peasant Initiative

Rodrigo A. Medellin E.

This chapter offers a special view of the Mexican Food System, that of a particular peasant region. It describes the experience of a peasant movement that, in the course of its struggles, met with SAM. The peasants responded actively to SAM's proposals, fundamentally because SAM was the state's response to their struggles. The program achieved results, but often in spite of the institutional apparatus that implemented SAM. The results might have been more efficient if the state had left many aspects of SAM's implementation in the peasants' hands. But this would have required substantial changes in the state's political behavior and in the dynamics of Mexican society. In retrospect, the historical moment had not yet arrived, but the vigor of the peasant movement nationwide indicates that it may.

Strategic Propositions

The Initial Questions. SAM's strategy regarding the peasants, as enunciated by its national director, was the following:

The productive strategy of SAM depends on three principles which correspond and reinforce one another. First, the crystallization of the state-peasant alliance, with the state sharing with poor producers the risks that are inherent to their productive co-investment, guaranteeing to them a minimum of income and of food if their crops are lost. Second, it is imperative to respect and promote democratic base organizations and those derived from these at higher echelons. Finally, it is necessary to induce—largely, massively, and continously—technological change in the agricultural, livestock, and fishing activities that are most backward in terms of the control of the peasant organizations over their working conditions. This stems from the recognition that it is necessary for producers to retain the surplus generated by their ac-

148

tivity as a basis for their productive self-support. However, the proposition should be extended to other links in the agrofood chain, by fostering a technological-productive sequence for the whole of production and marketing of food.[1]

In both production and marketing, the peasants are main actors in the system. Strictly speaking, therefore, rather than being SAM's target population, they are its primary subject. SAM's results would depend to a large extent on the behavior of the peasants, on their capacity to respond to the state's propositions. Thus the first question is, What is this peasant response? Furthermore, what is the nature of this capacity to respond to the state's various stimuli? And even more fundamentally, Who are peasants and what role can they play in a process of social transformation?

Methodological Approach. An adequate answer to these questions requires a shift in the theoretical and methodological approach to social research.[2] Specifically, it implies the clarification of these questions from a particular theoretical vantage—conceptualizing society as a whole from the structural position of the peasantry. It also implies clarifying the peasants' role in social change, not through conceptual analysis, but rather through participation in their organized social action, that is, on the basis of their concrete struggles.[3]

1. The quotation continues: "It is essential for peasant organizations to be the seat of the strategy, in an alliance with the state, because in order to have access to a higher share of the surplus there must be genuine organizations based on the full control of their own productive-reproductive process and of the technical and financial sequences it may generate. This is the only way in which the share of primary producers' income will increase all through the food chain. The excessive fragmentation of plots and the scarcity of resources being the main obstacle to increased production in rain-fed areas, one cannot separate the impulse to production from the policies that simultaneously allow producers to withhold the surplus of their own activity. A substantial improvement in productivity may be reached by applying land-concentrated technologies, where fertilizer and seeds, combined in simple processes, are fundamental. One must keep in mind that a process of rapid modernization would cut short producers' control over productive and working conditions, and would accentuate the concentration of income. Hence the proposition that peasant organizations be in charge of controlling the productive process, so they can promote the creation of stable relations—in terms of generation and appropriation of value—with other agents in the food chain, like those within the phases of industrial processing, and distribution. Each of the 25,000 *ejidos* of the country, and their union, are a key to this basic strategy of alliance with the state in search for self-sufficiency. Likewise, this would be the point of departure that is essential for the process of creating integrated peasant agroindustries, and a response to the progressive denationalization of agriculture." Cassio Luiselli Fernandez, "Qué es el SAM? Ojectivos y Programa del Sistema Alimentario Mexicano," *Nexos* 3(32) (August 1980): 32–33.
2. See Orlando Fals-Borda, "Una perspectiva para las ciencias sociales del Tercer Mundo," *Comercio Exterior* 30(7) (July 1980): 671–674.
3. Throughout this chapter the terms *struggle* and *peasant struggles* are used. I do not simply refer to an *effort*. In the case of peasants, I mean a fight with somebody or against somebody. As is clearly seen in the text, there are always specific social actors whose interests oppose those of the peasants, and who are usually accountable for the looting of the peasant economy. These actors, with their interests, react when peasants try to defend their rights and

When SAM was launched, I had been working for five years as an adviser to peasant organizations—successively at the community, microregional, and regional levels—confronting theoretical propositions with concrete realities, with a research methodology of participatory action. For this reason, I learned of SAM not only as an academic researcher of the problems of the Mexican countryside, but also from concrete involvement in specific peasant struggles.

Hence there are certain limitations to this discussion. It is not a general description of the peasant reaction in the country as a whole; rather, it is the description of an example, but with sufficient theoretical and methodological support to advance propositions and derive lessons beyond narrow geographic boundaries. I attempt confirmation of these lessons in the last section with an analysis of the general characteristics of Mexican peasant movements.

Theoretical Premises. Before studying the specific case of peasant barley producers and their meeting with SAM, it is worth spelling out some theoretical assumptions about Mexican peasants that might begin to answer the initial questions posed above: Who are peasants and what is their role? What are their struggles and how do these relate to state initiatives such as SAM?

A single standard answer is provided, curiously enough, by two widely divergent schools of thought: peasants will inevitably disappear in the process of social transformation; their struggles are their death throes. The *modernization* school thinks of peasants as traditional, inefficient producers and speaks of a change in the sectoral composition of the economy, with agriculture producing an increasingly smaller percentage of the GNP and employing an even smaller proportion of the labor force, thanks to highly productive technology.[4] Orthodox Marxism considers the peasantry to be a transitional class based on a precapitalist mode of production that is gradually being destroyed by the expansion of the dominant capitalist mode of production. Only when capitalism is fully

---

fight for their interests. Of course, the term *struggle* does not refer to violence or aggression. The latter is what is often used against peasants (aggressive violence); they, in turn, do not usually resort to it, except in extreme cases (defensive violence).

The theoretical framework for this approach is not based on the analysis of modes of production and their articulation, as in Roger Bartra, *Estructura agraria y clases sociales en México* (Mexico City: Editorial Era, 1974); various authors, *Modos de producción en America Latina* (Mexico City: Ediciones de Cultura Popular, 1978). Rather it is based on an analysis of the circulation of social class struggles (see Harry Cleaver, *Reading Capital Politically* (Austin: University of Texas Press, 1979).

4. In literature about development a strong current speaks precisely of the *modernization* of the rural sector in these terms. See, for example, Rasnan Weitz, *De campesino a agricultor* (Mexico City: Fondo de Cultura Económica, 1973).

developed and has turned all peasants into proletariat can they contribute to a radical transformation of society.[5]

Both schools seem to err in their application of a social analysis—one derived from processes that took place in countries now fully industrialized—to countries with totally different characteristics and international contexts. At least in a country such as Mexico, peasants are not likely to disappear as a class or social sector in the foreseeable future. The national economy, as dependent capitalism, cannot absorb them into the labor market. The destruction of the peasant economy would create a politically unmanageable problem of massive unemployment in Mexico and increased illegal migration to the United States. Peasants know from experience that there is no alternative for them in society, and they fight fiercely to maintain and improve their status. A possible prospect is for prolonged agony in the present social structure. But is this inevitable?

The conditions of poverty and relative helplessness that peasants face today cannot be denied; neither can the enormous difficulties in overcoming them. The peasants are often referred to as the most underdeveloped sector of underdeveloped countries. How are we to understand this underdevelopment?[6]

Underdevelopment is not a static state in which rural communities might have traditionally languished, and from which they are rescued when modernization is brought to them from the outside. Nor is underdevelopment essentially the *lack* of certain elements present in "modern" sectors, such as legal security, minimally economical plot size, adequate inputs, advanced technology, credit, communication and transportation facilities, literacy and technical training, modern attitudes, adequate nutritional level, and health services, although a merely empirical description of rural communities would emphasize that they lack many of these. Consequently, underdevelopment cannot be considered overcome when rural communities are provided with these elements.

On the contrary, underdevelopment is a process. The so-called underdeveloped rural community is actually *underdeveloping.* The process of underdevelopment has a clear historical beginning and is the result of the type of relations imposed on rural communities by other sectors of society. In the case of countries such as Mexico, the expansion of capitalism results in the loss of economic surplus and the loss of means of production

5. See, for example, Sergio de la Peña, "Los prejuicios campesinistas," *Nexos* (2)(February 1984), within a series of articles that reopened the agrarian discussion in Mexico's academic circles at the end of 1983. For a very clear rejection of this position, see Ann Lucas de Rouffignac, *The Contemporary Peasantry in Mexico: A Class Analysis* (New York: Praeger, 1985).

6. See Rodrigo A. Medellin, "Sanctorum: The Recovery of a Wrecked Campesinos Community in Mexico" (Ph.D. diss. Harvard University, 1978), pp. 25–28.

by rural communities.[7] In other words, peasants are subjected to a process of surplus extraction and *exo-accumulation*.[8] As a consequence, rural communities as a whole are increasingly less able to satisfy the needs of a growing population.[9] The role of providers of cheap food, raw materials, and manual labor for the public or private sector has been imposed upon them. Yet Mexico's economy is not strong enough to absorb the peasant sector and provide employment for all individuals expelled from its labor force. Consequently, it offers the peasants no real alternative to their own economy, and in fact rural communities maintain a strenuous struggle for survival.

It is difficult to speak of the Mexican peasantry in an abstract or a general manner. The heterogeneity of Mexican peasants is enormous, as a result of their ethnic heritage, cultural diversity, and the characteristics of the regional environment. There is practically no peasant sector in Mexico not subjected to this process of exo-accumulation. In every case, however, modalities are specific. Therefore, any further analysis must be concrete and particular, such as the case presented here of peasant barley producers from the highlands (*altiplano*) of Mexico's central plateau, in the states of Tlaxcala, Puebla, Hidalgo, and México. Nevertheless, one must not think that barley producers are an atypical sector of the peasantry. A similar analysis, with similar results, can be made in the cases of

7. One of the best-documented topics of research on peasant reality in Latin American is the exploitation of peasants by public or private capital and the deterioration that such exploitation causes in their economy. Bartra, for example, analyzes the exploitation of peasant labor on the part of capital in terms of the relationship of unequal exchange in the markets of products, money, and labor; Armando Bartra, *La explotación del trabajo campesino por el capital* (Mexico City: Editorial Macehual, 1979). A pitfall that researchers often encounter is the consideration of capital as the only active party of the process and the peasant as helpless before this capitalist Goliath—to the degree that the peasant has no possibility of surviving if capital does not "spare him." See Sergio de la Peña, *Capitalismo en cuatro comunidades rurales* (Mexico City: Siglo XXI, 1981). In this sense, there is often a unilateral interpretation of the processes of agrarian reform in Latin America, as if they were the exclusive initiative of private or state capital, aimed at further exploiting the peasants, and not as a peasant conquest—albeit deficient and limited—in which capital must redefine its strategies to continue its exploitation; see, for example, M. Gutelman, *Capitalismo y reforma agraria en México* (Mexico City: Editorial Era, 1974). There is rather a clearly dialectic relationship in which both parties, the peasants and capital, are dynamic actors.

8. The term *exo-accumulation* is used to refer to the process through which economic surplus is extracted from a peasant productive unit to be appropriated by an alien actor in a process of capital accumulation. As can be seen, when defining the peasant unit we refer to different levels: family, group, community, microregional, or regional organization. In this sense, the prefix *exo-* has no geographic connotation, but rather refers to social class. *Exo-accumulation* may occur among actors physically settled in a peasant town but belonging to a different social class. The opposite is *endo-accumulation,* which implies a social process of accumulation at the various levels of a peasant unit. In a deeper sense, there can be processes of exo-accumulation in every peasant unit or in every peasant, as accumulated resources are used against the common interests of the peasants as a social class. In this sense, there is a constant struggle inside peasant organizations themselves to eliminate exo-accumulation and to promote peasant endo-accumulation.

9. This inability is one motive for migration from rural to urban locales.

coffee, tobacco, sugarcane, honey, cocoa, and fruit producers.[10] Even in the case of maize, the basic staple for peasant subsistence, the mechanisms of surplus extraction and exo-accumulation appear.[11]

Our case study has three parts: a general description of the producers and other actors, an analysis of their interaction prior to SAM, and an examination of their experience during SAM.

The Peasant Barley Producers and Other Actors

There are two important barley-producing regions in Mexico.[12] One is El Bajio, with low (1000 meters) plains of good, often irrigated, land midway between Mexico City and the northern part of the country. Here barley is a winter crop. The second region is located in the high valleys on Mexico's central plateau. Here barley is mostly a rain-fed, spring-summer

10. The case of peasant producers of agroindustrial raw materials is well documented, though at times there may be different theoretical interpretation of the peasant condition. Coffee: C. Deverre, "La production de l' indien: Les relations de production agrarie dans l'état de Chiapas" (dissertation, Paris V, 1976); D. Early and J. Capistran, *Las consecuencias de la dependencia las condiciones de los cafeticultores en la Sierra de Zongolica, Ver.* (Mexico City: INMECAFE, 1976); Jacques Félix, *Le café* (Paris: Presses Universitaires de France, 1968); Luis Fernandez, and R. Wasserstrom, "Los municipios Alteños de Chiapas, sus relaciones con la economía regional: Dos estudios de caso" (Mexico City: Meeting of the Comisión de Estudios Rurales del Consejo Latinoamericano de Ciencias Sociales, June, 1976); Margarita Flores, and Arturo León, "Elements pour l'analyse de l'organisation des paysans producteurs de café" (Ph.D. diss., Université de Paris I, Pantheon Sorbonne, 1979), and "La politica del INMECAFE y la Sierra Mazateca (1973-1976)," *Comercio Exterior* 29 (7) (July 1979): 676-778; S. Gallardo, *Los efectos socioeconómicos de las organizaciones campesinas promovidas por el INMECAFE* (Mexico City: Escuela Nacional de Antropologia, 1971); Arturo Leon, "Las comunidades indigenas y un cultivo comercial: El café. Estudio de caso, Region Ch'ol del norte de Chiapas" (diss., UNAM Mexico City, 1976). Sugarcane: Susan Kaufman Purcell, "Politics and the Market in Mexico: The Case of the Sugar Industry" (manuscript, n.d.); Luisa Pare (coordinator), *Ensayos sobre el problema cañero* (Mexico City: UNAM, 1979). Tobacco: Jesús Jáuregui, et al., *TABAMEX: Un caso de integración vertical de la agricultura* (Mexico City: Nueva Imagen, 1980).

11. Even in the case of peasant units that produce for self-consumption, there are well-patterned mechanisms for the extraction of economic surplus. For example, given the subsistence level at which most peasant units operate, at harvest time (November-December) there are whole series of debts to repay, to the bank or to local moneylenders. Consequently, peasants have to sell off part of their maize crop immediately. Later, during the sowing season (January-May), peasant families consume some more maize and sell some in small amounts—not because they might have a surplus of grain, but because they need to buy other essential staples. As a result, by May most of the peasants' maize is in the hands of hoarders or CONASUPO. Now it so happens that precisely in May each year the agriculture and livestock authorities announce the new guaranteed prices for the following year's crop. Automatically the stocked maize is worth more. Thereafter, until the next harvest, peasants have to buy back maize for their own consumption at much higher prices. To pay for these purchases, peasants have to abandon their communities and work as wage laborers—ordinarily below the minimum wage.

12. For a longer description of the issue of peasant barley producers, see Rodrigo A. Medellin, "Los campesinos cebaderos y la industria cervecera en Mexico," *Comercio Exterior* (9) (Sept. 1980): 927-930.

crop, exposed to frequent frosts. In this latter region there is a sharp stratification among producers, ranging from *ejidatarios*—with plots anywhere from half a hectare to 8 or 10—to relatively large private landowners with up to 700 hectares.[13]

Some barley is locally consumed as animal feed, but most of it is sold. This leads to the other end of the market. The breweries, as part of their production process, buy barley to manufacture malt, which is then used as raw material for beer.

There are three large beer manufacturing corporations in Mexico, and each is part of a large economic group: Cervecería Cuauhtemoc, Cervecería Moctezuma, and Cervecería Modelo.[14] Each has its own malt-manufacturing plant. However, neither the breweries nor their malt plants buy barley on their own. The three corporations have established a subsidiary company, Impulsora Agrícola S.A. (IASA), which is responsible for organizing the purchase of barley and for promoting its cultivation.

Finally, the international barley market must be considered, especially U.S. firms that produce, handle, and export the grain. If a shortage of the product ever develops in Mexico, the breweries must import it from across the border, after obtaining an import permit from Mexico's Ministry of Trade.

Official Institutions Related to Barley. Several official institutions take part in the production and sale of rain-fed barley. The National Rural Credit Bank (BANRURAL) is the most important; it grants credit for production (seeds, fertilizers, herbicides, and equipment). For its part, the National Crop and Livestock Insurance Agency (ANAGSA) insures the amount of credit in case of total or partial loss of the harvest. The National Institute of Agricultural Research (INIA) develops new and improved varieties of barley for beer production; the National Seed Producer (PRONASE) authorizes and supervises the reproduction of INIA improved seeds, and the extension services of the Ministry of Agriculture and Water Resources (SARH) offer technical assistance. In the commercialization of barley the role of CONASUPO is marginal, almost nonexistent, with IASA taking its place. Finally, SARH is in charge of organizing small-scale producers. SAM was to operate through these same institutions.

The Network of Intermediaries in the Commercialization of Barley. Not many peasants deliver their barley directly to the malt plants. Most of them, especially the smaller producers, sell their harvest to one of the many local or regional middlemen, some of whom are officially commissioned by IASA. (For a general view of the actors involved, see Figure 8.1.)

The Price of Barley. Every year, the board of directors of IASA (which

---

13. In the past few years, areas have been opened for the cultivation of rain-fed barley in the states of Zacatecas, Coahuila (the Saltillo area), and Chihuahua.
14. Small or regional factories have already been purchased by the three large ones.

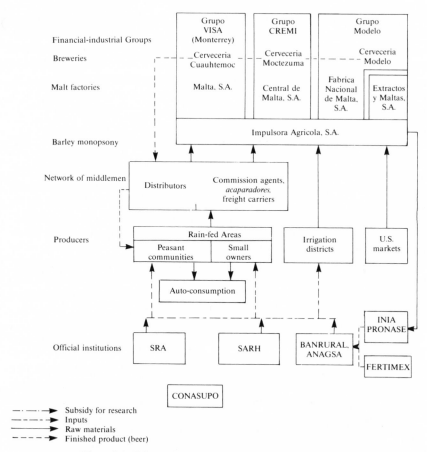

Figure 8.1. Schematic illustration of the malt barley market

represents the brewers) sets the price that malt plants will pay for barley—in accordance with government authorities—generally above the support price for wheat set by CONASUPO. Ordinarily, the price is announced when the rain-fed barley harvest is well advanced, and it is applicable to the winter season as well. It is actually a price range—determined by the quality of the barley—to be decided on according to the quality standards of the plants (percentage of usable grain, humidity, impurities, useless grains, germination percentage, variety mix).[15] The price of barley in Mexico increases constantly due to inflation, as opposed to the price in the U.S. market, where it fluctuates in response to supply and demand.

15. For further details on the requirements of factories, see Rodrigo A. Medellin. *Las relaciones de los compesinos productores de cebada con intermediarios, cervecería y dependencias públicas en Mexico* (Mexico City: COPIDER editions, 1979).

In the countryside, the prices of barley are set on the basis of a certain margin beneath the price paid by malt plants; they are set through the interaction of a small number of purchasers with a large number of sellers. Nevertheless, the outstanding element is the real demand of the factories, in other words, how much barley they really need and where they can get it.

The First Struggles and the First Conclusions

If the initial hypothesis on the extraction of economic surplus and exo-accumulation were correct, the fundamental development strategy in the barley-producing rural community would be to strive to change the terms of exchange (trade) in order to retain as much of the surplus as possible, and once retained, not to consume it all, but to accumulate at least part of it in the community.

With this hypothesis and its implicit objective as a guide, an experimental commercialization program was carried out for two years (1975–1977) in a community of the state of Tlaxcala. The idea was for the *ejido* to become the organizational structure through which *ejidatarios* would market their product. The aim was to deliver directly to factories.[16]

During the first year, on a pilot scale, results were encouraging, but during the second year there were many problems. IASA and malt factories often obstructed the delivery of barley; the price in the countryside fell, and in the 1976–1977 season much malt barley remained in the countryside. Those who had sold early to middlemen were better off. Nevertheless, time and effort were not wasted. The peasants formulated a clearer picture of the structural relations among the various actors, and of the strategy to follow in the future. It is worth analyzing these two aspects.

What Motivates Brewers and Peasants? The beer industry wants a large production of barley and a reliable supply system. It has three sources of supply: the high valleys (Tlaxcala included), the central Bajío region, and imports from the United States. Barley is convenient for peasants of the high valleys because it alternates well with maize, its cultivation requires less work, and its timing is compatible with that of maize. It is a cash crop. It lets the land rest. Its biological cycle is shorter, so there is less risk of frost. No real functional alternative is available in the short run. The two interests seem compatible. The beer industry is interested in excess supply to insure sufficient raw materials without any, or only small, price increases. Individual peasants want large crops and high prices.

The Brewers' Monopsony. The beer industry has always constituted an oligopsony with great economic resources and substantial leverage over the state's decisions. In the past, however, there had been some competi-

16. Ibid.

tion among the three major beer corporations in dealing with the peasants. Buyers from the three beer producers would approach the peasant communities and bid for available barley. The individual *ejidatario* had access to buyers and had a visible representative with whom to bargain. If there was an excess of supply, however, the situation would change, and it would be the peasants who would have to compete with each other to sell the barley.

The organizational tendency of the two sets of actors ran in opposite directions. The three beer giants soon realized that it was to their advantage to reach an agreement and "rationalize" the barley market. In 1959, they established IASA to handle all the different aspects of the barley market—to stimulate production, to guarantee adequate supply to factories, and so on—thereby creating a monopsony that virtually eliminated competition among the buyers. IASA was the only institution that could authorize the purchase of barley on behalf of the three big beer corporations. Without a delivery order from IASA, a factory could not receive a single grain of barley nor could anyone sell barley to any of the three beer companies. A delivery order became an indispensable passport into a tightly controlled market.

The Intermediaries. These rules of the game strengthened the role of the network of intermediaries. On one hand, the rural information channels aided the middlemen: the peasants were kept as ignorant as possible of the procedures required to deliver barley to the factories; the difficulties and problems involved in the delivery were magnified. On the other hand, the factories preferred to deal with the middlemen, who were reliable suppliers in bulk and who were assigned the dirty work of dealing with individual producers. The more important middlemen were officially recognized as suppliers by IASA, *comisionistas,* and were awarded commissions per ton delivered, supposedly to cover costs of storage, handling, transportation, and financing.

In this context, the middleman could present himself as a benefactor to the peasant, one who could help him face a mysterious and hostile market. Within this system, the peasant increasingly contributed to rationalizing the market for the beer factories. The factories' storage space is limited compared to the bulk of the barley crop or even their own needs. They cannot receive barley at the same rate that it is harvested, nor can they keep a supply on hand that sits around very long. Consequently, an important feature of market rationality is to maintain a steady flow of delivery from field to factory. The combined action of IASA and the middlemen assures this steady flow. Middlemen strive to deliver to the factories as much barley as possible. IASA controls the flow from the middlemen to the factories by granting or withholding delivery orders. This chain necessarily goes back to the peasant, who must provide storage space and tend to the barley until it is needed at the factory. In the early stage of the crop, middlemen are eager to buy. After that, they close the purchase deal but ask the peasant to wait until the barley can be transported. For this reason, the peasant is eager to sell as early as possible, even at a lower price; otherwise he will have to act as depository and caretaker of his own barley on behalf of the beer factories, while at the same time running all the risk if something goes wrong (e.g., germination).

To a large extent, the peasant also has to finance part of the whole marketing operation. IASA does not actually buy barley and consequently has no funds to pay the sellers. The malt factories buy barley, process it, and sell the malt to their parent companies, the breweries. The malt factories pay the barley sellers out of what they are paid by brewers. Malt factories usually pay the seller three days to a week after the barley is delivered. Middlemen, however, always buy barley from the peasants on credit, and they take anywhere from three weeks to a month and a half, or more, to pay them. The peasant is then not only temporary caretaker but even temporary financier—at no interest, with no documents—and seller in a very disadvantageous position.

The Bank. As the peasants' economic situation becomes more precarious and their production capacity declines, IASA must see to it that the supply of barley does not fall below demand. Consequently, it must do something to increase the peasants' production, or at least to halt its decline. For this purpose, a high-level agreement has been established between the brewers and top government officials, an agreement that is implemented through the government's technobureaucratic apparatus, specifically BANRURAL and SARH's extension program. IASA, through INIA, breeds improved varieties of barley seeds in its irrigated experimental fields. The seed is sold to the bank, which in turn delivers it on credit to the peasants. The bank is thus acting as an intermediary between IASA-INIA and the peasant in providing the latter with the type of seed malt factories require. Naturally, the credit is provided in kind, never in cash. Along with the seed, a certain amount of chemical fertilizer and herbicide, determined by the technobureaucrats, is given. The varieties of barley seed provided by IASA-INIA are obviously more expensive than the seed the peasants could save from the previous crop. Though it yields more, it means a greater risk. And it so happens that the grain produced is not fit for other traditional peasant uses, such as feeding animals—it is too hard.

On the other hand, BANRURAL requires the loan to be repaid in cash and charges interest in cash—normal 10% interest until December 31 and 12% thereafter (in 1976). By no means does the bank accept payment in kind (in barley). Nor is the bank at all effective in getting delivery orders from IASA for the peasants. The concerted action of IASA and the bank forces the peasant to sell—cheaply and quickly—to the middleman.

The IASA-BANRURAL Alliance. Is there a contradiction between BANRURAL and IASA in their effort to promote production and their market policies? Can production increase be sustained when the peasants are systematically losing much of the economic surplus they are able to produce? Can the peasants increase their productive capacity if they are systematically prevented from accumulating resources?

In a capitalist enterprise, credit is one means of accelerating the process of capital accumulation. For the peasant, credit plays several roles. For a peasant deprived of all his resources, credit makes up for the resources a normally developing agriculture should have. This would be fine if credit provided the impulse needed to begin or accelerate a process of resource accumulation, a kind of mini-Marshall Plan. But it does not. Credit is needed to prevent a total collapse of an already badly damaged productive system. It can best be conceptualized as a resource injection necessary to make further surplus extraction possible in a socioeconomic system that has imposed an exploitative relationship on the peasant class. Consequently, there is no contradiction, but rather a perfect congruence, in the IASA-BANRURAL alliance, one that first promotes production through improved seeds on credit and then facilitates the extraction of surplus by forcing *campesinos* to sell to the middlemen.

The Impact of the International Market on Peasants. It is not possible to place the barley growers' situation in a wider theoretical framework, that of the internationalization of capital and of the rural sector in Third World countries.[17]

Although the breweries in this study are Mexican corporations, they have close ties with international capitalism; through them even the smallest barley producing *ejidatario* is linked to the whole international process of capital accumulation. Specifically, the price of barley in U.S. markets has a direct impact on the real price of that grain in the Mexican countryside. *Official* IASA prices do not follow the upward and downward fluctuations of barley prices, present and future, in the international market. The *real* price in the countryside, however, fluctuates up and down. The behavior of market prices in Tlaxcala in 1976 is an example. The factory price was 2200 pesos per ton; in the field, the price fell from a high of 1850 pesos, to the point where peasants could not sell it even at 1500 pesos. In fact, substantial amounts of barley remained unsold that year. The *ejidatarios* had no idea why, except what the middlemen told them: "the factories do not want it."

Whenever the price of U.S. barley falls sufficiently below domestic prices—some experts say 10% is enough—it is more profitable for Mexican malt factories to import rather than to buy locally. Imports may be legal or illegal. The transportation costs from Tlaxcala to Monterrey are not lower than from Texas, for example, to Monterrey. The quality of the grain, however, is always better in the United States.

17. On this subject, see Carlos Rozo and David Barkin, "La producción de alimentos en el proceso de internacionalización del capital," *El Trimestre Económico* 50(34) (July-Sept. 1983).

The peasants learned through the 1975–1977 commercialization experiment that it was the price set by malt factories, rather than the actions of the intermediaries, that prevented their getting an adequate income from their crops. The risks they ran in trying to bypass the intermediaries did not compensate for the relatively small additional income they could get. Only an increase in factory prices would change their economic situation substantially.

But there was a double constraint on their possibility of bargaining to get a price increase—assuming that they could unite in a sellers' organization. Malt factories can always dispense with rain-fed, highland barley by waiting for the immediately following winter crop from the irrigated Bajio, where costs per ton are lower anyway. The second constraint is the price of U.S. barley. The government is always inclined to grant import permits if there are real, operational, or alleged barley shortages, since the revenues it receives from an alcoholic beverages tax on beer are very high. Moreover, government officials may be bribed to import some barley without a permit.

The First Conclusions. Though a case study cannot prove or disprove a general hypothesis, it can contribute to its conceptualization. What light does this first phase of the struggle shed on the debate regarding the theoretical status of peasants?[18]

The poverty of the peasant economy and its low agricultural yields cannot be accounted for adequately by its intrinsic characteristics; the disadvantageous relationships imposed on it (the process of exo-accumulation) must be considered. Peasant barley producers are small producers, but they are not independent, because they lack full control over their own means of production and over their production process. They do not produce in response to market signals (prices), nor can they market their harvest at a sufficiently profitable price. Once their true condition is unveiled, peasant barley producers appear as *workers* in an industrial process of beer manufacturing. Peasants process, in their fields, raw materials that

18. Some of the more important theoretical propositions on peasants can be found in the following works: Henry-Lansberger, *Rural Protest: Peasant Movements and Social Change* (New York: Barnes & Noble, 1973); Eric Wolf, *Peasants* (Englewood Cliffs, N.J.: Prentice-Hall, 1966); Héctor Dias Polanco, "Análisis de los movimientos campesinos," *Nueva Antropología* no. 2 (1975); Samir Amin and Kostas Vergopoulos, *La question paysanne et le capitalisme* (Paris: Editions Anthopos-Idep, 1974); Roger Bartra, *Estructura agraria y clases sociales en México* (Mexico City: Editorial Era, 1974); Bernard Lambet, *Les paysans dans la lutte des classes* (Paris: Editions du Seuil, 1970); Robert Redfield, *The Little Community and Peasant Society and Culture* (Chicago: University of Chicago Press, 1979); Arturo Warman, *Y venimos a contradecir* (Mexico City: Ediciones de la Casa Chata, 1978); Gerit Huizer, "Resistance to Change and Radical Peasant Mobilization: Foster and Erasmus Reconsidered," *Human Organization* 39 (4)(Winter 1970): 303–312; Theodor Shanin, *Peasants and Peasant Societies* (Baltimore: Penguin Books, 1971) and "The Nature and Logic of the Peasant Economy," *Journal of Peasant Studies* 2 (2) (January 1975): 159–182; Ann Lucas, "El debate sobre los campesinos y el capitalismo en México," *Comercio Exterior* 32 (4) (April 1982): 371–383.

brewers need to produce malt. Peasants receive on credit the seed specified by IASA, at the price set by IASA—seed developed in accordance to the needs and specification of malt-producing factories. Their crops are grown according to the indications of IASA technicians, or their deputies—the bank or extension workers. Once the grain is harvested, it is sent to IASA—directly or through intermediaries—at the price set by IASA or intermediaries. From this standpoint, one may say that the peasants are working *for* the factory, even if they do not work in the factory. They are workers *of* the factory, but without such advantages as fringe benefits or other labor rights. Peasants produce in their fields (the factory invests nothing in the land), with their own resources (the factory gives them no tools), with their own labor and on their own time (the factory pays no wage), and at their own risk (the factory risks nothing). If the harvest is lost through a frost or drought, the peasants lose everything; the factory loses nothing. If the harvest is good, the factory and the middlemen make a profit; the peasants get very little benefit. Thus peasants are factory workers who work under worse conditions than wage laborers of the factory.[19] Their other economic activities complement the income derived from their work for the factory, and they contribute to the fiction of the independent agricultural producer.[20]

A Lesson: A Broad Organization. If the hypothesis derived from the first cycle of struggle is true, peasants cannot organize effectively as small independent producers (producers' association). Though they are affected by unequal exchange relations in a commercial transaction, this does not define the nature of their main relationship with purchasers (the factories). It is true that they may attempt to associate in order to commercialize their harvests as producers in an anonymous market, in order to have a stronger position and get higher prices. But they will have little success; they cannot achieve an advantageous position in the market.

Peasants would have a greater chance of successfully organizing themselves if they were recognized as a unique type of workers for capital, and if, as such, they were to negotiate, with capitalists, the terms of their productive labor. They must, first and foremost, eliminate intermediaries, so that their relationship with capital is clear and direct and their payment is received in full. Moreover, they should secure technical support for production, adequate prices, and other benefits.

19. The term *worker* is used in the context of a sociological analysis that tries to find out the social role that peasants actually play as barley producers, under the specific conditions mentioned. The term is not being used in a narrow labor-law definition: "a worker is the individual that works personally and subordinated to another physical or moral person" (Federal Labor Law, art. 8). *Subordinated* is understood as far as place, time, activity, and technical direction is concerned. In this sense, no labor lawyer would accept that peasants are *workers* of a company.

20. Later, by estimating their production costs and comparing them with prices paid by malt factories, the peasants realized that they were actually working to subsidize the breweries.

The basic building blocks for a broad class organization must begin at the group or community level, and with time it must consolidate at higher levels. The seeds for such an organization are there, in the capacity of peasants to mobilize and pressure the government for support in their dealings with IASA, the breweries, and BANRURAL.

Peasant Barley Producers and SAM

The Barley Producers' Struggle. In 1980 Mexico had to import over 12 million tons of basic grains to meet national demand. In view of the logistic and financial problems caused by these massive imports, it was very unlikely that the government would authorize the import of barley for brewers. Thus the climate seemed favorable for a broad organization of peasant barley producers to propose negotiations with IASA and the brewing companies about the various aspects of their relations: timely inputs, appropriate seeds, technical assistance, direct sale without middlemen (and, to this end, the creation of collection centers in producing areas), the setting of prices on the basis of production costs, the revision of quality standards and testing procedures.[21]

A team of rural-development professionals led by the author had already participated for several years in this area. In July 1980 the general sense that the bargaining situation with brewers was favorable triggered a regional peasant movement that involved, in its most active phase, over one hundred communities.[22] The movement was promoted and directed by the peasants themselves, who took charge of inviting more and more communities, holding weekly meetings with community representatives, and setting their objectives and strategies. The professionals were assigned an active but subordinate role as advisers. All along, the peasants were careful not to lose control of the movement. For example, they decided not to elect individual leaders or to allow them to emerge, experience having shown that they could be bought or eliminated by outside adversaries. Rather, they operated through assemblies to share information and to make broad resolutions and through ad hoc commissions to carry out specific tasks.

Peasants put forward their demands to IASA, which rejected them. After intense peasant mobilization, including use of mass media, the president instructed the minister of agriculture to urge the directors of brewing companies to negotiate with the peasants. The brewers had to accede but

21. For a complete set of policy proposition that could benefit both peasants and factories, see Medellin, "Los campesinos cebaderos y la industria Cervecera en México," *Comercio Exterior* 30 (9) (Sept. 1980): 934–935.
22. The history of this peasant movement, accompanied and advised by the author along with a team of professionals of the Centro de Investigación y Capacitación Rural (CEDICAR), is carefully documented though not yet written.

stipulated that only a small number of peasants, and no advisers, were to participate.

A small increase was achieved—from 4.45 to 4.65 pesos per kilogram of barley, a bonus, as the brewers called it—as well as the promise to install eleven collection centers near the communities and to revise quality standards under the arbitration of government. These promises were not kept fully and satisfactorily. From the peasants' point of view, the increase in price was totally insufficient: "Not even enough to pay for the expense of carrying out this negotiation." Estimates indicated that the brewers' price was well below barley production costs. Peasants ended up subsidizing the brewers by over 100 million pesos that year alone—enough for each one of the one hundred communities to have bought a good-sized tractor. Early in the second year of negotiating with the brewers, the peasants discovered SAM.

The First Encounter with SAM. For the Tlaxcala peasants, SAM represented an opportunity, since it changed some of the conditions under which their production units operated. Traditionally this region produced barley as a commercial crop and maize basically for household consumption.

Although SAM was highly publicized by the mass media, and although the region has many radios and television sets, information about SAM programs was still incomplete and confusing. Initially SAM was thought of as "benefits that the government was giving to the peasants," for instance, cheaper fertilizers and seeds, and sharing of risks. The first reaction of the peasants was to ask the government to include barley under SAM's coverage and to grant them the same subsidies given to producers of other basic crops. They thought of the "shared risk" as even more important, since Tlaxcala's climate ruins harvests every two or three years. Although there was agricultural insurance, what ANAGSA actually insured was the loan from BANRURAL. Without quite understanding just what shared risk was, the Tlaxcaltecan peasants asked that it be applied to their risks in planting barley. Specifically, they asked that ANAGSA insure not only the bank loan but also their crop's value, that is, what they would have obtained if the crop had not failed.

Of course, none of this had been envisaged by SAM. Barley is not considered a basic crop that needs to be encouraged, nor did the shared-risks program have anything to do with the problems faced by the Tlaxcaltecan peasants. Once this was explained to them and they were convinced that barley would receive no support, they began to analyze the benefits that SAM really offered and how they could take advantage of them.

A Change in Crops? There were many discussions about the usefulness of the incentives SAM offered for the Tlaxcala region. The peasants decided to seek discounts on fertilizers and seeds for planting maize, which they already planted for their own consumption. A limiting factor was that SAM offered this benefit only to those who received their inputs through the bank. Many peasants had found BANRURAL to be a heavy burden; they had struggled to free themselves from it and save sufficient resources to buy their own inputs. Other factors also weighed heavily: Who was the peasants' representative to the bank? Was that person also the bank's representative to the peasants? Did the representative belong to the community that requested credit? Although the discussions were conducted in groups, it was basically an individual decision whether to apply or reapply for credit to the bank in order to take advantage of the benefits offered by SAM. The case of a more drastic decision was very different: whether to change crops from barley to wheat.

From Barley to Wheat. SAM's incentive package promoted the cultivation of basic foodstuffs, which include wheat but not barley. In times past, wheat had been planted extensively in Tlaxcala, not only during the spring-summer cycle, but also in the winter. This was known as "gambler's wheat," since planting it was always a gamble that depended on the *cabanuela* rains (in the first weeks of January) and on not having a freeze. However, the climate of the high Tlaxcaltecan valleys had changed, rains had become scarcer, in part a result of the fierce deforestation and consequent erosion of the land. Hence wheat planting had ceased. Gambler's wheat was abandoned first, and then the spring and summer wheat was replaced by forage barley and later by malt barley.

Considering the obstacles to marketing the peasants' malt barley, and SAM's incentives to plant wheat, a return to wheat seemed attractive. There were additional influential factors: first, CONASUPO promised to buy their wheat, although this awakened the peasants' acquired mistrust of official institutions; second, a change from barley to wheat, with a new technological package (SARH and BANRURAL offered seeds of specific variety and recommendations on fertilizer formula and dosage), did fall within the definition of SAM's shared-risk program; moreover, both federal and state agencies in Tlaxcala were carrying out a major campaign to convince the peasants to plant wheat extensively. The state government hoped to score a political point by presenting itself as a model of SAM's implementation.

The case for wheat, unlike that for maize, was not basically an individual decision. In a peasant community, a shift in crops is largely a communal decision.[23] Unless a large number of community members take the same course together, those who venture out on their own run many risks: a scarce crop always risks being harvested by others, and it faces major difficulties in marketing; invasion by livestock and irregular planting times can also be special problems.

A good number of the peasants decided to devote at least part of the land normally planted with barley to wheat. Since the communities did

---

23. It was not that there was an authority structure that had to clear the change, as is the case for many Indian communities and some collective *ejidos*.

not have seed, and SAM's benefits were accessible only through the institutions, they had to resort to BANRURAL loans and take out insurance with ANAGSA. The bank provided the seed, a little late, as usual—of an *improved* variety, the peasants were told—and the fertilizer—without any explanation or discussion about the formula or dose, as usual. Insurance coverage was extended only after the seed had germinated, again as usual.

That year the rainfall was good all during spring and summer. There were no early frosts. The wheat harvest seemed to be coming along nicely. Once the wheat began to spike, however, black spots began to appear. The peasants know this phenomenon quite well. The wheat had the *chahuistle* fall on it. This fungus (*Tilletia caries*) attacks the grain during its maturation phase and ruins it completely. In addition, its spores can remain in the area and attack later crops. In other words, the zone can become permanently infested. The harvest, therefore, was totally lost, and the possibility of consolidating wheat as a permanent crop was in danger. The peasants concluded that experimentation should have been carried out to ensure that the wheat variety to be distributed was in fact appropriate for the region. Because of the government's haste in promoting the crop, and because of SAM's benefits, they had agreed to follow a procedure that was not their own.

During the discussions on barley, it had been explained to the peasants what shared risk meant, and why it could be applied to wheat but not to barley. When the wheat crop was lost, the peasants expected the benefits of this program. The answer was typically bureaucratic. What had not been explained was that for the program to operate in a particular zone, a contract in which the technological package to be applied was clearly stipulated, and through which the state agreed to run the risk jointly with the peasants, first had to be signed. This contract had not been drawn up, and therefore the peasants had no right to the corresponding compensation. After further inquiry, it turned out that the risk-sharing trust fund had begun operations in only a few regions. The only thing the peasants got was the insurance agency's coverage. "So the bank wouldn't lose," they said. They felt no obligation to repay, and "where from?" they rightly commented. But whatever income they could have obtained, had they planted barley, was lost.

Some Lessons. One of the issues debated by students of the peasantry is the capacity of peasant productive units to respond to market signals and, more broadly, their capacity to respond to the various economic or political stimuli of the state and of other social classes. Based on the peculiar conceptions of the peasant economy, which emphasize production for household consumption to the extreme of a supposed autarky, there are those who maintain that the peasant production units are immune to market signals, that their economic behavior is motivated by traditional behavior patterns and not by any "economic rationality," that is, not by an orientation toward the maximization of profits at the end of each economic cycle. Alternatively, in Chayanov's reasoning, these production units never accumulate, since once their needs have been met at a level considered equivalent to the effort invested, the work stops. This explains their poverty and their lack of drive and entrepreneurial spirit. The prognosis is that these production units will be eliminated as a capitalist society modernizes or as a centrally planned economy takes over. Frequently this model is the result of a static analysis of peasant reality based on a research methodology that described phenomena without attempting to explain their causes.

The theoretical model derived from a historical and structural analysis of the peasants' condition is quite different. It is true that peasants do not accumulate; it is true that they do not seem to possess initiative or an enterprising spirit, and that they do not seem willing to assume risks. But, in fact, the peasant economy operates in a context of generally adverse economic and political forces, forces which systematically close off the opportunities supposedly available to anyone competing in a free-market economy. These forces remove, through different extraction mechanisms, the economic surplus generated; they keep the peasants under constant threat and oppression.

In this case, peasants *seem* to be apathetic, without any possibilities of accumulating, not very willing to take advantage of opportunities, incapable of responding to market stimuli—and these features appear as part of their personality. But these are in fact all adaptive forms of behavior that allow them to survive. Their behavior is perfectly rational within the specific constraints imposed upon them. The same is true of peasant organization. Far from being disorganized and requiring the services of an external agent, the peasants *are* organized—although their organization is usually quite battered and is basically oriented toward survival.[24] Whenever an opportunity is offered, when some endogenous or exogenous factor changes the correlation of contextual forces, peasants and their organizations have an enormous capacity to respond, especially when they have been waging an active struggle. What has been observed recently in Mexico is the peasant's capacity to transform the contexts and change the correlation of forces that determine their fate.

Conclusion and Aftermath. One could propose the hypothesis that, in general, the peasants did have a capacity to respond to the strategies proposed by SAM, because in reality SAM and its strategies responded, to a

24. This does not mean that external supports are not needed. On the contrary, they are needed very much, if they follow the peasant rationality. Often these external supports are the ones that open opportunities to peasants: they can help change the correlation of forces, and this change allows the peasant organization to express itself and to act.

large extent, to earlier peasant struggles and initiatives. Seen from this point of view, the question is not whether the peasants had a sufficient capacity to respond to SAM's stimuli, but rather whether SAM and its stimuli responded adequately to the proposals of the peasant movement. In general, if there were limitations, they would have to be attributed to SAM and its implementation rather than to the Mexican peasants' capacity to respond. The reason is simple: SAM was, at least in part, the state's response to peasant demands; it was not a program designed and put into practice by Mexico's peasant organizations themselves; they were still not sufficiently strong.

SAM confirmed the peasants once more in their ancient mistrust of official institutions. The decision to plant wheat, though induced by SAM and its incentives, was actually a response to the peasants' growing awareness that they needed to find ways to value their own work against their exploitation by the brewing industry. SAM opened a road they had been looking for. It did not carry them very far, but it allowed them to continue exploring new paths for their future.

The following year (1982), the area cultivated with wheat in this region was significantly decreased, but wheat did not disappear. Since then, it has continued gaining ground. The fungus infestation was controlled, and other varieties were tried. Gradually the regional ratio of wheat to barley changed.

Given the limited achievements in their negotiations with factories, the peasants decided to explore ways of decoupling themselves from capital. Producing wheat to sell to flour mills would lead to the same situation. Consequently, at present the Peasant Union of the Altiplano is studying a project to install a wheat mill in the area, with appropriate technology. The flour could then be directly sold to other popular organizations, both peasant and urban.

Another strategy was to use barley to feed dairy cattle. Thus, instead of having to sell their grain as raw material, the peasants could add some value. From this initial concept of an alternative use for barley they moved to the idea of an alternative use for the soil; they could plant a different crop, one that would constitute a better cattle feed. In this way, surplus family labor would be put to use and income would be stable throughout the year, allowing family members to live permanently in their communities. And besides milk, cattle produce manure that can be used as fertilizer.

Though cattle and their daily care belong to each family, other services are communal or even microregional. With this form of organization, peasants have access to modern technological elements that can only be handled at a level larger than that of the family, such as, a mixer to make animal feed rations, prophylactic services, breeding centers, and joint commercialization. And in order to avoid again falling into capital circuits by delivering fluid milk to pasteurizing plants, they began to make cheese that could be sold directly to consumers.[25]

Thus the peasants' initiative—in this case that of barley producers—was alive and well before SAM. It took from SAM the elements it could use and then continued forward when SAM—by a governmental decree—was extinguished. Thus we may say that peasants' initiative is more important than temporary state programs. Let us now take a look at this initiative on a broader level.

The Peasant Movement in Mexico

Throughout their struggle, barley producers and their advisers came in contact with many other similar peasant organizations. Gradually an image of the organic nature of the peasant movement began to emerge. SAM had been an event, in some cases important, but temporary. Their struggle was deeper and more stable.

One often hears in official, political, and academic circles of the need to organize peasants. Actually, it is clear that there is no such thing as disorganized peasants—they could not survive! Peasants are quite organized for whatever is possible at a given time, in order, fundamentally, to survive in a very hostile society. However, under favorable conditions their organizations revitalize and acquire new dimensions. An enlightened policy should permit, and even encourage, autonomous peasant organization. Peasants should not be repressed as soon as their organizations acquire some strength.[26] When this happens, peasants retreat, but the pressure increases while they await a more opportune moment.

Another phenomenon in the peasant movement is the cooperation of urban actors with peasant organizations. Ties like those I and my colleagues made with the peasant barley producers are repeated time and

25. In order to support these peasant initiatives, a network of professional institutions has lately been created in the areas of financing, technical assistance, commercialization, supply, managements, training, and so on. These are nongovernment organizations, self-financed and nonprofit. They constitute the network called ANADEGES (Análisis, Desarrollo y Gestión), in Mexico City.

26. The problem of repression is an important trait in Mexico's rural reality. It seems that the basic objective of the official policy toward the countryside is political control over the peasants to facilitate the extraction of their economic surpluses. When peasants build their organization, the state follows this evolution closely. If any of these acquire strength, state forces are at once mobilized to control them. The strategies are varied and gradual, as peasants advance. They range from cutting off all supports, buying the peasants out with benefits, dividing their union, corrupting their leaders, and killing the leaders that cannot be purchased or controlled, to generalized repression through police forces (*guardias blancas*) and the army. In spite of this state attitude, peasants continue to advance. They too have their strategies to face this official policy. Eventually, the struggle strengthens the organizations that overcome it.

again in other areas. It is rare to find a peasant organization that does not have the support of urban actors—or at least of sons of peasants who have succeeded in arriving at higher levels of education and experience.[27] Thus a bond of great potential is created, because a new subject emerges from the integration; It is no longer, so to speak, the peasants in their original situation; It is no longer the urban actor with his or her own culture. It is the integration of the two approaches that gives shape to a subject with greater potential than the sum of its parts. Peasant organizations also attempt to reorient the apparatus of the state, fighting against adversaries and striving to win allies. And the impression is that there are increasingly more officials, professionals, and technicians within the state that ally themselves with peasant struggles.

What are the peasants struggling for? For land, natural resources, productive resources, the fruit of their labor, the control of commercialization, to add economic value, to withhold surpluses, and be able to accumulate, in a context of their own democratically controlled organizations. The whole country is alive with these peasant struggles. This forceful initiative proves wrong those who consider peasants to be traditional, passive, careless, or resistant to change. On the other hand, the orthodox Left, which does perceive these struggles, maintains that they are merely *reformist,* and not *revolutionary,* demands; that peasants lack any proposal for the transformation of society.[28]

A deeper analysis allows us to draw from peasant struggles the elements that, extrapolated to a macrosocial level, shape a project for a future society that will no doubt have influence during Mexico's current times of crisis and transition (see Table 8.1). These elements can achieve deeper results than SAM achieved in its time.

Another characteristic of peasant struggles is their spiraling nature. These are not random struggles, nor do they follow a linear undifferentiated process. For a given group or organization there seem to be *struggle cycles.* This means that (1) the objective to be pursued is set (for example, a land grant or its restitution), (2) it is discussed until an agreement is reached, (3) the struggle begins and continues as long as it is feasible. When, finally, (4) the goal is achieved, the cycle for that struggle is closed; on the basis of that conquest, another cycle is opened for a higher objective. In every case, the conquests of the previous cycle constitute the bases for the struggle of the next cycle. There is no endogenous limit.

Similarly, there is also a *circulation* of the struggle, that is, a gradual extension of the struggle to other groups. At present, for example, the struggle for good prices for their products in the commercialization pro-

27. It could be a state promoter committed to the organization, or nongovernmental, civil, or church organizations.

28. See the overview of Marxist literature by Ann Lucas de Rouffignac, "The Contemporary Peasantry in Mexico" (New York: Praeger, 1985).

Table 8.1. Peasant struggles and the macrosocial project

Object of peasant struggle	Characteristics of a macrosocial project
For the land to be granted or returned. ("The land belongs to those who work it.")	The means of production in the hands of organized workers.
For the means to make the land produce—inputs, credit, technical assistance.	Appropriation of the factors that complement or increase the productivity of labor.
For the fruits of the labor invested in the land.	Distribution of income that fundamentally rewards labor.
For food, clothing, housing, education.	A reorientation of the productive apparatus toward the satisfaction of the real needs as defined by the majority of people.
For control over the production process, with labor-intensive technologies.	Full productive occupation of the population as the basis of social life.
For the diversification and increase of value-added in the productive complex.	Integration of the productive processes by popular organizations.
For the retention of the economic surplus and its endo-accumulation.	Endo-accumulation by the social sector of the economy, instead of exo-accumulation by private capital, technocracies, or bureaucracies.
For mutually beneficial exchanges between organizations, which tend toward class self-sufficiency.	Intensification of horizontal relations that integrate civil society.
For the democratization and autonomy of their organizations.	Integration of the democratic process starting from grassroots social organizations.
For the elimination of relations of exploitation, oppression, repression, and external political control.	Establishment of a macrosocial structure of just and supportive relations.
For the revalorization of ethnic and regional cultures.	Sociocultural pluralism in national unity.

cess is extending from the very few groups that started it to more and more organizations. On the other hand, there are convergences: (1) when several organizations coordinate themselves to support and strengthen one another's struggle—for example, the "Plan de Ayala" National Coordinating Committee, or (2) when the same struggle requires the linkage of several peasant organizations—for example, the struggle for class self-sufficiency through the exchange of products between peasant organizations or with urban slum dwellers.[29]

29. The most important target of this struggle for class self-sufficiency is maize. Being the basis of the peasant diet, as well as that of the urban poor, it is of strategic importance for peasants to be well supplied and not to depend on middlemen, *acaparadores,* or official agencies (CONASUPO) for their supply. At present, important experiments are being made among peasant organizations to achieve this end.

The peasant movement strategy has two steps: first, the coupling with capital through negotiation with private or state captial to arrive at more advantageous terms; second, the decoupling, through development of their productive capacity and the control over the terms of trade. These are not, of course, two independent spheres of action; rather they are dialectically related. The advance of one is a precondition for the advance of the other, and so on.

Final Conclusions

SAM died at the end of the López Portillo administration. The problem of food and its production in the countryside, however, continues. The new administration has already created the National Food Program (PRONAL), which again proposes, as one of its basic strategies, to support peasant organizations and to find support in them (see Chapter 15). However, the problem of contradictory policy directions continues. In the past administration, SAM was inconsistent with the Law on Agricultural and Livestock Development. In this administration, PRONAL is inconsistent with the amendments to the Federal Law on Agrarian Reform, whose basic sense is to give further power to the state bureaucracy at the expense of peasant organizations. The latter continue to struggle to influence national policies within a context of global planning, and especially for rural development. In order to succeed, national planning has to take into account the strength of the peasant organizations, leaving in their hands the solution of the countryside's major problems: productive processes, land tenure, credit, inputs, commercialization, production increases, employment, and a rural life viable enough for rural people to stop migrating. Otherwise we see no constructive way out of the present crisis, nor a feasible solution to Mexico's food problem.

It seems safe to say that the fate of Mexico, as a society, is linked to the fate of Mexican peasants; that Mexico will have food if the peasants have food; that the country will be self-sufficient if they are too; that she will be a free, strong, and democratic nation if her peasant organizations advance in that direction.

CHAPTER 9

Feeding Mexico City

Cynthia Hewitt de Alcantara

The Mexican food system is distinguished by its extraordinary concentration. In 1980, when SAM began to function, almost a quarter of the entire population of the nation lived within 550 square kilometers of the Valley of Mexico, and 50% or more of effective demand for certain selected food products was exercised within the same very limited area. The system was, in other words, a macrocephalic one—one in which distortions of all kinds in the use of resources for food provisioning had been introduced in the course of the formation of a metropolitan area that was on its way to becoming the largest single settlement of urban dwellers in the history of the world.

Although SAM's strategy was aimed largely at the countryside, it did single out the urban poor as a priority target group. At least a million people in Mexico City were deemed to suffer from caloric deficiencies. SAM was linked to the cities through consumption subsidies and through the government's CONASUPO distribution system (described in Chapter 4). But, of course, it takes a lot more than CONASUPO to supply Mexico City. An analysis of the implementation of SAM would be incomplete if it failed to examine how the private sector of Mexico City's food distribution system affected and was affected by SAM. Accelerating urbanization is one of the most fundamental structural transformations occurring throughout the Third World. Effective food policy must be based on a

This article draws on material gathered collectively by members of the Project on the Food System of Mexico City, carried out with the support of UNRISD, UNDP/CIDER, CONACYT, SAM, and the International Development Research Centre of Canada. Much of the information provided on the central wholesale markets of La Merced and Iztapalapa is taken from reports of Hector Castillo Berthier, and much of that on small retail commerce in low-income neighborhoods from the work of Carol Meyers de Ortiz. The way the material is interpreted remains, however, the sole responsibility of the author.

clear understanding of how the massive urban centers affect and are affected by the structures and dynamics of the food system. In this chapter I strive to add to that understanding in the context of Mexico City and SAM.

The rapid spatial redistribution of the demand for food, which occurred within Mexican society during the period of roughly sixty years separating the end of the revolution from our own times, can be illustrated quite simply by reviewing the course of demographic growth of the capital. In 1920 only 900,000 people, or slightly over 6% of the population of the nation, lived in Mexico City and had to be fed from local house lots and from the surrounding countryside. By 1940 the figure had grown to 1.8 million, or something over 9% of the national population; by 1960, to 5.2 million (15%) and by 1980, to 14.5 million (22%). When one remembers that urban growth implies, not only the physical movement of food consumers from rural villages and provincial towns toward the metropolis, but also in most cases a qualitative change in the nature of their diet, the significance of the phenomenon in shaping the parameters of the broader food system should be apparent. Although there are notorious deficiencies in all estimates of the total volume of food consumed in metropolitan Mexico City, it seems likely that residents of the capital around the turn of this decade had at their disposal some 1,200,000 tons of maize per year, 220,000 tons of beans, 3,400,000 tons of fresh fruits and vegetables, and 300,000 tons each of fresh beef and eggs. The figures for fruits, vegetables, beef, and eggs represent between 40 and 50% of all marketed national production.

Meeting this kind of urban demand on the overall food resources of the nation has required both the constant intensification and extension of production throughout the Mexican countryside and the progressive adaptation of all forms of economic activity (finance, transport, commerce) intervening in the transfer of food products from rural to urban areas. In social and political terms, there has been a necessary refashioning of the network of relations among groups located at any number of points along the food chain, implying concomitant change in the degree to which any group may benefit from participation in the activity. The result appears to the contemporary observer as a kind of "model" of food supply incorporating actors into relatively fixed patterns of interaction that determine the terms of access to food available to each. At the risk of oversimplification, it is my purpose in this chapter to trace the development of such a model, which has come to underlie the daily existence of almost a quarter of the nation's population, as well as to suggest some of the difficulties encountered by SAM due to the nature of the model. To understand the interaction between SAM and the food distribution system it is essential to first trace the evolution of the structure and dynamics of that system.

Traditional Pattern of Commercial Control
over Small Producers

When one looks at maps that register the geographic origin of the principal food products consumed in Mexico City over successive decades from the time of the revolution onward, one cannot help but be struck by the physical expansion of provisioning boundaries that has accompanied the growth of the capital. At the beginning, for example, grain consumption in the city was guaranteed through shipment by railroad from no farther away than the Bajío, Jalisco, or Nayarit. But by the 1960s and 1970s, urban dwellers had come to depend in good measure for their wheat on the distant farmers of Sonora and for their maize on the farmers of Tamaulipas. And in bad years, when inclement weather decimated national production, residents of the capital were fed to an ever greater extent with imported grains, brought overland from the port of Veracruz or freighted thousands of kilometers south from the border with the United States. Similarly, by the 1960s the exploding urban demand for beef outpaced supply from regions like Huasteca, long linked to the market of the Federal District, and encouraged the growth of new areas of supply in Chiapas and Tabasco. Tomatoes began to be shipped to the capital in volume from as far away as Sinaloa (as well as being exported to the United States), and oranges from Nuevo León.

The long-distance expansion of food trade, on which the population of Mexico City now partly depends, has generally been associated with modernization of cattle raising and agriculture, and the insertion of newly opened irrigation districts, within the boundaries of an international capitalist system. But is has also been spurred by the extension to the farthest regions of the nation of the same kind of parasitic commercial relations that underlay the earlier development of the food system of the capital. In contrast to the process characteristic of the participation by large capitalist enterprises in the food system, the resources of innumerable small cultivators have for many years been drawn into the sphere of influence of the Mexico City market not through response to the stimulus of favorable prices but as a necessary result of increasing involvement in local commercial networks. This has involved the exchange of foodstuffs for manufactured products, needed not only to permit continued agricultural production on overworked or poor-quality soils but also to satisfy a growing demand for consumer goods among rural families at all levels of income. At best, this kind of market incorporation has tended to imply voluntary emphasis on the cultivation of commercial crops on small holdings and a consequent exposure to the greater risks of market involvement; and at worst, the progressive subjection of the weakest producers to a network of control based on usury. In either case, the flow of food products toward urban destinations has been speeded, and local self-sufficiency lessened. Thus, SAM encountered a situation in which trying to satisfy the ever-expanding appetite of Mexico City was like running up a down escalator.

It is important to understand the nature of the socioeconomic and political relations that have permitted the remarkable expansion of control by commercial agents of the capital city over an ever-increasing number of smallholders in the Mexican countryside and have thus contributed to a constant broadening of the supply possibilities of the former. In some regions, and during some historical periods, these relations have approached the textbook model of "primitive accumulation" in which goods have literally been appropriated at token prices under threat of force. This has particularly been the case in areas of sharp ethnic differentiation in which traders culturally equipped to deal with the outside world have been able to work within the confines of an archaic and extremely repressive structure of regional political power.[1] With far greater frequency, however, those who serve as brokers between isolated small producers and the market for food products centered in the capital have been able to control vital elements of the economic life of villagers, which permits relegation of the threatened use of force to the background of commercial relations, although by no means always entirely eliminating it.

One of these elements of control is transport. One might speculate that if the population of Mexico City had not grown so spectacularly in the 1930s and 1940s, at a time when the system of road transport beyond the confines of the capital was still in its infancy, the imbalance between urban demand for foodstuffs and existing transport capacity in the central area of the country (not to mention those regions farther removed from the hub of the market) might have been less extreme and the opportunities for carving out local and regional domains of power through control over the physical mobilization of goods less tempting. As it happened, however, newly built roads permitting the commercialization of local products—and thus an entirely new level of access to money—were from the moment of their construction a productive economic resource to be controlled through political negotiation. To obtain governmental assistance in the building of roads required the forging of political ties; to obtain official permission to operate trucks on the roads required further exploitation of those ties; and with great frequency, the purchase of trucks was also directly or indirectly associated with the kind of preexisting distribution of power that permitted either a previous accumulation of sufficient capital to be able to make the investment or privileged access to credit.

There was a long-term tendency, then, for smallholders who needed to obtain money from the sale of foodstuffs to be subjected to conditions imposed by the relatively few people—whether local residents or outsiders—

1. See Ricardo Pozas, *Chamula* (Mexico City: INI, 1959); and Alejandro Marroquin, *La ciudad mercado (Ilaxiaco)* (Mexico City: UNAM, 1957).

who could transport their goods.[2] And this situation was very often complicated by the likelihood that those who controlled transport were also in a position to offer agricultural credit or had formed an alliance with those who would. With a certain frequency, the latter were village owners of drygoods stores, who were accustomed to financing the necessary consumption of their neighbors in return for a lien on their crops and were therefore natural agents of collection of a local surplus that was later channeled by larger merchants toward consumers in urban areas, very often including Mexico City. But there were other kinds of local moneylenders, including relatively large landowners and politicians with the ability to obtain funds from modern lending institutions and then to invest a part of their capital in assuring control over some of the output of their smaller neighbors.

From the 1930s onward, a number of the more enterprising merchants, landowners, and truckers of productive agricultural areas in central Mexico responded to the incentives of mushrooming demand for foodstuffs in the capital, not only by forming alliances with urban wholesalers, but at times by becoming urban wholesalers themselves. At the opposite end of the commercial chain, the more successful merchants of Mexico City's central wholesale market, La Merced, began to extend their operations backward toward the countryside, utilizing family connections and ties of ritual kinship (*compadrazgo*) to assure the direct delivery of foodstuffs from particular regions in which they not only monopolized credit and transport but also served as benefactors and *caciques* (rulers) of entire communities. As this process advanced, they not surprisingly were likely to become relatively large landowners and thus to utilize production from their own fields to round out and balance the supply of goods that they received from other sources. At the same time, their direct incursion into local production often permitted them to encourage technological change among neighboring clients through the lending or renting of agricultural machinery as well as eventual renting of the latter's fields.[3]

One of the clearest indicators of the force with which this process of commercial reorganization swept over the Mexican countryside is provided by data reproduced in the national agricultural censuses of 1940 and 1950. Between these two years, the proportion of all agricultural production sold for money increased from 50 to 82%, tipping the balance between local self-sufficiency and participation in the market quite decisively toward the latter and weaving the livelihood of most rural families directly into a trading equation in which urban consumers of all income strata were ultimately included as well.

2. Lourdes Arizpe discusses a very interesting case of this kind in Chapter 7 of *Indigenas en la Ciudad de México: El caso de las Marias* (Mexico City: SEP-Sententas, 1975).
3. This process is analyzed in Luisa Pare, "El capital comercial en la agricultura mexicana," *Historia y Sociedad,* no. 4(1974): 81–91.

A central element in the web of socioeconomic relations through which private entrepreneurs supplied Mexico City with food from the 1940s onward was thus oligopolistic manipulation of credit and transport, through which they assured themselves a steady supply of saleable goods at an extremely low cost. Indebted producers, or those who could not independently ship their crops out to market, could not bargain seriously; they were, as later students of peasant life were eventually to point out, little more than the cottage laborers of the intermediaries, producing goods so cheaply that at times the remuneration they received did not cover the cost of their own labor while concomitantly absorbing a great deal of the risk for any crop that failed. The increasing complexity of agricultural technology and marketing over the years tended to exacerbate this situation. To participate in commercial agricultural production, one needed to spend more money on the most modern inputs and one therefore incurred more debt; and to market crops successfully, one needed extremely sophisticated knowledge of market conditions, which isolated smallholders could not hope to attain.[4]

In such a situation, the great wholesalers of Mexico City became, in effect, the central planners of a large part of the food system of the country. It was ultimately their capital, or ability to guarantee loans, their knowledge of the market, and their influence within the transport sector that stood behind the activities of a host of intermediaries operating at regional and local levels. And given the relative cheapness of the goods that were under their control, as well as the extensiveness of the network of supply that they managed, the wholesalers could afford to ignore certain imperatives of efficiency that, under less oligopolistic circumstances, might have been required to ensure an adequate return on investment. They tended, for example, to be uncautious concerning the careful handling of their product and unwilling to invest in costly forms of packaging and conservation. Postharvest spoilage on such staple items of urban consumption as oranges, onions, and tomatoes, shipped under makeshift conditions and rarely refrigerated, claimed between 10 and 20% of the volume of goods entering the central wholesale market; animals transported haphazardly over long distances before being slaughtered in the capital were similarly wasted.

Even though a scarcity of food might exist in the market in relatively peripheral areas of the Mexican food system, a surplus of goods could be drawn toward the center with relatively little concern for the cost of spoilage or even of transport itself. By the 1960s the control of Mexico City wholesalers over the food resources of the nation was so great that they

4. For an excellent discussion of the relations between small agricultural producers and potato wholesalers of La Merced in the mid-1970s, see Ursula Oswald, "El monopolio de la central de abastos y sus efectos en la sociedad campesina," in Oswal, ed., *Mercado y dependencia* (Mexico City: CIS-INAH, 1979), pp. 171–200.

could insist on bringing perishable commodities many hundreds of kilometers overland from their point of origin to the metropolitan area, only to be sold there to clients who would ship these same goods back to the region where they were grown, incurring additional damage and added costs. The logic of such maneuvers was grounded, not in calculations that would offer competitive prices to local consumers, but in the need to draw closed boundaries around the available supply of goods. The great wholesalers built their success upon their ability to orchestrate the arrival in Mexico City of complementary streams of products from Mexico's various zones, with their particular climates and growing seasons, and thus to ensure, not only that their own warehouses would never be empty, but also that no one else could successfuly challenge the structure of prices that they imposed.

Secondary Urban Intermediaries

As the node of the urban fresh-produce distribution process, the wholesalers of Mexico City depended on two things from the moment their businesses were established: (1) successfully obligating a sufficient number of farmers, cattlemen, or fishermen to deliver particular commodities exclusively into the hands of a given merchant (or his agents), so that the volume could be carefully maintained and (2) cultivating the kinds of social relations that would assure an expeditious sale of perishable goods. Being unable to get rid of accumulated merchandise was even more ruinous for a wholesaler than being unable to obtain it. For that reason, the complex structure of volume provisioning that accompanied the explosive growth of postrevolutionary Mexico City included an increasing number of secondary urban intermediaries who operated at times as "half-wholesalers," buying a sufficient amount from larger wholesalers to be able to distribute to a number of retailers at a higher price, and at times as sales agents paid a commission by wholesalers for assuring the rapid distribution of perishables to retail outlets.

The remuneration received by these additional actors in the food chain was, of course, ultimately incorporated into the final market price of perishables, along with that attributable to every other participant in the process, from the producer to the consumer. And it was inherent in the form of operation of a supply system based on oligopolistic control that, as demand expanded ceaselessly, obliging wholesalers to handle ever greater volumes from all over the nation, the existing structure of power could only be maintained through incorporation of a still larger group of subordinate agents in the chain, including those whose function it was to manage information concerning the daily availability of goods not already contracted by large wholesalers from the countryside.

These subordinate agents, known in commercial circles as *coyotes,* at times owed direct allegiance to certain wholesalers and played the secondary role of buying up unattached merchandise on its arrival in the environs of the central market and passing the goods on to the patrons who needed them to complete existing supplies or to defend prevailing prices by holding further production off the market. A similar role, although one formally considered to serve entering independent producers rather than wholesalers, was played by a growing number of commission agents, *comisionistas,* who offered their knowledge of the market to representatives of producers when their goods arrived unconsigned in the wholesale district. In fact, the *comisionistas* accepting goods on consignment from the producer were likely to utilize their unrivaled familiarity with the market to serve their own interests and those of wholesale buyers. They offered producers a low price that, given the closed oligopoly confronting them, they could either accept or reject (and return home with a truckload of unsold produce).

This intermediate structure of control over uncontracted foodstuffs tended to expand largely because both *coyotes* and *comisionistas* were ready to inject ready cash into a commercial process otherwise heavily dependent on credit. Established wholesalers paid their contracted suppliers only after goods were sold, thereby husbanding their resources to be utilized more for investment in preharvest stages of provisioning than in postharvest liquidation of obligations. Such a strategy was, as already noted, in fact the essence of their success in enforcing the delivery of goods by indebted producers. But independent producers, arriving at the wholesale district, required immediate payment; and here the field was open for virtually unlimited speculation on the part of *coyotes* and *comisionistas* with money in hand or with sufficient knowledge of the market to know where to obtain it.

The large wholesalers and their secondary intermediaries survived and thrived on the imperfections in the marketplace: inadequate market information, low mobility of products and people, limited access to credit and services, few buyers. By providing services and inputs in these arenas of imperfection, they performed functions necessary to the operation of the marketplace. However, the resultant oligopolistic structure concentrated bargaining power in their hands, enabled them to capture a relatively larger share of the economic surplus, and led to a less efficient system than would have prevailed under a more competitive market environment.

As a result of the numerous kinds of wages, "commissions," and profits that were generated in the passage of fresh produce from the countryside to the metropolis, the price of food products was likely to double in transit from producer to wholesalers and to double again between the latter and the consumer. Margins at each stage of the transaction depended on the product, the kind and costs of intermediation associated with it, and the nature of producers and consumers involved. Calculations from a 1976 study of margins for three basic products in the diet of Mexico City families (oranges, bananas, and tomatoes) indicated that producers whose output was promised before harvest to a particular buyer were likely to receive only 5–15% of the total retail sales value of their goods, and producers who sold at harvest time, 10–30%.[5] The remaining 70–95% of the price paid by metropolitan consumers went to intermediaries of various kinds to cover their costs and profits. Table 9.1 presents the margin structure for perishables at each level in the marketing system from producer to retailer. In the worst cases, the final prices might be twenty times higher than those received by indebted farmers; and one-half of that increase was attributable to the wholesale sphere. It was generally estimated that, even after discounting all costs of operation, some 30% of the wholesale price of fresh fruits and vegetables was attributable to profit.[6]

The Position of Small Retail Merchants
in Low-Income Neighborhoods

Even the most cursory examination of statistics such as those in Table 9.1 makes clear not only the high cost of intermediation associated with the handling of fresh produce but also the great weight of remuneration to retailers in the overall process of price formation. In fact, as it is often pointed out in justification of programs to modernize the small retail trade, there is no part of the marketing chain that absorbs as great a part of all value-added, with as relatively small a part of all apparent business expenditures, as the retail merchant, who is generally responsible for almost one-half of the total increase in price confronted by urban families in the course of the daily process of provisioning.

The retail sector is, of course, extremely heterogenous; it contains supermarkets, specialty stores, and large meat markets, which can make extraordinary profits by selling perishables at an average market price that may be double the wholesale level. But in general, fresh produce reaches metropolitan consumers, not through such channels, but rather through the efforts of tens of thousands of small merchants, grouped in fixed public markets, street markets, or neighborhood shops, who fulfill the vital function of collecting perishable goods from the wholesale district, transporting them throughout the hundreds of square kilometers of the

5. Data are from Secretaria de Industria y Comercio, Oficialia Mayor, *El sistema de centros de abastecimiento de alimentos no elaborados y semi-elaborados en el Distrito Federal* (Mexico City: 1976).
6. See Richard Anson et al., "Overview of a Developing Country's Fruit and Vegetable Production and Domestic Market System: The Case of the Role and Prospects of the Mexican Small Farmer," draft of a report for the World Bank, November 1977, Table 4.7. Data for 1976.

Table 9.1. Average variations in the price of perishable agricultural products in various phases of intermediation

	Variation (%)		Interphase increase (%)	
	Min.	Max.	Min.	Max.
To producer (futures sales)	100	100	—	—
To producer (at harvest time)	185	270	85	170
Rural market	220	350	20	30
Prewholesale	290	525	30	50
Wholesale	360	740	25	40
Half wholesale	430	1030	20	40
Retail	650	2060	50	100

Source: Secretaría de Industria y Comercio, Oficialia Mayor, *El sistema de centros de abastecimiento de alimentos no elaborados y semi-elaborados en el Distrito Federal* (Mexico City: 1976).

metropolitan area, and making them available within walking distance of the homes of millions of consuming families who depend on fresh produce for by far the greatest of all their nongrain food consumption. For residents of better-off zones within the city, such a service is generally considered a convenience, as well as an opportunity to receive the personalized attention of a specialist who can procure particular products, of particular quality, not necessarily offered in the impersonal environment of a supermarket.

For most families constituting the bulk of the labor force of the capital, however, the existence of small produce retailers is not just a convenience; it is a necessity. In the vast working-class neighborhoods that have grown up around the periphery of Mexico City during the past decades, average family income is not sufficient to provide the kind of demand required to justify the establishment of modern supermarkets from which an acceptable rate of profit is expected. Under- and unemployment of residents is widespread; income is likely to be sporadic, requiring the expenditure of small sums on a day-to-day basis; and food purchases center around basic staples rather than the kind of luxury products on which supermarkets count for profit. The commercial structure of these neighborhoods is therefore composed of a variable number of CONASUPO's state-run supermarkets that were part of SAM's distribution channels, supplemented in some areas with other kinds of middle- or large-scale self-service outlets sponsored by labor unions or by the government of the Federal District, on one hand, and of a myriad of small family enterprises on the other.

Strictly speaking, none of these forms of food commerce serving homogenous working-class areas of Mexico City can be characterized as operating within the limits of a capitalist rationale. The public- and union-run stores are managed as modern enterprises, but their function is to provide a service at a cost that will not exceed income—in other words to break even. This they generally can do by buying manufactured and dry goods in great volume, on credit, and selling at a price usually 10% cheaper than that prevailing in private supermarket chains. With the partial exception of beef and eggs, these government distribution channels used by SAM have not been successful at handling perishables on a scale anywhere near that required to supply inhabitants residing within their areas. Fresh produce requires a kind of attention, at both wholesale and retail levels, that surpasses the capacity of modern nonprofit retail outlets, and existing operations in that field are likely to be characterized by low quality, high wastage, and financial loss.

Given such a situation, the buying pattern of most low-income families of the metropolitan area, when they can count on a sufficient amount of time and cash with which to implement a maximizing strategy, includes regular patronage of state- or union-run stores following receipt of a paycheck in order to stock nonperishable items. Thus SAM's public outlets were meeting some of the needs. For daily provisioning of perishables, however, the consumers must have recourse to nearby small merchants; and if the precariousness of family livelihood precludes investing time in traveling to a state-run store, or money in transport, or if the size of daily purchases is so limited that no expenditure on transport can be justified at all, then even staple goods must be bought from small merchants no matter how much lower the prices of larger and more modern public outlets might be.[7]

The great dependence of residents in homogeneous low-income parts of the city on nearby provisioning and the consequent creation of a peculiar kind of oligopolistic advantage of small merchants over their clientele is often cited as the root cause of high prices for most basic foodstuffs within poor neighborhoods. This is indeed an element of the problem in relatively isolated areas with little more than an incipient commercial structure. But it is certainly not the principal reason why small retailers are likely to account for such a very large part of all value-added in most working-class neighborhoods, where the commercial structure is highly developed, often to the point of involution. The problem lies elsewhere, not in lack of competition but in the very frailty of most small commercial enterprises and in the requirement that they support a family under the most inauspicious of socioeconomic conditions.

When one remembers that the retail handling of food by even the most modern forms of (state- or union-run) self-service outlets in low-income areas is clearly not a profit-making activity, it should not be surprising to learn that small family outlets are also very seldom viable business operations in the sense supposed by a wider capitalist society. Rather, the

---

7. These observations were taken from the field notes of Cristina Padilla Dieste and Carol Meyers de Ortiz for the Project on the Food System of Mexico City.

latter represent a peculiar kind of organization for subsistence, designed to remunerate otherwise unemployed or underemployed labor by providing a service that no other form of economic organization has yet proved able to displace. Small food retailers in low-income areas survive by substituting family resources for business investment, utilizing parts of their living quarters as storefronts or for storage, diverting small amounts of money from family consumption funds toward the purchase of merchandise, and above all, providing endless hours of labor on the part of men, women, and children with which to meet demands otherwise satisfiable only through the allocation of capital. Remuneration from this activity, shared by the entire family, often is not likely to reach the equivalent of a minimum wage for time invested by each of its members, much less to produce real profits; but it must of necessity provide sufficient income to meet certain minimum standards of family livelihood. Given the extremely low volume of sales generated within a pauperized environment, the margin between wholesale and retail prices must consequently be substantial.

The problem of large retail margins is thus in part a function of the requirement that family income be generated from an especially restricted number of sales, but that does not exhaust the explanation of the small merchants' dilemma. To a great extent, retail food outlets in low-income areas are constrained by the nature of relations surrounding their procurement of merchandise; for in their role as poor and dispersed buyers of necessary products, they are equally at a disadvantage with wholesalers as low-income clients may be with small merchants. The fact that the latter must buy very little at a time and must sell most of their original purchase before being able to invest the proceeds in another small quantity of goods makes it very difficult for them to supply their stores or market stalls at true wholesale prices. As a rule, in fact, only the better-off among the small retail establishments manage to obtain "half-wholesale" prices from intermediaries charged by wholesalers with distributing fresh produce or manufactured goods in medium-sized lots. Poorer merchants are forced to buy their merchandise at retail or near retail prices at the great public market stalls of the central warehouse district (in the case of perishables) or from grocery distributors (in the case of manufactured products). When supply problems are extremely grave, small retailers have even been known to buy goods in public or private supermarkets, or in government-controlled street markets (*mercados sobre ruedas*), in order to obtain the foodstuffs that form the basis of a part of their family livelihood.

In summary, then, the same limited bargaining power that impedes small merchants from lowering retail prices also prevents them from challenging the premises of the existing wholesale and "half-wholesale" structure. For this reason, one might conclude that the tens of thousands of small family enterprises that distribute foodstuffs to consumers living in homogeneous low-income zones of metropolitan Mexico City are in a very real sense functional elements of a system characterized by the prevalence of oligopolistic practices from the moment in which agricultural products are gathered up from small, dispersed, and relatively powerless peasant enterprises until they are sold, in natural or processed form, through equally small, dispersed, and powerless retail enterprises in the vast working-class neighborhoods that make up an important part of the national demand for food. Just as peasant producers have been integrated into the system as captive sellers faced with the alternatives of selling at a low price or being excluded from the process of circulation of commercial capital on which part of their livelihood depends, so too small retailers have been integrated as captive buyers willing to pay almost anything in order to ensure continued access to one of the few economic activities open to their families in a sluggish and oversaturated urban labor market. Without both forms of economic coercion, the extraordinary expansion of oligopolistic intermediation characteristic of traditional wholesale activity in the Mexico City market would literally have been impossible. During SAM, CONASUPO addressed this problem by expanding its wholesaling activities directly to the small private retailers, giving them credit and lower prices in return for their commitment to sell the goods at official CONASUPO prices.

Elements of Diversification in a
Traditional Oligopolistic Strategy

The form of operation characterized above, constituting the dominant pattern of socioeconomic interaction within the food system of the nation's capital (in terms of number of people involved) from the early postrevolutionary period onward, has been repeatedly challenged not only by governmental programs designed to lower the cost of food provisioning in metropolitan Mexico City but also by development of a host of new food-related activities within a rapidly modernizing private sector. In both cases, emphasis has been placed on reducing the length of the chain linking producers to consumers, eliminating unnecessary intermediation, and increasing that part of all value-added attributable to producers at one extreme and manufacturers or retail outlets at the other. Such efforts have quite clearly benefited relatively more powerful sectors within national society, to the extent that the latter have been granted privileged access (as urban consumers) to an ever larger stream of high-quality produce or (as modern agricultural entrepreneurs or industrialists) to a lucrative urban market. Even so, however, there are few signs that the traditional oligopolistic structure of supply involving the poorer sectors of both countryside and city has been fundamentally transformed. On the contrary, this structure would seem rather to have been adapted to new possibilities and constraints, often utilizing the particular advantages inherent in buying from and selling to the poor in order to bolster a parallel process of relative modernization permitting the diversification of operations and the enlargement of profits. Let us illustrate this complex development within overall food supply patterns for the metropolitan area during recent years by turning first to activities within the private sector of the economy.

Beginning around the time of World War II and greatly increasing in importance from the late 1950s onward, private capital—including a growing volume of foreign capital—has been invested in areas of the Mexican food system susceptible to the kind of technification that would produce a relatively high rate of return and would likely be associated with supplying food needs of middle- or high-income consumers. These areas have included intensive livestock and poultry breeding, fresh fruit and vegetable cultivation for winter export to the United States, the manufacture of beer, wines, and liquors for an expanding national market, as well as the development of a modern food-processing industry making everything from vegetable oils and canned fruit and vegetables to processed meats and dairy products, sauces, seasonings, and the most nutritionally unsatisfactory kinds of sweet and salty snack foods. The production requirements of such modern economic activities have been met frequently through imports of foreign agricultural practices and industrial equipment, while the commercial requirements have been met through a combination of direct export, extablishment of powerful producers' associations operating at national and international levels, coordination with expanding new supermarket chains, as well as exploitation of the possibilities inherent in the traditional retailing structure described above.

When one examines the nature of links between modern food-related enterprises and agricultural producers, it is interesting to note, one finds that except at the level of the very largest farmers, differences between traditional and modern forms of operation are not particularly great. The ideal kind of supplier for a modern industry or export concern, as for a traditional rural intermediary, is a sure one—obligated to deliver a specified amount at a specified time and, if at all possible, obligated to produce a particular type or quality of good. Therefore, despite the fact that industries and exporters often make every effort to establish direct links with producers, thus bypassing what they consider to be dysfunctional structures of intermediation, the quality of their relation with small commercial farmers is often quite similar to that between the latter and old-fashioned representatives of urban fresh-produce wholesalers (see the Chapter 8 example of small barley producers). Through production contracts, farmers operating within the area of influence of a food-processing industry or meat-packing plant, or tied to an exporting enterprise, are financed in return for committing their production before the harvest. When this proves to be a profitable arrangement for farmers above a certain minimum size, renting of land increases; and in many cases, large agricultural operations begin to predominate as suppliers for industry and export, even in regions where expansion beyond a legal maximum must take place through subterfuge.

The growing importance of export and food-processing enterprises as buyers of agricultural products and livestock in the Mexican countryside has no doubt permitted some small farming families with sufficient resources (falling into a category that has come to be called a "stationary" or "surplus-generating" peasantry)[8] to increase their bargaining strength—at least to the extent of choosing the buyer with which they make preharvest arrangements. Nevertheless, this does not constitute a definitive challenge to the system of oligopolisitic control that serves as the principal provisioning mechanism of the nation's capital. Regions or producers that, because of very limited resources or association with particular crop, have not been drawn into modern commercial networks are still objects of traditional forms of debt and patronage, financed by a variety of wholesalers in the central market of Mexico City, most particularly those who occupy the medium or lower-level rungs within the wholesaling hierarchy. At the same time, the elite among the wholesalers have integrated into their operations the very kind of participation in export activities and food processing that permits them, not only to maintain control over an extraordinary part of all perishables consumed in the Valley of Mexico, but also to play a significant role in food provisioning of other kinds and in other places.

In fact, the modernization of the Mexican food system as a whole has fostered an increase in the range of activities of great wholesalers and in the geographic extension of their influence. The advisability of delivering a given volume of perishables to the central market of Mexico City is now weighed against advantages to be obtained from export or from shipment to some other part of the country. In addition, surplus quantities of produce, in danger of spoiling, can be channeled toward food-processing plants in order to avoid saturating the lucrative fresh market. Only when export markets are closed, or when oversupply surpasses demand in all other channels and regions, must prices be drastically lowered or produce destroyed.

Here the clear demarcation of differing kinds of demand, satisfied through distinguishable forms of retail outlets within the metropolitan area, provides an additional element of flexibility in an oligopolistic strategy. At moments of oversupply, prices of fresh fruit and vegetables, as well as meats of all kinds, can be be maintained steady (or raised) for better-quality merchandise destined to be sold within public markets, meat markets, and supermarkets serving middle- and upper-class buyers, while lower-quality goods are sacrificed through channels reaching a vast

8. See CEPAL, *Economía campesina y agricultura empresarial: Tipología de productores del agro mexicano* (Mexico City: Siglo XXI, 1982.)

but poor clientele characterized by an extremely great elasticity of demand. Small family retailers in low-income neighborhoods become, in periods of threatened price collapse, particularly central supports for an oligopolistic system, distributing inferior produce efficiently and at minimum cost to wholesaler. When more lucrative, alternative uses of such produce appear, of course, prices for low-income buyers rise as high as the market allows.

For the distribution of processed and manufactured foods, as distinct from perishables, the existence of an endless number of small neighborhood grocery stores can be equally useful to certain kinds of modern industrial and commercial establishments. In the context of the economic expansion characteristic of much of the past decade, demand for many processed products (and most particularly for those assiduously promoted through advertising) tended to surpass supply; and there was ferocious competition among retail outlets for access to an adequate volume of goods. But at the same time, industries were likely to gamble on some products for which the Mexican market proved weak, and thus to find themselves faced with surpluses of particular items for which an outlet had to be sought. In such a situation, rapidly growing supermarket chains received privileged treatment, for they not only bought in great quantity but also offered access to the kind of buying public that large national or international food industries were established to reach. With the injection of international capital, these great retail enterprises began to exert considerable control over the manufacturing sector, receiving very favorable terms of credit, imposing promotional conditions, and encouraging the production of house brands to the obvious benefit of a middle- and upper-income clientele.

Small grocery merchants in low-income areas, in contrast, received sporadic deliveries of less prestigious products on very much more onerous terms. They paid cash on delivery, through dry-goods wholesalers who seldom offered prices more than a few cents below the suggested retail level, and were also subject to the condition that in order to receive the most solicited products, they must purchase a given amount of seldom-solicited ones. They served, in other words, as a kind of alternative market that permitted grocery wholesalers—and the food-processing industry itself—to avoid losses on unpopular or overstocked products and to receive immediate payment from an important part of the retail sector, so that credit conditions imposed by supermarket chains could be balanced. In addition, for the specific distribution of beer and bottled drinks, as well as sweet and salty snacks, small grocers constituted a vital element in a network that systematically placed these relatively expensive and nutritionally inadequate commodities within easy reach of virtually every family residing within the metropolitan area of Mexico City.

It is hardly an exaggeration to say, as a growing number of observers have done, that by the 1970s the majority of all family-run grocery outlets in the nation's capital could be less adequately characterized as independent enterprises than as a disguised form of company outlet for breweries, soft drink distributors, and the snack food industry.[9] The number of families with little alternative except to earn a living through commerce was so great, and the difficulties of competing for the existing supply of the few manufactured products regularly utilized in the preparation of the daily meals of most people so severe, that many small merchants opted to dedicate the limited space at their disposal almost completely to displaying nonessential items, which were delivered assiduously, on credit, by beer or snack food distributors, displayed on stands or in refrigerators provided by the latter, and sold at prices that were fixed at the factory level. The function of the merchant under such an arrangement was simply to receive the customers' money and turn it over to the distributor, retaining a small commission.

It is one of the painful ironies of the food system that such activities, channeling marginally valuable food products at a very high price toward segments of the population receiving extremely limited incomes and sometimes physically undernourished, should absorb millions of man-days of labor time—quite likely as many as those invested in supplying the staples of a nutritious diet. But then the phenomenon is merely illustrative of a much broader tendency toward the association of the superfluous with profit and the necessary with loss, which can best be understood within the context of a discussion of the policy of the state toward food issues, and the way in which this policy has shaped—and been shaped by—the general outlines of the network of social relations that determines the terms of access to food for all competing groups within the population of the nation's capital.

The Role of the State in the Food System of Mexico City

Attempts by public authorities to impose limits on private speculation with foodstuffs, and to assure an adequate supply of basic goods to all inhabitants of the principal urban centers of Mexico, have run parallel to the growth of those cities and to the expanding activities of private intermediaries. Just as private entrepreneurs began in the 1930s to elaborate new networks permitting them to channel food commodities from ever more distant parts of the countryside toward Mexico City, so too the newly incorporated postrevolutionary state took measures to establish its own lines of supply in order to counteract cyclical scarcity in the capital and to lower the price of provisioning for an ever-increasing number of residents. By the turn of the 1940s, the broad outlines of an official policy had begun

9. See Jorge Alonso, ed., *Lucha urbana y acumulación de capital* (Mexico City: CIS-INAH, 1980).

to emerge, and during the following decades, certain patterns of interaction between the public and private sectors appeared repeatedly, reflecting the institutionalization of relations that have changed relatively little over almost a half a century.

These relations, centered around food-related issues, were grounded in the socioeconomic and political imperatives of life in postrevolutionary Mexico. Armed conflict among contending factions ended in 1917 with no clear mandate to permit the reorganization of the national economy on entirely capitalist or entirely socialist lines; the "mixed economy" that appeared as the only viable alternative to further strife was therefore to be sustained in coming years through careful balancing of the demands of all sectors of the population within a corporative structure permitting interaction between each contending sector and the state, but not directly among sectors in the absence of the state. The state became both arbiter and benefactor, and in this extremely ambiguous role it was required to intervene in the food system in a way that would maintain political stability while providing a certain minimum level of physical sustenance for the population of the nation as a whole.

No doubt because of the central importance of basic grains in the diet of most families, official policy was first concerned with limiting private speculation in maize and wheat. The ability of great grain merchants to withhold large volumes of goods from the principal urban markets of the country was for hundred of years of critical weight in determining the stability of governments in Mexico; in the immediate postrevolutionary period, it was obvious that state control over that sector of the food system was a prerequisite to political survival. From the time of the revolution itself, provisional governments began to experiment with programs to provide rationed grain to hungry urban families, but a combination of rural devastation and lack of control over the principal elements of the commercial structure made those efforts largely symbolic. It was only in the late 1930s, with the institutionalization of political power within a corporate party structure, that real progress began to be made in designing a program that was eventually to permit effective state intervention in the grain market, at least insofar as the latter affected the daily livelihood of families in the capital of the republic.

The strategy that emerged during this period, one that was strengthened with SAM and that still underlies food policy today, embodies all of the contradictions generated by conflicting interests within the postrevolutionary Mexican food system. The production of basic grains, as well as their sale, could not be removed from the sphere of private enterprise, but alternative sources of supply, on land-reform holdings receiving credit from official institutions, could be encouraged. Intermediaries could not be prevented from driving rural prices down, but they could be challenged through the elaboration of a network of public receiving stations where grain would be bought at a guaranteed price (see Chapter 4). Intermediaries were also free to speculate by driving urban prices up, but if they did so they would confront official measures to defend a maximum limit through importation. And as a last resort, urban consumers would be protected through recourse by the state to subsidies, which would assure a certain minimum level of consumption through fiscal transfers ideally designed to redistribute resources from higher- to lower-income groups within the country, but in fact increasingly dependent on international borrowing.

Such a program successfully defended—and indeed greatly improved— the terms of access of metropolitan consumers to such dietary staples as tortillas and white bread. But the very incompleteness of official control over the sector, as well as its obvious bias toward that part of the population settled near the seat of power and constituting the principal labor force of national industry, encouraged a widening in the gap between life chances in the countryside and in the city. The grain market in more isolated rural areas continued firmly under the control of traditional intermediaries, whose monopoly over credit and transport often gave them privileged access to government-run grain reception centers. The largest part of the national population, whether in rural villages or provincial towns, paid far higher prices for grain-based staples than the population of Mexico City, which even as late as 1980 was estimated to absorb approximately 50% of all federal funds destined to subsidies on maize and 40% of all funds earmarked for subsidies on the consumption of white bread.[10] SAM attempted to address these weaknesses, but its focus was more on the countryside than on Mexico City. CONASUPO's grain-purchasing program (PACE, see Chapter 4) attempted to give greater access to the small farmers, while its retail store network (DICONSA), with lower prices, expanded greatly in the rural areas rather than in Mexico City. As mentioned before, CONASUPO did expand its urban wholesaling operations to small private retailers.

Nonetheless, the political imperative to maintain the balance of power between public and private sectors impeded thoroughgoing control by the state over the process through which subsidized grains were converted into tortillas and bread and implied reliance on private manufacturers who received maize and wheat from public agencies on the understanding that products for popular consumption would be sold at officially established prices. It is not surprising that this arrangement gave rise to both an onerous requirement for constant inspection of participating private enterprises and to noteworthy opportunities for corruption. Thus maize sold by the government to be used for tortillas could be resold at a very much higher price as an input for whisky or breakfast cereal or starch; and wheat supposedly destined for low-cost bread could more profitably

10. Estimates were made by Eduardo Cifuentes Guzman in the Subsidies Subproject Report of the Project on the Food System of Mexico City.

be transformed into expensive pastries or cakes. The strength of this black market during and after SAM continues to provoke heated criticism of the program and spurs demands that the subsidized tortillas and bread on which low-income residents depend so heavily be produced entirely by state-run, rather than private, industries.

The constant strain toward deviation of basic grains from use in the manufacture of staple items of the popular urban diet toward more lucrative uses in relatively superfluous food products finds counterparts, in one form or another, in virtually all other aspects of relations between the public and the private sector concerned with definition of the legitimate uses to which edible commodities may be put. As the state tries to limit areas of profit making from food in response to the insistent demands of low- and middle-income groups that make up by far the largest part of the total constituency of the government, there is a concomitant displacement of areas of speculation toward nonregulated parts of the food system, where private enterprise is completely unhindered, accompanied by an immediate shift of available resources toward new fields of activity. The process is perhaps nowhere better illustrated than in the field of price control.

For over four decades there has been some form of state control over the prices of selected basic food products in Mexico City, although the number of products concerned and the degree of vigilance exercised has varied considerably. Throughout that period, the threat of scarcity has always accompanied the attempted limitation of profits. In part, the disappearance of particular items on which price margins are low has been the work of wholesale or retail merchants who speculate when possible by holding goods off the market until public officials have no choice but to raise the price. In part, a relative decline in the availability of basic goods with controlled prices reflects the production decisions of farmers and industrialists who have sufficient resources to determine how their capital can most profitably be allocated, and who consequently choose to produce the most lucrative foodstuffs. Thus, when milk prices are controlled, dairy farmers and food processors turn to cheese; and when certain popularly consumed types of cheese are brought into the price control program, production is oriented toward luxury products like yogurt or gruyere. The list could be extended. What the exercise suggests is a long-term trend toward dedication of the best-quality food resources of the nation to the least-necessary elements of an adequate, but inexpensive, diet.

Public authorities have responded to this dilemma in a number of ways, including the proposal of a package of economic incentives that, during the latter 1970s, were designed to stimulate the participation of private industry in production of basic and semibasic items, and a much broader effort to extend subsidies on agricultural inputs to farmers and animal breeders who agreed to keep farmgate prices low. SAM continued the policy of subsidizing actors at various points along the food chain in an effort to reduce consumer prices of eggs, milk, poultry, ham and certain cheeses, rice, beans, and a limited number of canned products including chilies, tomato paste, and a variety of processed fruits and vegetables. But the rationale of private enterprise is only transitorily compatible with this kind of program, the problems of which are legion.

Aside from importation, the only other measure available to the Mexican state, given the inability of all of the urban (and some of the rural) population to produce its own food, as well as the patent incapacity of millions of low-income families to pay prices as high as a completely unrestricted market would demand, has involved direct competition with the private sector in the food system through establishment of state commercial and manufacturing enterprises (see Chapter 4). The possibilities of this strategy were first tested largely in the field of commerce: intermittent public activity as direct wholesaler or retailer of basic food products in the postrevolutionary period began in the late 1930s, was institutionalized in the late 1950s with the establishment of a network of fixed and mobile retail outlets in Mexico City, and gained new impetus in the 1970s, when grocery commerce supported by CONASUPO ceased to be a service restricted to basic goods in low-income areas and expanded into the field of full supermarket activities.

Much of the early effort of CONASUPO (or its predecessors) in retail commerce rested on control over the capacity to import and was concerned primarily with basic grains and milk. But as the program expanded into the provisioning of certain basic manufactured products, publicly supported outlets were dependent for supply on private industry, which tended to grant priority to private supermarkets, not CONASUPO stores. Therefore the 1970s witnessed unprecedented state investment in food processing, including the stepped-up manufacture of maize and wheat flour, vegetable oils, pastas, and canned and powdered milk, and, toward the end of the decade, a renewed effort to promote small private industries that would take the place of the principal food-processing corporations in the provisioning strategy of state-run retail outlets. As mentioned previously, under SAM, CONASUPO expanded its direct operations with thousands of relatively small retailers in the metropolitan area, providing goods at low wholesale prices, on credit, in exchange for the promise to sell basic staples at official prices. This program was significantly expanded under SAM through CONASUPO's subsidiary, IMPECSA.

One can conclude, after reviewing the general outlines of public policy and the SAM strategy toward provisioning of basic grains, milk, eggs, and basic processed foods like pastas and oils, that certain limits to private speculative activities have been set and maintained by the state, whatever the difficulties of enforcement, and that the result has been of immediate benefit to low-income consumers of Mexico City. But what of the role of the state in the remaining spheres of food commerce, most particularly those concerned with the provisioning of fresh meats and fish, and fresh vegetables and fruits? With processed luxury and snack foods, including soft drinks and alcoholic beverages, these are the edible commodities that remain most firmly in the hands of the private sector and that constitute, in effect, the primary profit-making zones of the metropolitan food system. SAM made some effort through nutrition education to counter the consumption of what it labeled "junk food."

Attempted state regulation of fresh meats, vegetables, and fruits has certainly not been entirely lacking. The history of food provisioning in Mexico City is replete, for example, with efforts to control the supply and price of fresh meats through establishment of limitations on the kind of facilities legally available for the purpose of slaughtering. At one point, only the publicly run municipal slaughterhouse could serve as the point of introduction for fresh meats into the city; at another, that monopoly was lifted. Oscillations of this kind, combined with de jure price control on fresh beef, have never been of more than token value, however, in imposing criteria of public utility on the great intermediaries in control of the beef supply of the city. And in the case of fresh fruits and vegetables, the domain of great wholesalers is equally unchallenged. During the 1970s a marketing agency within CONASUPO was established for the purpose of competing with the fresh fruits and vegetable dealers, buying up uncommissioned produce in the countryside and selling it through publicly run stores in the Federal District. Under SAM the agency continued to operate and supplied some 24,000 tons of perishables to CONASUPO outlets; but the impact of that volume upon the overall supply mechanism of the metropolitan areas was negligible.

Since the oligopolistic strength of the great wholesalers of perishables was developed over the course of many decades with their physical control over warehouses and reception centers concentrated in the downtown neighborhood of La Merced, and since the extremely congested conditions of the central market implied extraordinary expenditures of both time and money on the part of buyers and sellers dependent on motor transport, there was growing agreement among public officials from the 1960s onward that new facilities outside La Merced must be provided for the wholesale market and that such facilities should be designed to stimulate competition. If the state could replace private merchants as owner or administrator of that part of the metropolitan territory from which the latter directed their provisioning activities, then limits on speculation might be enforced and the rights of independent producers to participate in the market on fair terms guaranteed.

This kind of rationale, in fact similar in many respects to that justifying earlier state control over the municipal slaughterhouse, stood behind the construction during the SAM period of the largest modern central marketing complex in the world and the forced transfer of wholesale activities for perishables from La Merced to the new installations in the municipal delegation of Iztapalapa. The possibilities subsequently opened for urban renewal in the congested downtown area contiguous to La Merced were indeed great. However, even though the geographic location of the market was changed, the impact of the transfer on power relations within the fresh-produce sector was relatively minimal. If anything, the great wholesalers strengthened their oligopolistic position, for only the very wealthy could afford at the outset to buy warehouses in the Iztapalapa complex. Some small wholesalers disappeared, taking up other aspects of their activities as agricultural producers or merchants in other parts of the country; or they began to set up independent warehouses on the outskirts of the city, where they could supply their clientele without paying the price required for entry into the Iztapalapa market.

The political skirmishes associated with the Iztapalapa project, like those periodically surrounding the supply of fresh meat, illustrate the underlying currents of socioeconomic and political interaction that continue to make state control over the perishable-food sector so problematic. The new central market may legally have been declared a national property, but it is the wholesalers who manage the supply process from its inception in the countryside to its completion in the thousands of small stands and stores of the metropolitan area. The simple threat of a supply boycott on the part of the fresh fruit and vegetable wholesalers (or the great beef dealers, in the case of meats) is therefore sufficient, in the present as in the past, to give intermediaries an immediate advantage over urban planners. The concentration under official auspices of the supply function in one single urban space paradoxically means that public investment in the monumental installations at Iztapalapa, temporarily covered by the state through foreign borrowing, must be repaid through leasing warehouses (for a period of ninety-nine years) to intermediaries. The latter, if sufficiently pressured, might consider the option of reneging on their payments and moving operations out to areas (particularly in the contiguous state of México) not controlled by authorities of the Federal District, just as beef wholesalers have done when unhappy with conditions at the municipal slaughterhouse in the past.

At present, the cost of the new infrastructure at Iztapalapa is being passed on to the Mexico City consumer to an extent that would hardly be possible under less oligopolistic conditions. The retail price of the most frequently consumed fresh fruits and vegetables has risen approximately 400% between March 1982 (six months before the new central market began to function) and March 1984; this supasses even the rate of increase for fresh beef, 250% in the same period.[11] When compared with a general index of inflation estimated at around 180% for the economy as a whole, the entrenchment of fresh-produce merchandising as the single most speculative type of activity within the food system should be obvious.

11. Statistics were provided by the Instituto Nacional del Consumidor. The fresh fruits and vegetables considered are tomatoes, onions, chilies, squash, oranges, bananas.

Problems at Iztapalapa are clearly related not only to the economic advantage of great wholesalers within a provisioning structure historically dependent on extraordinarily massive injections of private know-how and capital but also to the political requirements of the Mexican state. Important intermediaries can be vital political allies, just as (in a different way) the smallest retail food merchant of low-income areas, when grouped in corporative interest organizations, can contribute to the maintenance of the existing political balance of power. It is this truism that underlies a host of contradictory policies on the part of the state within the food system, perhaps most clearly manifested, because of the extensiveness and complexity of the groups involved, within the sphere of food retailing.

The handling of food as a commodity is contested, at the retail level alone, by a truly bewildering number of metropolitan residents, equipped economically and politically in a variety of ways to assert their claims to a part of the surplus generated within the food system, as a precondition for their continuing support of the government. This is particularly the case in times of economic crisis, when food becomes one of the only marketable items for which demand continues to be high and toward which large number of the under- and unemployed therefore gravitate in search of remunerated activity. Here competition is not limited simply by any clearly drawn lines of antagonism between large capitalist enterprises and small subsistence ones, or between the public and the private sphere. Every square meter of remunerative commercial space is likely to be contested by small merchants intent on enforcing their domain to the exclusion of others of similar means: permanent merchants of public markets protest against the periodic appearance of rotating street markets (*tianguis*) and defend nearby sidewalks from ambulant vendors; established neighborhood storekeepers protest against the appearance of new public markets; and all feel themselves threatened when government-sponsored markets on wheels (*mercados sobre ruedas*), selling in volume and at fixed prices, appear nearby. Then, of course, there is the clearly delineated conflict between all these merchants and both publicly and privately run supermarkets, and the final confrontation within the supermarket sector itself. Thus, all of the government distribution channels used in the implementation of the SAM strategy encountered resistance from the food distributors.

The activity of policy makers has over decades been determined by the need to capitalize on political benefits that can accrue to the recognized arbiter in such a contest, requiring shows of political allegiance in return for permission to operate or assurance that others will not be allowed to operate at any given point in the system. This may reinforce the stability of the government, but at the same time it contributes visibly to undermining the coherence of food policy. And such all-encompassing, yet diffuse, official intervention in the structure of food supply also lends itself quite readily to manipulation for personal political ends, or as a means to accumulate money for private individuals. In that case, permission to participate in any given way in food commerce may be granted even though the activity in question may not necessarily be congruent with more broadly defined public policy goals, or even with prevailing legal norms. Thus carried to its logical extreme, such a rationale permits the marking out of privately appropriated sections of the system, the use of which implies payment of certain amounts of money to the "brokers" who stand between the economic activity of an individual entrepreneur and the acquiescence of public authorities.

Policy Implications

This very brief discussion of certain key relations among agents intervening in the process of producing and distributing food for Mexico City consumers by no means exhausts the topic at hand; but it does provide grounds for characterizing the food market of the capital city in terms that make policy options somewhat easier to understand. In the first place, it is abundantly clear that food products travel from the countryside to the metropolis through the working of oligopolistic mechanisms that greatly restrict the ability of both low-income rural producers and low-income urban consumers to bargain for the best price or to challenge the authority of the great intermediaries who in effect fulfill the function of central planners, not only in the perishables sector but also in much of the grocery sector as well. This function has been assured through consolidating control over rural credit and transport, knowledge of the urban market (as well as export and manufacturing alternatives), and access to an urban half-wholesaling and retailing network. It has also been won through the elaboration of useful political ties. Such a structure cannot be challenged simply by forcing those who integrate it to change their domicile from La Merced to Iztapalapa or by attempting to impose controls on wholesale or retail prices. Up to the present, the structure has not been effectively disciplined by opening alternative channels for direct entrance of rural producers into the Mexico City market. Until the kind of sure, though exploitative, outlet for goods provided in the traditional system can be bettered, small farmers will continue to be tied to agents of large intermediaries; similarly, until small retailers can be absolutely certain that someone other than the established wholesalers of the central market can supply them constantly, they cannot afford to risk losing their ties to the source of the only goods that daily stand between them and unemployment. CONASUPO's procurement and wholesaling efforts during SAM attempted to address these problem areas, but the traditional structure remained largely intact.

A second point to be made is that one confronts a speculative market in which private capital is totally free to seek out the most lucrative areas within the Mexico City food system and to concentrate a disproportionate share of all economic resources where profits are greatest and the satisfaction of popular needs least. For a number of decades, the key profit-making zones of the system have been fresh fruit and vegetables, fresh meats and fish, luxury specialty products, snack foods, and alcoholic beverages. If returns on investment are limited too drastically in any of these areas, capital flows elsewhere, causing scarcity that may in some instances be highly salutary in nutritional terms, although a motive for protest that could at the same time prove politically volatile. In the specific case of animal and vegetable perishables, which constitute an important part of most programs for a nutritionally adequate popular diet, the problem is a particularly thorny one.

Precisely because speculation has been limited in those areas of the food system concerned with the provision of grains, milk, and several other basic products, their provisioning (both under SAM and now) remains largely dependent on the allocation of subsidies that drain public coffers while assuring low-income families important nutritional benefits. One way to make transfers fiscally sound, as well as socially just, would be to finance the subsidy program with funds obtained from taxing speculative activities within the food system heavily. Yet the great wholesalers have up to the present almost entirely avoided taxation, on threat of leaving the capital unsupplied for a number of days; and great manufacturers can often obtain significant tax benefits in return for their own pledge to continue token production of certain basic products, while concentrating most of their resources on more profitable items. As has often been pointed out, the subsidies program constitutes at least as much of a benefit to private enterprise as to the low-income families of Mexico City, for, through public expenditure to maintain the population above the hunger level, many of those who hire labor are relieved of the obligation to provide wages sufficiently remunerative to permit their work force to pay unsubsidized prices for the basic elements of their daily diet.

A fourth characteristic of the food system of Mexico City is the inefficiency with which perishable food products tend to be handled and the consequent likelihood that a significant part of all goods destined for consumption within the nation's capital will be discarded before ever reaching their final destination. It is particularly ironic that the strength of the Mexico City market systematically draws foodstuffs away from local populations, toward the center of national demand, only to waste in transit as much as a fifth or more of the volume extracted, and then to expose the remainder to further loss as it passes down the half-wholesale and retail chain toward the consumer. As long as oligopoly makes more efficient management of the perishable food resources of the nation economically unnecessary, such a situation is not likely to be remedied; but at least in the case of basic grains, a thoroughgoing restructuring of the rural storage and transport system has long been on the public agenda.

High rates of spoilage are of course concomitant upon a fifth characteristic of the food system of Mexico City, its long-distance nature, made inevitable not simply by the rapid growth of the metropolitan population but also by an equally rapid abandonment of nearby sources of provisioning. The physical expansion of the metropolitan area has converted large areas of former farmland into urban settlement at the same time that a noteworthy increase in water requirements for domestic and industrial use implied the progressive drying up of irrigated parts of the Valley of Mexico, once considered among the most productive agricultural systems in the world. The problem, however, is more than an ecological one. Even when rural land on the outskirts of the metropolitan area is physically suited to some form of agriculture, rising property values make speculation with idle land a remunerative alternative to farming. Given the relative cheapness of food products extracted by intermediaries from much of the countryside, transport costs have been absorbed without the kind of necessary reevaluation of the urgency of encouraging closer sources of supply that would immediately follow upon a readjustment of the cost structure of the urban food market.

In the last analysis, the food system of Mexico City has developed along lines that direct many of the best resources of the Mexican countryside, not toward feeding the inhabitants of rural villages or provincial towns, but toward supplying metropolitan consumers—either because the latter can exercise greatest monetary demand or because, despite being poor, they make up the immediate political constituency of the government. As a result, certain dietary advantages have come to be associated with migration to the capital. If not accompanied by new and vigorous action to assure similar levels of consumption in the countryside, programs to ensure adequate access to foodstuffs among low-income families in Mexico City are likely to have the contradictory effect of encouraging further migration, further increases in the volume of food consumed within the boundaries of the city, further extractions from an already burdened countryside—and then a closing of the circle once more as new families migrate toward the metropolis. The key to changing the course of events most obviously lies in improving the quality of life in the countryside and in refashioning the network of intermediation that now stands between many local populations and any meaningful attempt to reserve a part of their food-related resources for their own benefit.

SAM addressed some of the problem areas identified in this analysis: the economic revitalization of the small farmers enhances their quality of life; the increased direct grain purchasing from them by CONASUPO removed some layers of intermediation; CONASUPO's increased wholesaling of goods on credit to small retailers reduced their costs; and consumer subsidies at the retail level improved the access of low-income consumers in Mexico City to needed staples. However, the nature, magnitude, and scope of these efforts have been insufficient to change significantly the oligopolistic structure and behavior of the perishables marketing system for Mexico City. The resultant inefficiencies and inequities remain a critical item on the agenda of Mexico's food policy makers and the policy successor to SAM.

CHAPTER 10

SAM and the Mexican Private Sector

Carlos Guillermo Sequeira

It was unavoidable that the vast Mexican Food System (SAM) have a strong effect on that part of the private sector involved in Mexico's agricultural industry and its related areas. SAM, both as an institution and as a program, operated precisely along the interface of the public and private sectors; an evaluation of SAM's effect on economic and social life in Mexico clearly should consider the program's effect on the private sector.

The relations between the private sector and the various institutions of Mexico's public sector are usually characterized by mutual prejudice. Mexican entrepreneurs are generally quick to point out inefficiency and mismanagement in government projects and agencies, while those in the public sector are similarly quick to assume that the private sector lacks social values and a national spirit. Each often considers the other a necessary evil, one to be limited and controlled in order to guarantee the welfare of society. Though this occurs in other societies as well, the peculiarities of the Mexican political system and the unusual way Mexican economic and social affairs are handled accentuate these traits. Nevertheless, in practice the two sectors have learned to live with each other, developing working relations where their activities overlap.

In this chapter I try to shed light on the nature, orientation, and results of the private sector's involvement in SAM. The data and commentary are based on two main sources: secondary sources, such as journalistic, business, and academic publications, and a survey of different food-sector enterprises of varying size and specialization.[1]

The results of our research are based on a large number of entrepreneurs' perceptions of SAM's achievements of one of its strategic objec-

1. This survey was carried out in 1983 under the supervision of the author by a team of eighteen graduate students of the agribusiness course of the master program of the Instituto Panamericano de Alta Dirección de Empresas, IPADE, in Mexico City.

tives: increased production of basic foods through incentives and supports for the different agricultural and para-agricultural functions that affect production levels. Therefore, it is necessary to establish from the outset that this study does not conclude anything about private-sector perceptions beyond that expressed by private-sector representatives, nor does it evaluate SAM's actions outside of its support for private production. Our research does not purport to be based on a statistically representative sample or scientific survey. Our purpose is to reveal, at least partially, the perceptions of SAM of different private-sector agribusiness actors operating at various points in the food system. They were affected by and affected SAM's implementation, and so some appreciation for their point of view is important to understanding SAM.

Methodology

Within the limits of this framework, we tried to establish a sense of direction that would help confirm or reject the following initial propositions that guided the survey:

1. The demand for agricultural inputs increased in 1980–1982 (the years during which SAM effectively operated) because of an increase in the amount of credit granted to finance the purchase of such inputs.
2. The demand for inputs exceeded the supply of state corporations, thus increasing the demand from private suppliers.
3. Higher support prices and subsidized agricultural inputs were available to large as well as small producers.
4. The increase of agricultural insurance coverage by Aseguradora Nacional Agricola y Ganadera, S.A. (ANAGSA), which included 100% of costs and interest, reduced the producer's peception of risk.

These propositions were the initial guidelines in our contacts with the entrepreneurs interviewed, although they do not mention all the subjects covered.

In designing and carrying out the survey, we chose five industries related to agricultural activity: production, agricultural input supply, processing, distribution, and transportation. Financial institutions that supplied agricultural credit were left out because, when the survey was carried out, the Mexican banking system had already been nationalized and was not a reflection of the private sector, and because the expansion of agricultural credit during the SAM years was mostly channeled through the government agricultural bank (BANRURAL) and the discount mechanisms of the special development funds of the Banco de Mexico (especially FIRA), which operated as a second-floor bank and as the province of the state (see Chapters 4 and 5).

In each group of industries nine companies were chosen for the survey—to allow a cross-section that could differentiate between large, medium, and small enterprises. The differentation of the three groups depended mainly on the skill of the interviewers rather than on formal parameters. Three enterprises were included in each of the three size categories in each of the five sectors chosen.

Survey Results

Producers

The survey was carried out among producers located principally in the state of México and Veracruz.[2] The sample included two large producers participating in a variety of agricultural enterprises, four who cultivated from 20 to 60 hectares, and three small farmers working on *ejidal* land. The larger farmers produced sugarcane, coffee, citrus, avocado, cotton, milk, poultry, and hogs, while the small and medium producers grew maize.

From the interviews, we concluded that, during the three years in which SAM was in effect, use of and access to incentives varied widely across producer size categories. Both large and medium producers found fertilizer easily available and relatively cheap. Large producers said that they did not use SAM's special credit mechanisms, but rather the channels they had always used. On the other hand, medium producers stated that they had used the special mechanisms. Small *ejidatario* producers indicated that they found difficulties in using these credit mechanisms because the new agrarian reform law required them to secure credit in groups. However, the peasants' lack of education, their individualism, and their age-old mistrust often prevented the creation of borrowers' associations. According to BANRURAL officials, of every ten groups that formed, only four operated correctly and fully met their financial obligations.

All the producers said that the access to and availability of improved seed was highly positive. However, both large and small producers pointed out that the yields expected with improved seeds were uneven and that the quality of the seeds did not always meet the standards advertised by the program. Producers, particularly small ones, stated that crop insurance that covered the cost of inputs (such as fertilizers and financial expenses) definitely stimulated them to increase production, since it minimized risk. But for large producers it was the support price rather than crop insurance, that was the main incentive and risk-reducing factor, since it provided a greater sense of security and allowed the producer to

2. Emilia Magana, Laura Trejo, and Alvaro Romero were the survey team.

get adequate margins by using efficient techniques and production methods. However, small and medium producers saw support prices as insufficient to compensate for what they considered to be increasing production costs and the sanctions and discounts applied by the state purchasing entities for variations in quality. Many state that they would rather have larger financial expenses and wait for a private intermediary who would offer better prices, without sanctions, who would solve transportation problems, and who would allow them to keep part of the harvest for their own consumption. Some said that only when financial pressure was too great would they sell their harvest directly to CONASUPO.

All said that SAM had contributed to increased production levels, but they agreed that this was due to an increase in cultivated area and not to significant increases in yield per hectare. The increase in cultivated area, especially in the case of *ejidatarious,* brought about an increase in rural employment. In areas previously cultivated, there was no significant increase of employment per hectare. Producers, especially the large and medium, perceived that one of SAM's objectives was to reduce imports of inputs and agricultural implements, and they said that, indeed, they had not had to import any.

It is worth stating that many producers, when asked to indicate the benefits that stimulated production under SAM, said that many of the incentives for production attributed to SAM existed before it operated and continue after its demise. When asking producers in general terms to compare the incentive package before, during, and after SAM, the interviewers had the sense that producers definitely perceived SAM as having contributed to increased agricultural production.

Input Suppliers

Because of the nature and organization of the input-supply industry, the sample shows a predominance of large companies, such as Purina, Ciba-Geigy, and Massey-Ferguson, with some medium-sized according to the standards of the Mexican economy.[3] In this group, the interviews were more structured and followed the proposed guidelines more closely.

All respondents state that SAM did stimulate demand for agricultural inputs. Credit-support operations, especially through BANRURAL, were fluid and effective. In the case of the tractor and agricultural implement industry, it was stated that the entire stock was sold and that credits for machinery purchases were effective. Some said, however, that Mexico's economic boom was much more important than SAM in stimulating input demand. In the agrochemical sector, respondents said that, because of the introduction of contraband, sales did not increase in 1980 in spite of

3. Francisco Alonso Novo, Miguel León Patrón, Mario Ramírez Soto, and Fernando Sobernes Valenzuela were the survey team.

greater overall demand and the payment of higher bonuses for distributors. They indicated that in 1981, however, they enjoyed 18% increases in the sale of fungicides and 16% for herbicides. In the area of veterinary products and animal feed, respondents indicated that increases in demand not met by state industry led them to increase their production capacity and to extend their distribution networks. Private seed producers stated that they provided 90% of the seed for sorghum and that, of the total supply of seed, private industry had a 75% share. The remaining 25% was supplied by PRONASE (see Chapters 4 and 6).

All input suppliers said that support prices and production subsidies helped to stimulate agricultural production and that their benefits were effective for both large and small farmers. Respondents indicated, however, that when the subsidy policy was changed, the small farmer stopped using improved seeds, fertilizers, and other inputs. An input supplier pointed out that the transfer of subsidies became very slow and that he had still not received some outstanding rebates.

Crop insurance was not cosidered relevant by some suppliers. Others indicated that, in their experience, insurance did not always cover all of the farmer's costs, and that when it was applied, the insurance payment was excessively delayed. Another group saw insurance as a very effective mechanism, one that decreased the risk of suppliers as well as farmers. Being sure of collection allowed companies to work with more liberal credit policies.

As for the agricultural credit-support mechanisms available during the SAM period, some suppliers stated that, in fact, private banks did not see a decreased lending risk, since the rediscount mechanism established by FIRA left the private bank with the ultimate responsibility for loan recovery. Others indicated that the delays in delivering production credit had serious repercussions for dealers' inventory levels, with the consequent financial implications.

Price controls and competition with state companies were considered by all interviewed as negative effects of SAM. According to several suppliers, fertilizer distribution—initially done privately—had to find new channels in view of the sharp increase in demand. The state sector intervened in distribution by preventing intermediation in order to control costs, and the result—according to those interviewed—was cutthroat competition on the part of state institutions such as BORUCONSA, FERTIMEX, and BANRURAL, who worked with direct subsidies and considered profits irrelevant. They also felt that BANRURAL had taken an increased share of the industry. According to the respondents, it held 30% of total purchases of agrochemicals and intended to control the purchase of all imports. Some said that, although price controls affected them, they had tried to offset the effects by charging for handling and transportation services.

There was a generalized feeling that, while the quality of services that private suppliers gave to producers was better than that of state corporations, the producers would prefer private suppliers regardless of price controls and cutthroat competition. However, they also stated that commercial distributors were threatened with extinction. Fertilizer dealers said that their small commission payments prevented them from giving adequate service to their customers and reduced their role to that of simply providing storage. They indicated that there were only sixteen warehouses scattered around the country. Respondents also indicated that PRONASE's policies tend to cut the share of private seed distributors.

In general, they all agreed that Mexico's monetary policy of 1982 affected the private agricultural input industry, as well as large and small producers.

Agroindustrial Processors

The sample of those surveyed included large, small, and medium corporations involved in canning fruits, soups, and vegetables, producing jellies and mayonnaise, processing wheat flour and other foods, baking bread, making tortillas, and mixing balanced feed for livestock.[4]

According to the canners, although demand remained constant during these years, it could not be met because of a marked lack of raw materials. The limited availability of fruits and vegetables, they said was a consequence of the emphasis of production incentives for the basic grains. The problem of supply was sharpest for products such as pineapple, chilies (in slices), and strawberries for jelly.

On the other hand, flour producers and bakeries indicated a sharp increase in sales. According to these producers, the policy of government subsidies brought about an increase in the consumption of bread, and as a result the price of basic bread remained quite stable while that of other substitute products rose. They indicated, however, that subsidies kept the price of flour lower than the price of the equivalent of basic fodder, thus causing a displacement in the consumption of flour from human to animal. They also said that keeping the price of Mexican flour lower than that of neighboring countries fostered illegal exports. The groups involved in making tortillas and animal feed saw no substantial change in their sales.

When asked about production costs, fruit and vegetable processors stated that the shortage of raw materials caused their prices to rise so high that they were channeled only to the fresh-produce market, which paid considerably higher prices than the industrial market. As a result, some

4. Raul Sergio Jiménez, Jorge Hernáiz, Delfín Rubial, and Leonardo Contreras were the survey team.

processors were forced to work at well below capacity. One pineapple-canning plant produced at only 10% of its capacity in 1982. This substantially increased unit costs because the factory could not spread its fixed costs over larger volumes, and, of course, it decreased net margin profitability.

Flour producers indicated that their costs increased proportionally with the increase in support prices for wheat, given that it accounted for about 80% of their costs. At the same time, the price of flour remained constant and the difference was covered by subsidies. Bread producers and tortilla makers purchased flour and dough at subsidized prices, and the increases in other expenses, such as labor and energy, were offset by authorized price increases for the final product or by a decrease in the cost of flour.

The balanced-feed producers indicated that the cost of raw materials increased in direct proportion to increases in support prices of products such as sorghum, sunflowers, soybean, and alfalfa.

Nevertheless, it was found that the processing sector increased considerably its installed capacity between 1980 and 1982, responding to incentives to decentralize, as planned under the Global Development Plan. Flour producers increased their capacity by 30%, for example, to meet increases in demand. It is worth mentioning that, in all cases surveyed, these increases were extensions to existing facilities that already had basic infrastructure. Flour millers stated that no new mill has been built for the previous fifteen years.

In terms of government subsidies and price controls, some processing plants whose prices were subject to government controls held that during 1980–1982 their margins barely permitted them to survive. The case of the flour industry was another matter. Though its prices were controlled, it enjoyed subsidies to protect it from increases in production costs. Processors indicated that at first subsidies were given to wheat producers, while millers purchased cheap wheat from CONASUPO. Starting in 1980, however, subsidies began to be channeled to flour production. Nevertheless, when CONASUPO's warehouses were empty, subsidies were difficult to secure. Then the price set for flour was decreased from 150 to 137 and finally to 115 pesos per 44-kilogram bag in an attempt to maintain an adequate level of profitability for bakeries. These prices, as suggested earlier, made flour cheaper than balanced feeds, thus fostering the displacement of flour toward nonhuman consumption.

In order to solve this problem, the price of bread was increased (the price of buns was practically doubled), and a census of bakeries was carried out so that they would be the only ones with access to subsidized flour prices. The result of this measure was that, while bakers purchased flour at 300 pesos for a 44-kilogram bag, this same bag cost 653 pesos for the producers of pasta, crackers, and industrialized bread (loaves). The price difference encouraged the smuggling of flour from one industry to the other.

Tortilla makers indicated that their situation was different, because of the adjustments made to the final tortilla price. For balanced-feed producers, though their costs were affected by many variables, such as support prices for raw materials, the amounts produced, and the price of the various meats, government price-control policy was more flexible and allowed them to adjust better to the market.

In general, the processors were much more critical of SAM than were the groups discussed previously. The core of this criticism pointed to an alleged lack of coordination between incentives granted to production, such as support prices and subsidies, and consumer-protection mechanism such as price controls. According to this group, the lack of coordination caused many distortions, such as an excess of wheat production in 1981 and a shortage in 1982, the diversion of products destined for human consumption to animal consumption, and the transfer of the same product from one industry to another. As one respondent put it, "The SAM objectives of giving incentives for the consumption of basic products to achieve more balanced diets and increasing production in order to gain self-sufficiency in certain basic grains were analyzed and implemented with a short-term perspective, without all the pieces being in place."

Distributors

The sample of distributors included a chain of self-service supermarkets in Mexico City, a large wholesaler in the northwest, four medium-sized distributors of eggs, fruits, grains, vegetables, and groceries, and three small grain distributors.[5] Although they all recognized the label of SAM, most of them did not seem to know what the institution had done. Some indicated that SAM had little to do with trade; they thought it was mainly oriented toward producers through cheap credit and subsidies to inputs. It was evident that many individuals confused SAM and CON-ASUPO; some used the names interchangeably. Consequently, it is difficult to isolate this group's perceptions of how SAM was carried out. It is, however, worth mentioning some of their views. They believed that SAM was concerned mainly with grains and did not deal with perishables (a perception confirmed by the analysis in Chapter 9). This, in turn, meant that CONASUPO was not concerned with perishables, which opened the way for private distributors. Some distributors barely recognized DICONSA and BORUCONSA. The institution perceived to cover government intervention in commerce was CONASUPO, without any distinction between its parts. It is not surprising that this group (in contrast to the conclusions of Chapter 9) held that middlemen, known as *coyotes,* were very useful for the supply of grains, since they performed the func-

5. Edgar Zuniga León, Ener Escobar Aguirre, Francisco Burquez Valenzuela, and Raul Pavón Mortera were the survey team.

tion of collection, classification, logistics, and transportation, which CONASUPO has not been able to accomplish satisfactorily.

As to the actions of SAM proper, distributors stated that SAM was just another program devised by the government from time to time; it really did not produce anything new and had no follow-up. The distributors also stated that they received no effective means of support. Distribution systems stayed the same, following the same patterns they always had. (This perception coincides with the conclusions in Chapter 9.) They believed that SAM had no coherent and systematic policy, since credit and subsidies were granted only to producers. Furthermore, SAM had very little impact on production. Increases in 1980–1981 were believed to be due more to weather conditions than to SAM programs (see Chapter 11 for an analysis of this view). There was no timely action regarding transportation, and a good deal of the harvests were lost as a result. Finally, distributors' sales increases were the result of inflation. There was general ignorance of the real objectives of SAM, either because of their lack of content or ineffective publicity. Though this group was quite critical of SAM, its opinions were less coherent than the previous groups', and confusion and ignorance about the role of the various governmental institutions abounded.

Transporters

The group interviewed represented transportation firms of various sizes, though it was not possible to define clearly the difference between those considered medium and those small.[6] According to this group, SAM had an impact on agricultural production levels in 1981. However, a good part of this production increase did not get to the points of consumption, a complementary objective of SAM, because of the lack of storage centers and sufficient transport.

When SAM's programs were booming, it tried to deal with the problems of transportation and of efficiency in the storing, loading, and unloading of grains with medium- and long-term corrective measures that were ineffective for short-term bottlenecks. It was recognized that the deficiencies in logistics and transport that affect agricultural production in Mexico are age-old, and that perhaps it was too much to expect SAM to correct them from the start. The businessmen noted that SAM could not be blamed for inefficiency in grain storage and at loading and unloading centers, or for a poor road infrastructure. They did say, however, that SAM did not recognize in time that the transport system imposed serious restrictions on moving the product from production centers to consumption centers, and that it tried to apply long-term solutions to short-term problems. The

6. Daniel Borba Gavin, Mayolo Ballesteros Acosta, and Ricardo Quilantan Ortuno were the survey team.

group indicated that close to 25% of the grain crop in the state of Tamaulipas in 1981 was lost because of lack of storage facilities and transportation.

The transporters said that the transport industry itself did not benefit from the "flood of incentives released by SAM," and that there were not soft credits for the purchase of new units or help for the purchase of inputs and spare parts. They indicated that the number of companies organized for the transport of agricultural products did not increase significantly between 1980 and 1982.

Truck manufacturers, however, said that their production increased sharply during this period, though they were unable to say how much of the increase was due to SAM's effect on transport. As an illustration, one estimated that close to 3500 units were directly attributable to SAM; this figure could not be confirmed by other means. Some transporters indicated that they enjoyed some benefits, such as special tax relief and special concessions for free travel, but that these had already existed before SAM. Some mentioned that it was very complex for SAM to participate in the area of transportation, since access was very limited and was largely controlled by groups of influential politicians whose interests were difficult to affect. Others pointed out the need to target transportation by type of product in order to prevent specialized units with refrigerating equipment for meat transport from being used for grains, which risks damaging the equipment.

In general terms, there was a consensus that SAM had very little, if any, impact on the transportation system in Mexico.

Summary

It is clear from our interviews that the perceptions of private entrepreneurs of SAM's implementation vary from industry to industry. Agricultural-input suppliers and producers strongly believed that SAM had valuable benefits for their industries and that its mechanisms were capable of promoting and supporting their production. The perception of such benefits was less clear among entrepreneurs in the processing, distribution, and transportation sectors. From the whole range of opinions, we can extract some elements that are valuable for our analysis:

1. Credit Mechanisms to support production were available to large, medium, and small producers, but the procedures required to implement them hampered small producers. Moreover, these measures were often carried out very slowly, making the system less fluid.
2. The amount of inputs and agricultural machinery put into production was considerable, thanks to SAM.
3. Support prices and expanded crop insurance promoted production

by decreasing the perception of risk on the part of the three types of producers surveyed. The decrease of risk also allowed input suppliers to adopt more extensive distribution policies and more liberal credit policies.

4. SAM was perceived as helping to reduce dependence on imported inputs and to increase rural employment.
5. Production of basic grains increased as a consequence of the above. This was perceived, however, as due more to an increase in cultivated area than to an increase in yields per hectare. This, of course, is inconsistent with the recognized increase in the use of inputs such as fertilizer, improved seeds, and agricultural machinery. It was perceived that the number of hectares devoted to the production of basic grain was increased at the expense of land previously devoted to other crops that supplied raw materials to certain processing industries, consequently causing problems of supply in these industries.
6. The state procurement agencies were not sufficiently flexible, and the quality sanctions they imposed encouraged the channeling of part of the crop to private buyers.
7. The infrastructures of storage, transportation, and distribution were not ready to respond to production increases, thus causing bottlenecks in the system.
8. In some cases, input-demand increases surpassed the capacity of state industry to supply them, thus allowing private corporations to capture the difference and satisfy the demand by fully using, and even increasing, their installed capacity.
9. In many cases SAM was seen as the cause of cutthroat competition by the state against the private sector; it was also thought to be strangling some private companies with price controls that were adjusted without sufficient flexibility to compensate for the drastic cut in margins caused by inflationary cost increases.
10. Bread producers and tortilla makers, though they recognized large benefits promoted by SAM, also said that the state's lack of coordination between price policy and raw-materials supply policy fostered the transfer of these materials from one industry to another and abroad, thus detracting from the basic purpose of these policies.
11. In the group of distributors and transporters, condemnation of SAM prevailed.

In short, these groups stated—between the lines—that SAM was aimed totally at production and had neglected adjustments that the production increase required from its counterparts in logistics, transport, and distribution.

Some Afterthoughts

I close this chapter with some reflections synthesized from our survey results and from several conversations—held as in-depth interviews—with a small group of private entrepreneurs. These reflections may be taken as lessons, messages, or mere hypotheses that require more structured, empirical documentation. These are, however, genuine expressions and observations of a group of researchers who analyzed the SAM phenomenon rationally, devoid of vested interests.

In attempting to evaluate SAM's implementation, one immediately comes up against the problem that no undertaking of SAM's size and transcendence produces conclusive results in only three years' time. There are no doubts as to the validity and legitimacy of SAM's objectives, which encompassed a national aspiration of the highest priority. But one may question whether the mechanisms for implementation were efficient and effective. For practical and managerial purposes, it is impossible to prescribe "chemically pure" mechanisms. For the manager, the process of implementation is always—even under the best conditions—one of approximation and trial and error, of learning by doing.

The issue is, perhaps, not so much to evaluate SAM for what it did, but rather, for what it did not do. Its main deficiency was that it did not foresee and plan for its own continuity and preservation. Its conception and organization, instead of ensuring its institutionalization in Mexican society, assured it an ephemeral life by linking it with the end of José López Portillo's term. SAM did not manage to consolidate itself as an institution, because it was completely dependent on the office of the presidency. It never had a clear place within the structure of Mexico's public sector. Its hierarchical or work relations with other entities, such as the Ministry of Agriculture and CONASUPO among others, were never clearly defined.

SAM did not recognize, and consequently did not act within, a global view of the Mexican agroindustrial system. It did not give equal attention to the various functions that have to be performed concomitantly, though separately, in bringing the products of the countryside to the mouth of the consumer. SAM's action in the sphere of production and its support for credit, insurance, and input supply were notable, in contrast to its improvisation in the areas of logistics, transportation, and distribution. Because it was not organized coherently or efficiently, SAM never fully achieved its objectives.

SAM had at its disposal large amounts of the community's scarce resources. The use of these resources must be justified, in the final analysis, by the results and benefits for the community that gave them and diverted them from other uses. If we incorporate these cost-benefit considerations into our analysis, it is impossible to avoid questioning whether benefits to Mexican society measure up to the billions of dollars invested in SAM. Some believe that less could have been invested to obtain similar results. Others believe that such an investment could have produced greater benefits. Unfortunately, the answers lie in the realm of speculation, so the questions remain as the expressions of a real, though perhaps fruitless, concern.

PART IV

IMPACT

CHAPTER 11

SAM's Cost and Impact on Production

Armando Andrade and Nicole Blanc

In trying to assess SAM's success, we first need to examine its impact on production, which was, after all, its strongest focus. Beyond this, we need to weigh the benefits of the production increases that did occur against SAM's costs. Carlos Sequeria (see Chapter 10) found that the impression of many observers was that SAM delivered fewer benefits than its costs warranted; but as he further points out, these were merely impressions. In this chapter, we try to test this view in more quantitative terms.

The analysis of SAM's impact on production leads to three conclusions. First, it achieved a rapid increase in the production of priority basic crops, by surpassing its own goals in the area of yields; its self-sufficiency targets for maize and beans were surpassed in 1981, but slipped back in 1982. Second, this expansion was fundamentally due to the rise in productivity, attributable in part to the increase of basic crops in irrigated areas that replaced other crops, but also to a greater use of the new technological package in rain-fed areas. Third, SAM's incentives were widely utilized by "poor peasants" as well as by commercial farmers.

As to SAM's costs, our analysis leads to the conclusion that, during its existence, both total public and general agricultural expenditure grew at higher rates than did spending associated with SAM. This would imply that the spending on SAM reflected the general rapid expansion of public outlays prevailing at the time. The incremental costs attributable to SAM were large and may have exceeded the value of the production gains achieved. However, SAM's cost includes expenses that, strictly speaking, belong to other accounts; it could be that SAM did not cost much more than other previous strategies, given that some of the large increases in the agricultural, livestock, and food budget seem to be associated more with traditional programs than with SAM. The incremental costs attributable to SAM's consumption goals were similiar in size to the production costs.

215

Because of lack of data, we could not calculate a cost relationship to nutritional benefits.

In this chapter we present the basis for our conclusions. First, there is a description of what happened to output during the SAM years of 1980–1982, taking the three-year period 1977–1979 as a reference. In order to relate the analysis to the historical trends, we also compare the SAM experience to the longer period between 1965 and 1979. Next, some of the hypotheses employed to explain the behavior of production are presented, to help speculation about how much can be attributed to SAM. Finally, an analysis of SAM's costs is presented.

Changes in Production Levels

During the period 1980–1982 agricultural production showed clear signs of recovery, with positive growth rates in most indicators, thus reversing the decline observed since the late 1960s. In 1980–1982 annual average growth of agricultural output (GDP) was 5.20%, compared with 2.84% for the 1977–1979 period and only 0.83% between 1971 and 1976. However, this pace did not offset the fall in agriculture's share of GDP: from its level of 6.32% between 1971 and 1976, it fell to 5.63% in 1977–1979 and to 5.17% during the period of our analysis. This loss of relative importance can be attributed to accelerated economic growth associated with the oil boom of the last five-year period (see Chapter 13). Within the agricultural, livestock, and forestry sectors, agriculture improved its relative position slightly, from 58.63% of sectoral output (GDP) between 1977 and 1979 to 59.6% in 1980–1982.

In the sphere of production, agriculture reflects very clearly the change in historical trends, particularly in food crops considered basic. This fact is seen in harvested area as well as in the volume of production. Since 1965, the national harvested area stagnated at around 15 million hectares. Between 1970 and 1976 there was even a decrease of 0.12%, which was partially offset by a (0.38%) rise in 1977–1979. In contrast, between 1980 and 1982 the harvested area grew by an average rate of 4.96%, bringing the average level to 17 million hectares in spite of a severe fall (-15.76%) in 1982.

In national harvested area, the area devoted to basic products (maize, beans, wheat, and rice) stands out. Together, these accounted for 73.3% of the total area in 1965, but they systematically lost their relative importance, dropping to 53.4% in 1979. In 1980–1982 the area devoted to these crops increased slightly to 53.6% of the national harvested area. In terms of absolute area, this group of basic crops rose from 9.5 million hectares in 1977–1979 to 9.9 million in 1980–1982.

The changes in the area pattern of basic crops harvested was fundamentally due to the increase in the area sown with maize. The share of

Table 11.1. Production of basic crops, 1965-1982 (thousands of tons)

	Basic crops[a]	Annual change (%)	Maize	Annual change (%)
1965	12,323.9	—	8,936.4	—
1966	12,304.3	−0.16	9,271.5	3.75
1967	12,123.8	−1.47	8,603.3	−7.21
Average	12,250.7		8,937.1	
1968	12,346.6	1.84	9,061.8	5.33
1969	11,966.5	−3.08	8,411.0	−7.18
1970	12,889.2	7.71	8,879.4	5.57
Average	12,400.8		8,784.1	
1971	12,939.6	0.39	9,785.7	10.21
1972	12,304.5	−4.91	9,222.8	−5.75
1973	12,159.4	−1.18	8,609.1	−6.65
Average	12,467.8		9,205.9	
1974	12,009.6	−1.23	7,847.8	−8.84
1975	12,990.8	8.17	8,448.7	7.66
1976	12,583.8	−3.13	8,017.3	−5.11
Average	12,558.1		8,104.6	
1977	13,931.1	10.71	10,137.9	26.45
1978	15,065.3	8.14	10,930.1	7.81
1979	11,877.0	−21.16	8,457.8	−22.62
Average	13,624.5		9,841.9	
1980	16,539.8	39.26	12,374.4	46.31
1981	19,852.1	20.04	14,550.1	17.58
1982	16,045.6	−19.17	10,129.1	−30.38
Average	17,479.2		12,351.2	

[a]Includes maize, beans, wheat, and palay rice; figures are for crop years.
Source: Dirección General de Economía Agrícola, (DGEA). *Anuario Agrícola Nacional* (for 1965-1982) and *Agenda Estadística Agropecuario* (for 1981, 1982) (Mexico City: SARH).

total area devoted to this crop was 52.2% in 1965, falling to 37.3% in 1979 and recovering in 1980-1982 to reach an annual average of 38.6%. Nevertheless, in absolute terms, the average area devoted to maize in the 1977-1979 period was 6.7 million hectares and did not basically change during the 1980-1982 period. An aggregate analysis of all agricultural output is of little value.[1] It is useful, however, to do such an analysis of the basic crops, since they are relatively homogeneous. In the three-year period of analysis, this group showed significant increases in production. From an annual average of 12.5 million tons in 1970-1976, it expanded to 13.6 million in 1977-1979, and to an annual average of 17.5 million tons for 1980-1982 (see Table 11.1). This shift reflects an unprecedented change

1. The official source, SARH, neither publishes nor accepts any category implying the physical sum total of crops, because of their heterogeneity.

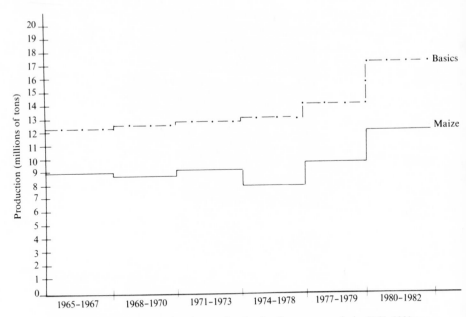

Figure 11.1. Averages of maize production by three-year periods, 1965–1982

in the level of basic food production and an exceptional contribution to increased per capita food availability, which reached one of its highest points ever.

Increased maize production played a central role in this change in the level of basic production. Its levels fluctuated around 8 to 10 million tons during the previous fifteen years and rose in the 1980–1982 period to 12 to 13 million tons (see Table 11.1 and Figure 11.1). Maize production increased 46% in 1980 compared to the previous poor year. In 1981 it grew another 17%, reaching a record high of 14.5 million tons and exceeding SAM's original self-sufficiency goal of 13.1 million tons (see Chapter 3). However, 1982 witnessed a 30% fall to 10.1 million tons as the effects of economic austerity measures and weather set in. Still, this level exceeded the average output of the 1977–1979 period or any of the annual production levels during the 1965–1976 period.

When comparing the evolution of harvested area and physical volume of basic crops, one can see that the latter grew more rapidly than the former; this was due to the accelerated increase in yields per hectare. For example, the average national maize yield per hectare stayed at around 1000 kilograms for many years. During the period 1977–1979, it reached 1464 kilograms but during SAM in 1980–1982 it rose 26% to 1840 kilograms. Wheat, with a traditional level of 3500 kilograms per hectare, reached almost 4000 kilograms per hectare during SAM.

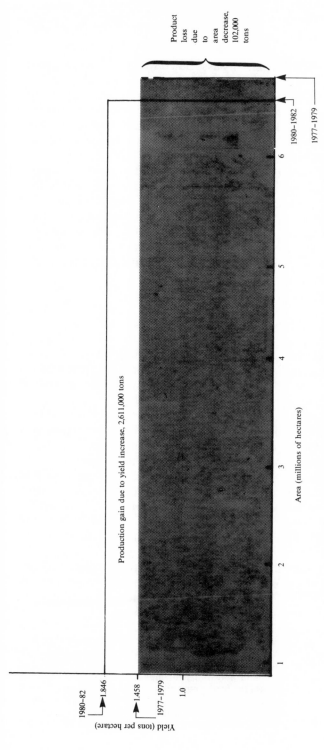

Figure 11.2. Maize production, 1977–1979 and 1980–1982: Changes in yield and area cultivated

This increase in yields was partially associated with the structure of harvested area, both for all crops and for the main crops, as well as its distribution between rain-fed and irrigated areas, which changed to favor the latter. In 1965, 14.7% of the total area destined to basic crops was irrigated and 85.3% was rain-fed. By 1977 this ratio had changed to 19.6% of irrigated land and 80.4% of rain-fed areas. The irrigated share of the area dedicated to growing maize increased from 13.8 to 15.4% between 1977–1979 and 1980–1982, even though total area planted in maize decreased slightly. This helps us understand why the increase in maize yields rather than an increase in area accounted for all the maize production gain (see Figure 11.2). As production grew, maize imports plummeted. From their 3.2 million ton peak in 1980 (to cover the 1979 shortfall), they dropped to 1.2 million tons in 1982. Inventories were built up and consumption increased. Imports as a percentage of apparent national consumption dropped from 27% in 1980 to 7% in 1982 (see Table 11.2).

Changes in Crop Structure

During the period 1980–1982, the share of total area devoted to basic crops was slightly higher than that of the previous three-year period; it increased from 60.10 to 60.67%. This improvement, however, did not reverse the historical trend toward reduction. In 1972–1976 the average share destined for these crops had been 64.02%, and 69.76% between 1965 and 1970 (see Table 11.3). On the other hand, the share of the harvested area devoted to other groups of crops clearly increased.[2] The category known as "other grains" increased from 0.80% between 1972 and 1979 to 1.27% during the period 1980–1982; fibers increased from 1.36% in 1972–1976 to 2.33% in 1980 to 1982; perennial fruit increased from 3.48 to 4.63% during the same period. The major change recorded was in feed grains, whose share increased from 10.92% in 1972–1976 to 12.17% in 1977–1979 and 12.23% in 1980–1982. The remaining groups maintained their relative positions.

When considering the value of production within the structure of crops, the horizon clearly changes. Basic foodstuffs increased their share from 30.4% in 1977–1979 to 39.11% in 1980–1982, and "other grains" went from

2. SARH classifies crops in the following groups: *Basic foodstuffs*: rice, beans, wheat, and maize; *other grains*: oats, green peas, broad beans, and lentils; *vegetables*: garlic, eggplant, onion, peas, dried chilies, green chilies, string beans, potatoes, tomatoes, green tomatoes, and sweet potatoes; *oilseeds*: sesame, peanuts, safflower, copra, linseed, soybean, cottonseed, olives, and palm oil; *fibers*: feather, cotton, and sisal; *feed*: alfalfa, beets, sorghum, birdseed, and chickpea; *fruits, short cycle*: strawberry, cantaloupe, pineapple, and watermelon; *fruits, long cycle*: avocado, capulin, plum, almonds, coconut, apricot, peaches, red pomegranate, guava, fig, lime, lemon, *mamey*, mango, apple, quince, orange, nuts, papaya, pears, bananas, tamarine, grapes, dates, and grapefruit; *industrialized products*: barley, cocoa, coffee, sugarcane, tobacco, and processed vanilla.

Table 11.2. Maize: Production, imports, exports, and apparent consumption by consumption-year, 1965–1982

Con-sumption year	Production (tons)[a]	Imports (tons)[b]	Inventory Jan. 1 (tons)	Exports (tons)	Inventory Dec. 31 (tons)	Apparent national consumption (tons)	Imports/ apparent national consumption (%)	Per capita consumption (kg)
1965	8,454,046	—	—	1,409,325	555,599	—	—	—
1966	8,936,381	—	555,599	878,694	699,688	7,913,598	—	186.7
1967	9,271,485	—	699,688	1,191,678	594,024	8,185,471	—	187.0
1968	8,603,279	—	594,024	907,184	495,898	7,794,221	—	172.4
1969	9,061,823	36,463	495,898	779,168	201,613	8,613,403	0.4	184.5
1970	8,410,894	732,356	201,613	—	667,454	8,677,318	8.4	179.9
1971	8,879,384	—	667,454	256,512	569,388	8,721,029	—	174.9
1972	9,785,734	190,698	569,388	428,596	342,759	9,774,465	1.9	189.6
1973	9,222,838	1,154,459	342,759	19,545	398,585	10,302,036	11.2	193.2
1974	8,609,132	1,318,873	398,585	—	311,975	10,014,615	13.2	181.6
1975	7,847,763	2,625,238	311,975	—	741,986	10,042,990	26.1	176.2
1976	8,448,708	955,127	741,986	—	416,780	9,729,041	9.8	165.0
1977	8,017,294	1,727,426	416,780	—	951,788	9,209,712	18.8	151.1
1978	10,137,914	1,465,180	951,788	—	889,473	11,665,409	12.6	185.1
1979	10,930,077	927,158	889,473	—	601,632	12,045,076	6.9	184.8
1980	8,457,849	3,167,262	601,632	—	631,000	11,595,743	27.3	172.1
1981	12,374,420	2,478,099	631,000	—	1,951,516	13,532,003	18.3	195.0
1982	14,765,760	1,176,841	1,951,516	—	813,935	17,080,182	6.9	239.0

[a]Data from DGEA (1965–1982); the figures differ from those in Table 11.1 because they correspond to consumption years, which have approximately a one-year lag compared to crop years.
[b]Data from "importación exportación y cambio de existencias: CONASUPO EN CIFRAS" (Mexico City: CONASUPO).
[c]To estimate yearly consumption, production of the previous year is considered.

Table 11.3. Changes in crop structure, 1972–1982 (percentage)

	Area			Production value		
	1972–1976	1977–1979	1980–1982	1972–1976	1977–1979	1980–1982
Basic foodstuffs	64.02	60.10	60.67	33.88	30.40	39.11
Other grains	0.80	0.80	1.27	0.31	0.39	0.61
Vegetables	1.80	2.00	1.80	8.99	10.75	8.92
Oilseeds	8.94	10.83	8.37	8.59	8.06	5.55
Fibers	1.36	1.17	2.33	8.96	8.15	4.78
Feed	10.92	12.17	12.23	11.34	12.03	11.85
Short-cycle fruits	0.38	0.50	0.40	1.70	2.21	1.62
Long-cycle fruits	3.48	3.93	4.63	11.83	12.47	13.68
Industrializable agricultural crops	8.30	8.50	8.30	14.40	15.54	13.88
	100.00	100.00	100.00	100.00	100.00	100.00

Source: Based on data of DGEA (1965–1982). The series was completed with 1982 preliminary data provided by DGEA.

0.39 to 0.61% during these periods. Other groups showed significant decreases: vegetables from 10.75 to 8.92%; oilseeds from 8.06 to 5.55%, fibers from 8.15 to 4.78%, and annual fruit from 2.21 to 1.62%.

In order to analyze this evolution, it is also necessary to review price changes. In the case of basic foodstuffs, a slight increase was recorded in domestic prices of all items. Vegetables and annual fruit were affected by a contraction in external demand that caused a severe fall in sales of tomatoes and strawberries. Domestic prices for oilseeds were stagnant or fell in real terms.

For all the above, during the three-year period analyzed, the structure of production was not significantly altered in terms of type of crop or harvested area. Commercial crops, despite popular belief, were not sacrificed to favor basic crops. The latter's increases are due to higher yields per hectare, not to an increase in area.

Impact by Type of Producer

Who actually benefited from SAM is one of the most controversial issues surrounding the program. A good part of this debate has been conducted in the absence of systematic information, and a great many generalizations have been based on fragmentary data or experiences or on a superficial analysis of available information. But it is not possible to examine thoroughly SAM's impact by types of producers, since none of the available statistics or surveys carried out during the period were based on a classification of the plots or producers or on variables that allowed for such distinctions to be made.

In the absence of such ideal data, we use an alternative approximation methodology. SAM's strategy aimed at the recovery of self-sufficiency in production of maize and beans by 1982 by supporting impoverished rain-fed peasants. Our hypothesis is that in the 1980–1982 period there was a significant increase in maize production in areas with a greater proportion of poor peasants. If this hypothesis can be confirmed, there would be a basis for defining the beneficiaries of SAM. In this regard, one must take into account the fact that maize is the country's most important crop, both in terms of area and the number of producers, as well as being the most typically peasant crop.

Since available information is broken down by states, our hypothesis was examined at this level of disaggregation. States were classified into two groups: those with basically maize-producing peasant economies, and those dominated by commercial agriculture, with a lesser share of peasant maize producers (see Table 11.4). In order to define the two groups of states, we took into account the percentage of rainfed maize-harvested area with respect to the total of harvested area and the percentage of producers falling into the categories of "infra-subsistence," "subsistence," and "stationary," which are considered to be the poorest peasants in the Economic Commission for Latin America's (ECLA) typology. These are the 80% who cultivate maize and beans in average plots of 50 or fewer hectares. Based on these data, maize-peasant-economy states were considered to be those devoting over 40% of their harvested area to rain-fed maize and having over 60% of their producers in the aforementioned categories. The remaining states, which met only one of the characteristics or none, were classified as commercial states with nonpredominant peasant economy. This does not imply that these states have no poor maize producers, but simply that the rain-fed maize area is not significant or that its specific weight does not define the type of state, or that poor rain-fed peasants devote themselves to other crops or activities. Such is the case of Chihuahua, an important livestock state, with 88% of producers within the aforementioned categories, but with only 29.9% of the harvested area devoted to rain-fed maize, or the case of Durango, which is a large wood-producing state. Jalisco, which is traditionally considered the grain silo of the country and which produces a significant share of maize, was considered a commerical state since more than half of its producers do not fall into the categories of poor peasants and, since, in addition, 19% of its significant maize area is irrigated. The categories of peasant and commercial states were further disaggregated into large producers (>150,000 hectares of rain-fed land), medium producers (50,000–150,000 hectares), and small producers (<50,000 hectares).

Three-year periods were analyzed by calculating the structural averages as defined by production increases, area, and yield, as well as by increases

Table 11.4. Maize areas and peasant populations in Mexico's peasant and commercial states

	Rain-fed maize area (%)[a]	Poor peasant population (%)[b]
Peasant States		
Large production states		
Veracruz	78.1	64.8
Chiapas	77.9	71.7
Oaxaca	70.0	91.0
Guerrero	68.4	85.0
México	67.2	94.4
Puebla	62.9	93.4
Hidalgo	54.1	92.5
Zacatecas	46.5	92.5
Michoacán	43.0	72.0
Medium production states		
Yucatán	84.9	91.4
Tabasco	79.5	61.7
San Luis Potosí	59.1	91.4
Tlaxcala	56.7	95.6
Aguascalientes	51.9	89.1
Querétaro	46.7	93.8
Small production states		
Quintana Roo	88.7	91.4
Campeche	69.2	86.5
Distrito Federal	52.3	95.4
Commercial States		
Large production states		
Jalisco	62.4	47.2
Chihuahua	29.9	80.9
Guanajuato	24.7	45.0
Medium production states		
Nayarit	36.9	31.9
Durango	31.1	95.4
Tamaulipas	11.3	62.3
Sinaloa	5.0	46.9
Small production states		
Colima	54.9	23.5
Nuevo León	28.6	87.4
Morelos	21.7	35.7
Coahuila	8.8	57.2
Sonora	0.1	33.4
Baja California Norte	0.06	5.5
Baja California Sur	—	48.2

[a]Percentage of harvested area of rain-fed maize with respect to total harvested area per state.
[b]Percentage of the sum of infrasubsistence, subsistence, and stationary peasants with respect to the total of producers as per classification of the Economic Commission of Latin America (ECLA) in *Economía campesina y agricultura empresarial* (Mexico City: Siglo XXI Editores, 1982).
Source: DGEA (1965–1982).

in consumption of fertilizers for all thirty-two states (see Table 11.5–11.10). The results of this analysis are as follows. The eighteen states with a predominantly peasant economy significantly increased their share of national maize production, from 59.05% in 1977–1979 to 64.28% during SAM's three-year period. Furthermore, they maintained their yield with respect to the national average and they slightly improved their position regarding harvested area. In contrast, the fourteen commercial states decreased in relative weight, in terms of production, from 40.94 to 35.72%, of area, from 32.99 to 32.03%, and of yields, from 1.26 to 1.08% with respect to the national average (Table 11.6).

The analysis of absolute increases in production reached by the two groups of states (Table 11.9) clearly shows the advance of peasant states, which increased their production by a total of 2,125,421 tons, accounting for 84.7% of the national net increase from the periods 1977–1979 to 1980–1982. Such a significant increase was due, on the one hand, to a net increase of 26,201 hectares in the harvested area, of which 24,398 were irrigated, and, on the other, to yield increases of 310 kilograms per hectare (see Table 11.9). In contrast, the commercial states reduced their total maize-harvested area by 80,825 hectares, mostly due to the reduction in rain-fed area and to an average net increase in yields of only 142 kilograms per hectare.[3]

In terms of the rate of growth, Table 11.8 shows that peasant states grew, on the average, more rapidly than commercial ones, both in production and yield (20.9 and 24.5% against 18.6 and 10.6%, respectively). In contrast (see Table 11.7), the latter's total harvested area grew more rapidly (8.7 against -4.2%), though peasant states grew at a faster pace in irrigated area (28.8 against 4.3%). In order to explain this contrast, one must take into account that the higher rate of growth in harvested area in commercial states is concentrated in some states that are not important in terms of the total area sown with maize.

As to the consumption of fertilizer, an important part of the same strategy, in peasant states there was an average growth of 89.3%, while in commercial states it reached 236.3% (see Table 11.8). The increase in the latter is accounted for by only one state, Baja California Sur, which showed a 3000% increase. This is a state that during the 1980–1982 period produced only 6428 tons on 2429 hectares, irrigated, and used little fertilizer. If this unusual case is eliminated because of its low relative weight, the average growth of commercial states decreases to 23.7% and, therefore, is one-fourth the average of peasant states.

As far as credit is concerned, coverage for maize production increased after 1980. The maize area financed went from an average of 34% in 1977–

---

3. Similar results are obtained, though with larger differences, if instead of comparing SAM's three-year period with the period 1977–1979, the comparison is made between SAM's best year, 1981, and the latter.

Table 11.5. Mexican maize production, 1977–1982

	Total area (ha)	Irrigated area (ha)	Rain-fed area (ha)	Production (tons)	Yield (tons/ha)	Fertilizers, nutrients, consumption (tons)
1977	7,469,649	979,251	6,490,398	10,137,914	1,357	1,089.5
1978	7,191,128	947,398	6,243,730	10,930,077	1,520	1,076.0
1979	5,581,158	856,516	4,724,642	8,457,849	1,515	1,225.3
Average	6,747,312	927,722	5,819,590	9,841,947	1,458	1,130.2
1980	6,766,479	1,115,492	5,650,987	12,374,400	1,829	1,319.5
1981	7,668,692	980,593	6,688,099	14,550,074	1,897	1,502.2
1982	5,642,893	1,000,768	4,635,524	10,129,083	1,795	1,727.5
Average	6,692,688	1,034,484	5,658,203	12,351,186	1,840	1,516.4

Table 11.6. Comparative maize production in peasant and commercial states, 1977–1979 and 1980–1982 (percentage)

	Area average		Production average		Yield average[a]	
	1977–1979	1980–1982	1977–1979	1980–1982	1977–1979	1980–1982
Peasant states						
Large production states						
Veracruz	8.31	7.51	7.62	6.64	0.92	0.88
Chiapas	6.33	8.57	7.64	11.50	1.20	1.34
Oaxaca	7.08	5.51	4.13	3.14	0.58	0.56
Guerrero	5.03	5.70	4.25	4.14	0.84	0.72
México	8.23	10.09	12.11	14.98	1.47	1.49
Puebla	6.58	7.14	6.46	6.95	0.98	0.95
Hidalgo	3.16	2.84	2.40	2.34	0.76	0.83
Zacatecas	5.69	4.95	2.53	2.26	1.06	0.45
Michoacán	6.16	6.68	5.15	6.18	0.81	0.92
	56.57	58.99	52.29	58.13	0.96	0.90
Medium production states						
Yucatán	1.70	2.17	1.10	1.12	0.66	0.52
Tabasco	0.81	0.52	0.67	0.55	0.83	1.05
San Luis Potosí	2.52	1.14	1.21	0.80	0.61	0.75
Tlaxcala	1.70	2.14	1.30	1.80	0.75	0.84
Aguascalientes	1.10	0.39	0.67	0.26	0.76	0.77
Querétaro	1.26	0.91	0.80	0.72	0.64	0.78
	9.09	7.27	5.75	5.25	0.71	0.78
Small production states						
Quintana Roo	0.54	0.86	0.33	0.29	0.63	0.34
Campeche	0.60	0.65	0.45	0.89	0.76	0.60
Distrito Federal	0.22	0.15	0.23	0.22	1.07	1.45
	1.36	1.66	1.01	0.90	0.82	0.80
Peasant state total	67.02	67.92	59.05	64.28	0.85	0.85
Commercial states						
Large production states						
Chihuahua	2.87	3.60	1.71	2.54	0.62	0.71
Guanajuato	3.90	3.73	4.21	3.12	1.07	0.68
Jalisco	12.78	11.75	19.52	16.29	8.51	1.38
	19.55	19.08	25.44	21.95	1.07	0.92
Medium production states						
Nayarit	1.57	1.17	1.93	1.70	1.21	1.45
Durango	2.38	2.29	1.72	1.36	0.72	0.62
Tamaulipas	4.95	5.18	6.92	6.39	1.46	1.26
Sinaloa	1.28	1.21	1.17	0.74	0.88	0.64
	10.18	9.85	11.74	10.19	1.07	0.99
Small production states						
Colima	0.56	0.59	0.65	0.66	1.15	1.12
Nuevo León	1.00	0.66	0.85	0.57	0.85	0.86
Morelos	0.70	0.63	0.75	0.66	1.08	0.98
Coahuila	0.58	0.46	0.62	0.40	1.12	0.91
Sonora	0.32	0.63	0.73	1.14	2.23	1.72
Baja California Norte	0.08	0.09	0.12	0.10	1.68	1.36
Baja California Sur	0.02	0.04	0.04	0.05	2.01	1.39
	3.26	3.26	3.76	3.58	1.44	1.19
Commercial state total	32.99	32.03	40.94	35.72	1.26	1.08

[a]Relative when national average equals one.
Source: DGEA.

Table 11.7. Increases in harvested area from 1977–1979 to 1980–1982 (percentage)

	Total harvested area	Irrigated harvested area	Rain-fed harvested area
Peasant states			
Large production states			
Veracruz	−10.35	70.41	−11.19
Chiapas	34.19	122.41	33.57
Oaxaca	−22.76	−32.52	−21.76
Guerrero	12.50	−29.87	15.44
México	21.59	36.17	18.47
Puebla	7.78	0.06	8.39
Hidalgo	−10.68	8.52	−15.36
Zacatecas	−13.58	6.85	14.86
Michoacán	7.67	−5.13	10.58
	2.93	19.66	5.89
Medium production states			
Yucatán	26.89	152.08	25.81
Tabasco	−36.29	−	−36.17
San Luis Potosí	−54.95	10.70	−63.89
Tlaxcala	24.64	43.66	23.01
Aguascalientes	−64.60	−45.79	−70.52
Querétaro	−28.34	8.01	−37.16
	−22.11	28.11	−26.59
Small production states			
Quintana Roo	57.23	98.55	56.46
Campeche	8.44	74.61	7.04
Distrito Federal	−34.02	−	−30.15
	10.55	57.85	11.12
Peasant state total	**−4.15**	**28.82**	**−4.03**
Commercial states			
Large production states			
Chihuahua	24.65	69.65	20.03
Guanajuato	−5.07	−12.24	−2.53
Jalisco	−8.78	41.90	−11.01
	3.60	33.10	2.16
Medium production states			
Nayarit	−25.77	−60.08	−22.65
Durango	−4.66	1.73	−6.61
Tamaulipas	3.90	22.20	32.62
Sinaloa	−6.32	25.89	−16.01
	−8.21	−2.57	−3.16
Small production states			
Colima	4.71	−18.62	11.99
Nuevo León	−34.42	−9.30	−48.68
Morelos	−10.46	3.85	−12.93
Coahuila	−21.85	−20.51	−24.04
Sonora	96.24	−95.42	114.04
Baja California Norte	21.48	22.27	−13.04
Baja California Sur	88.44	88.59	−
	20.59	−4.16	3.91
Commercial state total	**8.72**	**4.28**	**1.51**

Table 11.8. Increases in maize production from 1977–1979 to 1980–1982 (percentage)

	Maize production	Total harvested area	Yield	Fertilizer, nutrients consumption
Peasant states				
Large production states				
Veracruz	9.38	−10.35	21.1	36.2
Chiapas	88.84	34.19	40.8	79.0
Oaxaca	−4.36	−22.76	21.1	63.4
Guerrero	22.22	12.50	7.5	112.9
México	55.20	21.59	27.0	26.4
Puebla	34.95	7.78	22.6	60.7
Hidalgo	22.27	−10.68	37.4	38.5
Zacatecas	11.95	−13.58	−46.5	108.8
Michoacán	50.63	7.67	42.0	73.9
	32.34	2.93	19.2	61.1
Medium production states				
Yucatan	26.99	26.89	−1.0	666.7
Tabasco	3.27	−36.29	58.7	44.3
San Luis Potosí	−17.40	−54.95	53.5	107.8
Tlaxcala	74.04	24.64	41.4	27.2
Aguascalientes	−50.37	−64.60	27.7	54.5
Querétaro	12.96	−28.34	51.4	48.2
	8.25	−22.11	38.62	158.1
Small production states				
Quintana Roo	9.11	57.33	−33.4	−80.0
Campeche	8.58	8.44	−0.4	142.1
Distrito Federal	17.89	−34.02	70.1	47.0
	11.86	10.55	12.1	36.4
Peasant state total	20.90	−4.15	24.5	89.3
Commercial states				
Large production states				
Chihuahua	86.36	24.65	43.3	46.9
Guanajuato	−6.84	−5.07	−20.4	12.9
Jalisco	4.73	−8.78	14.7	46.8
	28.08	3.6	12.53	35.5
Medium production states				
Nayarit	10.37	−25.77	50.3	23.5
Durango	−0.54	−4.66	8.3	−5.8
Tamaulipas	15.90	3.90	8.7	34.3
Sinaloa	−20.96	−6.32	−8.7	14.3
	1.19	−8.21	14.65	16.6
Small production states				
Colima	27.50	4.71	22.6	19.3
Nuevo León	−16.04	−34.42	26.7	33.9
Morelos	9.78	−10.46	14.3	25.9
Coahuila	−19.75	−21.85	2.8	27.0
Sonora	96.54	96.24	−3.3	7.5
Baja California Norte	6.74	21.48	1.8	21.2
Baja California Sur	66.66	88.44	−13.3	3,000.0
	24.49	20.59	7.37	447.8
Commercial state total	18.60	8.72	10.6	236.3[a]

[a] If Baja California Sur is excluded, with an erratic behavior of 3000%, the average goes down to 23.7%.

Table 11.9. Absolute increases in production from 1977-1979 to 1980-1982

	Total production (tons)	Harvested area (ha)	Yield (tons/ha)	Fertilizer, nutrients consumption (tons)
Peasant States				
Large production states				
Veracruz	70,355	−58,058	0.284	28.6
Chiapas	668,214	146,101	0.716	23.4
Oaxaca	−17,697	−108,666	0.180	9.7
Guerrero	92,922	42,427	0.092	14.9
México	658,155	119,889	0.583	21.5
Puebla	222,384	34,538	0.324	24.7
Hidalgo	52,591	−22,752	0.417	3.5
Zacatecas	29,804	−52,113	−0.725	12.4
Michoacán	256,534	31,892	0.499	18.9
	2,033,262	133,258	0.260	157.6
Medium production states				
Yucatán	29,275	30,761	−0.010	10.0
Tabasco	2,144	−19,870	0.716	3.1
San Luis Potosí	−20,740	−93,288	0.479	9.7
Tlaxcala	94,615	28,337	0.454	4.9
Aguascalientes	−33,292	−48,112	0.309	6.0
Querétaro	10,271	−24,088	0.485	5.4
	82,273	−126,260	0.406	39.1
Small production states				
Quintana Roo	2,956	20,914	−0.310	−0.4
Campeche	3,811	3,409	−0.005	5.4
Distrito Federal	4,119	−5,120	1.098	3.1
	10,886	19,203	0.261	8.1
Peasant states total	125,421	26,201	0.310	204.8
Commercial states				
Large production states				
Chihuahua	145,508	47,658	0.394	23.2
Guanajuato	−28,314	−13,327	−0.320	15.3
Jalisco	90,913	−75,733	0.325	50.5
	208,107	−41,402	0.133	89.0
Medium production states				
Nayarit	19,703	−27,258	0.892	4.7
Durango	−911	−7,499	0.087	−1.9
Tamaulipas	108,197	13,036	0.186	12.5
Sinaloa	−24,142	−5,479	−0.112	16.7
	102,847	−27,200	0.263	32.0
Small production states				
Colima	17,693	1,784	0.381	2.3
Nuevo León	−13,386	−23,308	0.334	3.8
Morelos	7,264	−4,959	0.226	5.7
Coahuila	−12,093	−8,543	0.045	3.1
Sonora	69,015	20,554	−0.107	8.4
Baja California Norte	799	1,109	0.04	10.1
Baja California Sur	2,571	1,140	−0.392	3.0
	71,863	−12,223	0.076	36.4
Commercial states total	382,817	−80,825	0.142	157.4
National total	2,508,238	−54,624	0.452	362.2

Table 11.10. Absolute increases in harvested area from 1977-1979 to 1980-1982 (hectares)

	Total harvested area	Irrigated harvested area	Rain-fed harvested area
Peasant states			
Large production states			
Veracruz	−58,058	−4,080	−62,138
Chiapas	146,101	3,643	142,458
Oaxaca	−108,666	−104,436	−94,230
Guerrero	42,427	−6,579	49,007
México	119,889	35,426	84,463
Puebla	34,538	26	34,512
Hidalgo	−22,752	3,559	−26,312
Zacatecas	−52,113	1,551	−53,663
Michoacán	31,892	−4,042	35,935
	133,258	23,228	110,032
Medium production states			
Yucatán	30,761	1,501	29,260
Tabasco	−19,870	−106	−19,764
San Luis Potosí	−93,288	2,177	−95,465
Tlaxcala	28,337	3,979	24,358
Aguascalientes	−48,112	−8,164	−39,948
Querétaro	−24,088	1,331	−25,419
	−126,260	718	−126,978
Small production states			
Quintana Roo	20,914	659	20,256
Campeche	3,409	626	2,783
Distrito Federal	−5,120	−833	−4,287
	19,203	452	18,752
Peasant states total	26,201	24,398	1,806
Commercial states			
Large production states			
Chihuahua	47,658	12,547	35,111
Guanajuato	−13,327	−8,407	−4,920
Jalisco	−75,733	15,224	−90,958
	−41,402	19,364	−60,767
Medium production states			
Nayarit	−27,258	−5,293	−21,965
Durango	−7,499	649	−8,148
Tamaulipas	13,036	49,374	−36,338
Sinaloa	−5,479	5,187	−10,666
	−27,200	49,917	−77,117
Small production states			
Colima	1,784	−1,674	3,458
Nuevo León	−23,308	−2,280	−21,029
Morelos	−4,959	268	−5,227
Coahuila	−8,543	−4,974	−3,569
Sonora	20,554	19,481	1,072
Baja California Norte	1,109	1,124	−15
Baja California Sur	1,140	1,140	—
	−12,223	13,085	−25,310
Commercial states total	−80,825	82,366	−163,194
National total	−54,624	106,764	−161,388

1979 to 43.6% in 1980–1982, and working capital loans for maize rose, on average, from 17.1 to 28.7%. The amount of credit per hectare decreased in real terms, on average, from 1328 to 1141 pesos (1970 constant prices). It seems clear that a larger number of maize producers had access to credit, but, on average, they received less per hectare than in the previous three-year period (see Chapter 5).

This analysis by state shows that almost all states experienced growth, but that it was more significant in peasant states. In order to analyze this more accurately, it is useful to take into account the differentiation of states within each group (as indicated in the tables). Here we see that the behavior of the different variables fluctuates considerably from one state to another. The main average increases in production in peasant states were recorded for the group of large producer states, those with over 150,000 hectares of rain-fed maize land. The highest percentage increases in yield and consumption of fertilizers were concentrated in the medium producer states, those with 50,000 to 150,000 hectares. The small producer states, those with fewer than 50,000 hectares, show higher increases in area and smaller increases in yield and fertilizer consumption.

The fact that in some peasant states and in some commercial states there were decreases in the pace of growth, production, area, yield, and fertilizer consumption should support the relatively obvious conclusion that the response to SAM was not uniform. There were differences in the quality of the land, the weather, and the organization of producers, as well as in the intensity of support provided by the program.

The overall evolution of maize production in the different groups of states during the three-year SAM period reveals that its strategy reversed the trend toward stagnation or decline observed during the previous fifteen years (see Table 11.11). The average growth in maize production had been 17.1% for the group of peasant states (average growth rates of the group's states) and 25.4% for commercial states during the period 1970–1976; it fell to 7.8% for peasant states and 12.6% for commercial states during 1977–1979 and recovered—in both cases—to 28.1 and 30.4 during SAM. Growth rates of the previous periods had been higher in commercial states, but during SAM's three-year period, average growth in peasant states almost reached the same pace as that in commercial states.

Although information by states cannot be considered as equivalent to data by type of producer, the analysis allows us to infer reasonably that a significant portion of poor peasants received SAM benefits and that they were capable of translating them into higher production.

The Importance of Weather Changes

Since a large part of Mexico's agriculture is in rain-fed areas, weather variations significantly affect production results. This is even more signifi-

Table 11.11. Average growth of maize production, 1970–1982 (percentage)

	1970–1976	1970–1979	1977–1979	1980–1982
Peasant states				
Veracruz	2.7	1.0	−3.0	7.2
Chiapas	0.1	9.3	31.0	20.6
Oaxaca	−3.3	2.1	14.8	−7.0
Guerrero	1.3	14.0	32.9	−2.9
México	19.6	3.5	33.9	13.9
Puebla	18.2	17.9	15.6	8.0
Hidalgo	3.1	4.0	6.1	−1.1
Zacatecas	14.4	8.6	−5.0	35.3
Michoacán	3.6	3.3	−2.5	52.7
Yucatán	3.4	6.2	12.8	3.4
Tabasco	8.7	−1.8	−26.3	13.3
San Luis Potosí	8.4	2.0	−13.1	0.4
Tlaxcala	4.1	2.7	−0.7	56.1
Aguascalientes	−8.0	−4.3	4.5	−26.3
Querétaro	−7.9	6.4	−2.9	22.7
Quintana Roo	196.4	155.5	60.0	12.6
Campeche	18.3	8.3	−15.1	5.5
Distrito Federal	23.8	15.9	−2.7	291.6
Composite	17.1	14.1	7.8	28.1
Commercial states				
Chihuahua	−3.9	−4.3	−5.2	130.0
Guanajuato	24.0	12.0	−16.0	23.0
Jalisco	7.7	3.9	−4.8	7.1
Nayarit	3.6	14.9	58.1	16.8
Durango	24.9	150.9	8.5	11.0
Tamaulipas	4.9	8.8	18.1	17.8
Sinaloa	31.5	19.6	−9.1	43.8
Colima	−2.5	−4.3	−20.3	28.2
Nuevo León	30.9	18.4	−10.6	10.5
Morelos	19.6	14.0	0.9	0.2
Coahuila	4.2	1.7	−4.0	2.6
Sonora	−8.0	14.8	68.0	63.8
Baja California Norte	16.0	40.2	48.5	−13.2
Baja California Sur	15.2	23.7	43.5	84.1
Composite	25.4	22.4	12.6	30.4
National total	−2.5	−0.6	3.9	11.2

Source: DGEA.

cant in the case of basic crops, because a decisive share of production depends on the volume and the timeliness of the rains. The wide variations recorded yearly during the past decade—including SAM's period—have allowed for very different interpretations. According to many, weather changes are the prime cause for production results; the so-called SAM impact is biased, it is claimed, because in the middle year, 1981, the weather was exceptionally good.

Data on this issue are clearly insufficient to support a solid hypothesis one way or the other. Over long periods, a weak correlation between rainfall levels and agricultural production results has been observed, principally on the aggregate level, and there is no systematic information on the frequency and timeliness of rains and, especially, on the combination of these two elements (volume and timing), which would be essential to arrive at valid conclusions. Nevertheless, it is worth making some general observations. Over the longer run, there seems to be some statistical regularity in the distribution of good, average, and bad years. It seems impossible to forecast accurately the nature of any given year, or the specific distribution of their frequency in a given five-year period or a decade, but a general pattern appears to exist. By chance, frequency was absolutely regular during the three-year period of SAM, as well as during the previous period: 1978 and 1981 were good, 1977 and 1980 were average and 1979 and 1982 were bad. In other words, each of these three-year periods had a good, a bad, and an average year. At this admittedly gross level of generalization, therefore, one may discard the weather as a variable for comparative analytic purposes, since it would be an identical variable in both periods.

A more detailed analysis of weather conditions found that national averages for extreme weather variations (rainfall and temperature) did not differ during the SAM period from conditions prevailing in the previous fifteen years: 1979, the bad year, was no worse than other bad years, in spite of what has been stated otherwise, just as 1981 was no better than other good years.

SAM and Its Explicit Goals

One approach in analyzing SAM is to compare its concrete results with the objectives set forth when it was announced. SAM proposed to achieve self-sufficiency in production of maize and beans by 1982, and in production of other crops by 1985. It considered that these goals could be attained with different combinations of increased area or yields in both irrigated and rain-fed areas. The estimates for the two main crops were projected to meet the following production targets:

Maize. To produce by 1982 a volume of 13,050,000 tons, meaning an increase of 3,400,000 tons over the average of 1976–1978; an annual growth rate of 6.2%. This production was to be achieved by harvesting a total area of 7,673,000 hectares, of which 1,140,000 would be irrigated and 6,533,000 rain-fed, accounting for an annual cumulative increase in cultivated area of 1.5% in 1980, 2.6% in 1981, and 1.4% in 1982. Yields were to be of 2850 kilograms per hectare in irrigated areas and 1500 kilograms per hectare on rain-fed land representing an annual rate of increase of 4.4 and 4.5%, respectively.

Beans. To produce by 1982 a volume of 1,492,000 tons, meaning an increase of 685,000 tons over the average production of 1976–1978; an annual increase of 13.1%. This target was to be reached by harvesting a total area of 2,334,000 hectares: 344,000 irrigated and 1,990,000 rain-fed, with an annual growth rate of 9.2% in 1980, 11.0% in 1981, and 8.8% in 1982. Yields were to be 1302 kilograms per hectare on irrigated land and 525 kilograms per hectare in rain-fed areas, with an increase of 3.7 and 2.9%, respectively.

SAM's Production of Basic Foodstuffs program was to begin during the spring-summer cycle of 1980 with the following aims:

1. The program should develop rapidly, expanding from areas selected because of their productive potential during the initial phases to cover the whole country in the medium term. Efforts of the first three years would emphasize maize and bean production and would gradually incorporate other basic crops.
2. From the beginning, the program would concentrate on rain-fed districts and would foster production in rain-fed areas or those with small irrigation works. The target population of the program would be peasant producers on rain-fed land, with special emphasis on subsistence, semisubsistence, and stable strata that would have potential productive resources, though some policies and supports could be channeled to a wider spectrum of producers.

When we compare aspirations with results, we can see that SAM reached its goal of substantially increasing the volume of maize production in the term contemplated and even exceeded its specific target in 1981. It did not reach its targets for cultivated area, but after 1980 it greatly exceeded those for yield, both in irrigated and rain-fed areas (see Tables 11.12–11.13). In the case of beans, the volume target was surpassed, although the one set for area was not reached.

Results were achieved according to the terms originally foreseen by SAM, through a combination of increasing the irrigated and rain-fed areas producing maize and beans and of increasing yield in both types of areas. This combination relied on intensive production inputs. The irrigated area increased by 11.4% from one three-year period to the next (106,000 hectares), but rain-fed areas decreased by 2.8% (161,000 hectares). The increase in production was fundamentally due to yield increases, which allows us to assume that the technological package was effective. Nevertheless, one must take into account the higher increase in irrigated area—whose yields, as is obvious, are much higher than those of rain-fed areas—a factor that suggests that the bulk of the increase can be attributed to irrigation.

This information leads some analysts to state that the success of SAM was fundamentally due to the action of commercial farmers in large irrigated districts. It is argued that the attainment of quantitative goals was not due to the validity of the strategy itself, which committed itself to the impoverished rain-fed peasants. This implies that SAM was not a new strategy, but rather another turn of the screw of the prior model: only some more of the same medicine, administered at a very high cost.

This hypothesis is highly speculative, and there is not enough information available to either support or discard it. In particular, it is impossible to determine to what extent provisions of SAM were respected as to maximum size of irrigated plots (20 hectares) to which the strategy's incentives could be applied. It must be recalled that commercial farmers are concentrated in irrigation districts, but there are many small producers there as well. It may well be that the farmers on small irrigated plots were the primary source of the production increases.

In any case, there is no doubt that a large number of poor peasants in maize-producing rain-fed areas took advantage of the SAM incentives. It is clear that not all, or even only this group, did so, but this was what SAM proposed. Its objective was to start in certain selected areas and gradually cover the whole country. After it started, it admitted that part of the strategy covered a large spectrum of producers—not only peasants in rain-fed areas. In other words, SAM clearly arrived at its basic goals by following, fundamentally, the planned route. But there is insufficient evidence on this point. The rest remains speculative.

SAM's Food-Production Costs

It is virtually impossible to determine accurately the social cost of SAM. Several factors prevent it. First, during the period of operation, public spending was lavish due to the oil boom, which makes it especially difficult to distinguish precisely SAM's components. Furthermore, SAM's systemic approach and the way in which it was jammed into the Global Development Plan published only a few months after SAM caused the strategy to cut across administrative budget categories. It includes expenses of various sectors, and often under generic headings that are not sufficiently explanatory. Finally, some specific items that could be clearly associated with SAM appear in budgetary items that began in 1978, and for this reason it is not possible to compare their evolution with earlier periods.

In order to approach the problem analytically, we first examined public expenses channeled to the agricultural and livestock sectors (which include agricultural, livestock, forestry, fishing, and hunting activities in this budget classification), and then we analyzed expenses classified as "food-related." This category was typically multisectoral and is more clearly related to SAM.

Table 11.12. Maize: SAM objectives and results

	1980	1981	1982	Average	SAM 1982 target
Production (thousand tons)					
Irrigated	3,042	3,010	—	—	—
Rain-fed	9,333	11,540	—	—	—
	12,374	14,550	10,129	12,351	13,050
Harvested area (thousand ha)					
Irrigated	1,115	980	1,007	1,034	1,140
Rain-fed	5,651	6,689	4,636	5,658	6,533
	6,766	7,669	5,643	6,692	7,673
Yields (kg/ha)					
Irrigated	2,728	3,071	—	—	2,858
Rain-fed	1,652	1,726	—	—	1,500
	1,829	1,897	1,795	1,846	1,700

The agricultural and livestock share of total public spending went from 8.4% in the 1971–1976 administration to 8.0% in 1977–1979 and 7.7% in 1980–1982. As a share of sectoral GDP, public spending on agriculture and livestock accounted for 22.5% in 1971–1976, 29.9% in 1977–1979, and 45.6% in 1980–1982 (see Table 11.14). Here, total public spending is considered to be the total expenses made from the Federal Expenditure Budget (PEF), while public spending on agriculture consists of the expenditures of the Ministry of Agriculture and Water Resources (SARH), the Ministry of Agrarian Reform, and the Ministry of Fishing, as well as support for banks, funds, and insurance, and transfers to parastatal enterprises in the sector.

The evolution of sectoral public spending in constant terms (at 1970 prices, using the implicit GDP deflator) shows a slowing of its growth rate with respect to total public spending, with the former increasing at a proportionally smaller rate. Total spending grew at an annual average rate of 12.6% in the 1971–1976 administration and at 19.8% in 1977–1982. Sectoral spending increased, on the other hand, at an average rate of 12.6% in 1977–1982, against 34.5% during the previous six-year period. The three-year

Table 11.13. Maize: Harvested area

	1977–1979 (%)	1980–1982 (%)	Increment (%)
Irrigated	928 (13.8)	1,034 (15.4)	11.4
Rain-fed	5,819 (86.2)	5,658 (84.6)	−2.8
	6,747	6,692	−0.8

Table 11.14. Public expenditure for agricultural and livestock sectors, 1971-1982[a]

	1971-1973	1974-1976	1977-1979	1980-1982
Expenditures, as percentage of:				
total public spending	7.03	9.81	8.00	7.68
sectoral GDP	15.98	29.00	29.93	45.58
total public investment	14.52	16.30	18.30	16.00
Average growth[b] in:				
percentage of total GDP	2.38	1.86	3.81	4.25
total expense	11.40	13.90	11.43	28.17
sectoral percentage of total				
expense	55.45	13.53	10.07	15.08

[a]Total and sectoral expense, as well as total and sectoral investment, include the central and decentralized sectors.
[b]Real annual growth calculated at 1970 constant prices.

SAM period registered a 15.1% increase as compared to 10.1% in the previous three years.[4]

The levels of public investment in agriculture accounted for 16% of total public investment during the three-year SAM period, as compared to 18.3% in 1977-1979 and 16.3% in 1974-1976. If one considers that the growth in sectoral spending was almost the same (15.5%) as that of investment, the margin of increase in resources was channeled equally between investment, administration, and operational spending. (This classification of spending should be taken with serious reservations, however. For a long time, distinctions were based on a gross classification of the type of government agency.)

As a whole, these figures show that sectoral public spending during the SAM period recorded no exceptional increases; it was even below the average growth rate seen during the previous eight years, thereby losing relative importance within total spending. Nevertheless, the analysis of agricultural sectoral spending is insufficient to determine the real evolution of "SAM's budget," since, to a great extent, the strategy was channeled through agencies whose expenses and investments are not included in the items considered so far. There is one indicator that may illustrate the point: subsidies channeled to agriculture through the state-owned public-enterprise input producers increased in real terms at an average rate of 82.5% per year, accounting for 7.1% of the sector's GDP in 1977 and 33.5% in 1982 (see Table 11.15). In other words, aggregate spending accountable to the sector does not reflect what happened with the strategy and is not an adequate basis from which to compare its costs with other similar policies

4. In absolut terms (also at constant prices) average sectoral public spending for the 1972-1976 period was 16.952 billion pesos, whereas for the 1977-1982 period this average was 26.529 billion pesos and during the SAM period it was 32.482 billion pesos.

Table 11.15. Subsidies to the agricultural and livestock sector through inputs, 1977–1982 (millions of 1970 pesos)[a]

	Subsidies	Annual increase (%)	Subsidies (%)
1977	568.4	—	7.1
1978	537.8	5.3	5.7
1979	1225.7	127.9	11.4
1980	2308.5	88.3	11.3
1981	7113.5	208.1	27.8
1982	6634.9	−6.7	33.5

[a]Subsidies granted by parastatal industrial sector companies to the agricultural and livestock sector.
Source: SPP, *La política de transferencias y subsidios* (Sepafin y SHCP, 1982).

and programs, since it operated through allocations that are difficult to trace as such in public spending reports, such as subsidized inputs or preferential interest rates. We therefore attempt a more thorough analytic approach to the issue.

The analysis of the public food budget[5] involves examining all the programs and subprograms of the agricultural, livestock, fishing, commerce and industrial sectors that were considered basic to the implementation of SAM. The food budget (fiscal resources plus income of public agencies) was identified for the first time in the 1981 Federal Expenditure Budget (PEF); it is therefore not possible to make precise comparisons with previous periods. Another reason for reservations is that many public entities attached SAM's label to their usual programs, using the food budget as though it were synonymous with SAM; this would suggest that these programs are attributable to the strategy, but actually they are not.

In any case, the analysis shows that the food budget accounted for 8.5% of the PEF in 1978 (the first year specific budget items were identified) and decreased to 8.2 and 7.5% during the following two years (see Table 11.16). This means that during SAM's first year there was a decrease in food spending's share of the total budget. The most characteristic SAM year is, indeed, 1981; in 1980 it did not yet have the necessary momentum, and in 1982 it began to lose it. During the "stellar" year, the food budget increased to 10.8% of the total PEF, that is, an increase from 7.5% that amounted to over 100 billion pesos. The following year, it decreased to 9.9% of total PEF.

When considering the sectoral distribution of spending, we see that the agricultural component alone of SAM's spending in 1980–1982 had an average growth rate higher than previous periods, since the areas of fish-

5. This analysis is based on data published in SINE-SAM DGP, "Programa del gasto presupuestado en materia alimentaria" (México VIII, 1982).

Table 11.16. Sectoral evolution of Mexico's food budget, 1978–1982

	1978	1979	1980	1981	1982[a]	1980–1982
Total Federal Expenditure Budget (billions of pesos)	937.8	1273.9	2018.5	2575.2	3320.6	2638.1
Food budget (billions of pesos)[b]	79.6	104.8	151.9	277.2	329.0	252.7
Food share of total (%)	8.5	8.2	7.5	10.8	9.9	9.6
Composition of food budget						
Agricultural and livestock (%)[c]	41.0	42.3	50.1	43.7	43.0	45.6
Fishing (%)[d]	9.9	7.1	9.1	6.0	8.0	7.7
Trade (%)[e]	35.5	37.6	29.3	39.0	37.8	35.4
Industry (%)[f]	13.5	13.0	11.5	11.3	11.2	11.3

[a] Authorized budget.
[b] Food budget: Budget of combined sectoral programs that encompass the food chain stages, including the parastatals in each sector.
[c] Agricultural and Livestock food budget: Includes SARH, SRA and Paraestatales del Sector Agropecuario, PRONASE, CONAZA, ALBAMEX, BAN-RURAL, ANAGSA, FIRA, FICART, PRODEL, FEFA, FIRCO, FIOR, FOCC, FID Campaña Garrapata and FID.
[d] Fishing food budget: Includes SEPES and PROPEMEX; excludes BANPESCA for lack of data.
[e] Trade food budget: Includes CONASUPO, Sistema CONASUPO (affiliates), SECOM, PROPEMEX, SEPES, SARH, and SRA.
[f] Industry food budget: Includes industrial programs of SARH, SRA, SEPES, PROPEMEX and Sistema CONASUPO, SIDENA, PRONASE, ALBAMEX, and FERTIMEX.
Source: Elaborated on the basis of the Total Federal Expenditure Budget.

Table 11.17. Share of Mexico's food budget in the agricultural and livestock sector, 1978–1982 (billions of pesos)

	1978	1979	1980	1981	1982	1980–1982
Sector budget	46.3	59.7	134.5	196.5	256.4	195.8
Food budget	32.7	44.4	76.0	121.0	141.4	112.8
Food share (%)	(70.6)	(74.4)	(56.5)	(61.5)	(55.1)	(57.6)
SARH						
Total expense	27.1	32.2	91.0	126.9	169.9	129.3
Food expense	14.3	17.9	36.7	54.3	59.6	50.2
Food share (%)	(52.8)	(55.6)	(40.3)	(42.8)	(35.1)	(38.8)
SRA						
Total expense	2.5	3.4	7.9	8.5	9.7	8.7
Food expense	1.6	2.4	3.7	5.6	5.0	4.8
Food share (%)	(64.0)	(70.6)	(46.8)	(65.9)	(51.5)	(55.1)
SOEs						
Total expense	16.7	24.1	35.6	61.1	76.8	57.8
Food expense	16.7	24.1	35.6	61.1	76.8	57.8
Food share (%)	(100.0)	(100.0)	(100.0)	(100.0)	(100.0)	(100.0)

Source: Derived from the Federal Expenditure Budget.

ing and industry decreased and commerce stagnated. The sectoral comparison of the food budget share within total agricultural and livestock budgets shows a paradoxical fact: during the SAM period the food budget share fell from almost one-half to about one-third in 1982 (see Table 11.17). This may be partially due to problems of budgetary classification, but a more thorough analysis of the categories involved supports the hypothesis that central government agencies took advantage of SAM's momentum to pad their ordinary programs.

An analysis of the budget of the Ministry of Agriculture and Water Resources (SARH), the agency institutionally responsible for SAM's budget, confirms this impression (see Table 11.18). General expenses for administrative support increased continuously from 1978 to 1980, from 21.1% to 23.8 and 24.5% as a share of total SARH spending. In 1981 this figure decreased to 18.9%, which can be attributed only to the balance of total agency expenses for that year, as well as to the transfer of some items to other administrative budgets, as analyzed later. Finally, in 1982 it got back to its average level of the period, 21.0%.

In 1978 administrative expenses related to the operation of agricultural districts accounted for 1% of SARH's expenses. By 1979 they had already increased sharply to reach 5.5% of the total. During the SAM period, they reached an average of 11.9%. Only 28% of this item corresponds to the operation of rain-fed districts, more typically associated with the SAM strategy. In addition, in 1982—when SAM spending was already slowing down—allotments for this category increased faster, from 11.7% in 1981 to 14.4% in 1982, confirming the impression that ordinary programs were in-

Table 11.18. Allocation of SARH's expenses, 1978-1982 (percentage)

	1978	1979	1980	1981	1982	1980-1982
Administrative support[a]	21.1	23.8	24.5	18.9	21.0	21.5
Agricultural districts coord.	1.0	5.5	9.7	11.7	14.4	11.9
Agricultural[b]	66.9	55.0	49.2	48.2	41.6	46.3
Livestock	1.4	2.3	1.8	2.5	2.9	2.4
Forestry	1.1	1.3	1.7	1.8	—	1.8
Other[c]	7.5	12.1	13.1	16.9	18.3	16.1

[a]FF, FK, and 5A programs.
[b]Programs.
[c]Programs for trade, processing, research, and others not directly related to agriculture and livestock.
Source: Derived from the Federal Expenditure Budget.

flated in the name of SAM, and that the administrative payroll increased considerably, from 22.1% in 1978 to 35.4% in 1982, with respect to total agency spending.

One last confirmation of this hypothesis is that spending on agricultural programs shows greater concentration in infrastructure and other traditional agency programs than in items more clearly associated with the SAM strategy.

When examining the institutional distribution of the food budget, only SARH shows a tendency to increase its allocation to food resources. In the Ministry of Agrarian Reform, the food share decreases; in the parastatal corporations it remains the same (see Table 11.19).

Within the agricultural parastatal sector (SOEs), which is the largest item in the food budget, the budgetary dynamism of the SAM period was

Table 11.19. Institutional distribution of Mexico's food budget in agriculture and livestock, 1978-1982 (percentage)

	1978	1979	1980	1981	1982[a]	1980-1982
Agricultural and livestock food budget						
SARH food budget	43.9	40.2	48.3	44.9	42.3	45.1
SRA food budget	4.9	5.5	4.8	4.6	3.6	4.3
SOE food budget	51.2	54.3	46.9	50.5	54.1	50.5
Total SOEs						
Input production[b]	2.4	3.5	1.7	5.0	6.4	4.3
Financing institutions[c]	97.6	96.5	98.3	95.0	93.6	95.6

[a]Authorized budget.
[b]PRONASE, CONAZA, ALBAMEX.
[c]BANRURAL, ANAGSA, FIRA, FICART, PRODEL, FEFA, FIRCO, FOIR, FOCC, FID. Campaña Garrapata and FID Livestock Development.

due to the production of inputs, without even counting FERTIMEX's fertilizer, whose expenses are charged to food spending in industry and whose inclusion would make the rate increase by one-third. The financial sector, in turn, slightly reduced its share, though it continued to be the area with the greatest share.

It is useful to quote the comments of the Ministry of Planning and Budget about the Federal Treasury Accounts with respect to SAM's expenses for 1981:

> In 1981, SAM's decisive year, within the Federal Expenditure Budget, a total of 276 billion pesos is earmarked to support food as a priority. These are distributed among some of the programs and subprograms of the agricultural, livestock, and agrarian sectors, as well as fishing, trade, and industry. As basic for food strategy, the following sectors are considered: health, education, housing, and labor, stated as complementary; and regional promotion programs, PIDER, CUC, and COPLAMAR.
>
> It is worth stressing that these total resources, besides including the costs of measures stated above, are not in addition to the amounts of budget allocations traditionally given to these sectors, but rather correspond strictly to the priority share given to food through federal agencies spending.
>
> Because of the diversified nature of the actions of the agencies of the above mentioned sectors, they have been regrouped to differentiate budgetary allotments affecting the food chain in any of its phases from those related with the achievement of other minimum welfare levels. Thus, budgetary resources involved in production, commercialization, distribution, industrial processing, and food consumption reached 242 billion pesos, meaning 86% of the total budgeted as SAM's. Other minimum welfare standards—education, health, drinking water, and housing—absorbed the remaining 37 billion pesos.
>
> Besides fiscal resources earmarked with SAM's priority, entities of the parastatal sector managed huge volumes of their own resources, particularly the official banking system, and affiliates of Sistema CONASUPO. By virtue of this, total budgetary resources that were absorbed by food policy reached 409 billion pesos, of which 392 billion are reported in public accounts.
>
> Total expenses (fiscal resources plus income) of agencies and sectors that in 1981 were identified for the first time with SAM's priority reached 392,239 billion pesos. In view of the possible statement that the budget devoted to SAM was high, it is important to consider the following aspects:
>
> Resources identified in the Federal Expenditure Budget as SAM's acount for only 17.5% of expenses made by sectors involved.
>
> Resources identified as SAM's in total budget are reduced to 11.9%.
>
> These resources come down to 10.3% of the Federal Expenditure Budget if only food expenses are considered.
>
> The share of food resources with respect to the total Federal Expenditure Budget has remained relatively constant, around 8.9% for the five-year period from 1978 to 1982.
>
> The increase of 1.4% recorded by the share of Food Budget in 1981 (10.3%) with respect to the average of the five-year period (8.9%) is mainly due to

growth in financial support for CONASUPO and rural banking, as well as for the expansion of FERTIMEX. In the case of CONASUPO and the banks, these resources were destined to expand their operating plans, and for FER-TIMEX investments were to increase its medium-term production capacity. The 1.4% increase in absolute terms corresponds to 32 billion pesos and it was a lump sum expense.[6]

The spending analysis presented here is obviously insufficient to draw final conclusions as to the budgetary cost of SAM. It is evident, however, that items clearly associated with the strategy did not show unusual growth and were, in general terms, smaller than those recorded for other items of agricultural spending or for total public spending. One can infer from the analysis that, on one hand, traditional inertia of public agencies, particularly those of the central sector, might have blocked substantial modifications from being made in their programs and their spending. This might lead us to think that, far from diverting resources from traditional programs to SAM activities, for example from those that contradicted the strategy, SAM's momentum was taken advantage of to attract additional resources to these programs. The increase in agricultural and livestock outlays or food spending should therefore not be charged to SAM's account in a cost-benefit analysis, since its activities were not the specific items responsible for the most significant increases. On the other hand, the fact that SAM increased the bargaining power of agencies involved in food production implies that a mass of resources of various kinds was channeled to these entities. Even with strategic orientations different from those of SAM, or against it, this additional expense—reflecting a higher priority for food and for the agricultural and livestock sectors—may have had effects on overall results; therefore, they should not be ignored in the study of SAM's budgetary evolution, in terms of the costs as well as the benefits.

Incremental Costs and Results of SAM

Another way of approaching the estimation of costs incurred by SAM is to attempt to estimate the incremental expenses attributable to SAM, those in excess of the historical spending levels of the institutions that participated in the strategy. Once the incremental expense is determined, a comparison can be made with incremental production results obtained for the basic grains proposed as explicit targets of SAM, thus arriving at an indicator of the overall efficacy of the strategy.

6. Ministry of Planning and Budgets, "Cuenta de la Hacienda Pública Federal 1981, Programas Especiales" (Mexico City: SPP, 1981).

The incremental expenses attributable to SAM, that is, those costs that would have presumably not occurred without SAM, should be compared to the incremental production gains for the four basic crops: maize, wheat, beans, and rice during 1980, 1981, and 1982. Since SAM generated costs not only in production but also in industry, rural and urban commercialization, food processing, and so forth, one must consider—for the purposes of our comparison with production results—only SAM's expense item allotted to production.

Net Expense Attributable to SAM

The exercise of isolating SAM's total and production expenses was made for the years 1978 to 1982 because of the implementation of budgeting by programs. However, during this period it is difficult to estimate the food expenses that would allow one to calculate net expenses on a longer historical and statistical basis. Another factor that might distort the calculation is the inflationary pace, which would evidently underestimate the trend of expense and overestimate SAM's net expense.

With these considerations, we estimated SAM's net expense (ΔE_{SAM}) based on the sum of food budget expenditures (agricultural and livestock, B_a, and industrial inputs, B_i) for the period 1980–1982 and then deducted the trend of the food budget for the same years, which was estimated from the increase recorded between 1978 and 1979, since these are the only years that can be used as precedent. Thus

$$\Delta B_a + \Delta B_i = \Delta E_{SAM}. \tag{1}$$

The incremental agricultural and livestock food budget was obtained by deducting the estimated agricultural budget trend (T_a) from the real budget:

$$B_a - T_a = \Delta B_a. \tag{2}$$

The trend was estimated on the basis of the agricultural and livestock food budget increase for 1978–1979, which amounted to 35.8%. Increases in billions of current pesos, were as follows:

	B_a	T_a	ΔB_a
1980	76.0	60.3	15.7
1981	121.0	81.9	39.1
1982	141.4	111.2	30.2
Total			85.0

To estimate the incremental industrial food budget (ΔB_i), the trend for industrial growth was considered. Thus

$$B_i - T_i = \Delta B_i. \tag{3}$$

The tendency was estimated on the basis of the increase in the industrial food budget (inputs) for 1978–1979, which amounted to 26.7%. Increases, in billions of current pesos, were as follows:

	B_i	T_i	ΔB_i
1980	17.1	16.7	0.4
1981	31.5	21.1	10.4
1982	36.7	26.8	9.9
Total			20.7

Applying the estimated values for the incremental agricultural and livestock food budget and the incremental industrial food budget to Equation (1), we have

$$85.0 + 20.7 = 105.7:$$

The net increase of expenses attributable to SAM is 105.7 billion current pesos.

Value of Production Attributable to SAM

The value of production that can be attributed to SAM in net terms is obtained by comparing real production of basic foodstuffs (maize, beans, rice, and wheat) for the period 1980–1982 to estimated production based on historical trends (1963–1979) of the same products, and then by applying the support price for the corresponding year to the difference between real production and the trend (see Table 11.20).

In order to estimate the trend of production for the four crops, production for the period 1963–1979 is taken as the basis, and the least minimum squares principle is applied. In other words, a simple linear extrapolation is made. We then take real production data for 1980, 1981, and 1982 and deduct the values obtained on the basis of the estimated trend. The difference for every year is multiplied by the current support price in order to derive the incremental value of annual production for each crop. Finally, all incremental values that are obtained for all of SAM's years are added to obtain the total incremental value of production attributable to SAM. As Table 11.20 shows, an incremental production of 9.5 million tons valued at 73.9 billion pesos is attributable to SAM.

In addition, it can be said that the increase in production had a favorable impact on foreign exchange savings, mainly on account of reduced maize imports. These values are obtained from the calculation of the trend in national apparent consumption (consumption-years) and of the tendency of national production (see Table 11.21). The difference between the two results in a figure for the trend of imports that is to be deducted from the figure of real imports. Positive figures means a net savings in im-

Table 11.20. Value of production attributable to SAM

	Real (tons)	Trend (tons)	Difference (tons)	Support price (pesos)	Value (millions of pesos)
Maize					
1980	12,374,400	9,418,100	2,956,300	4,450	+13,155.5
1981	14,550,100	9,476,900	5,073,200	6,550	+33,229.5
1982	10,129,100	9,535,600	593,500	8,850	+5,525.6
			8,623,000		+51,910.5
Beans					
1980	935,174	694,700	240,474	12,000	+2,886.0
1981	1,469,021	667,900	801,121	16,000	+12,817.6
1982	943,300	641,100	302,200	21,100	+6,376.4
			1,343,795		+22,080.0
Rice					
1980	445,364	641,600	−196,236	4,500	−882.9
1981	643,550	667,200	−23,650	6,500	−153.4
1982	511,100	692,800	−181,700	8,600	−1,562.6
			−401,586		−2,598.9
Wheat					
1980	2,784,914	3,350,300	−565,386	3,500	−1,978.9
1981	3,189,402	3,482,500	−293,098	4,600	−1,348.3
1982	4,462,100	3,613,900	+848,200	6,930	+5,878.0
			10,284		+2,550.8
Total			+9,575,493		+73,942.4

Table 11.21. Maize: Net savings in imports

	Import savings (thousands of tons)	CONASUPO expense (pesos/ton)	Saved value (millions of pesos)
1980	−399	3,356	−1,305
1981	625	3,911	2,444
1982	165	5,209	859
Total savings			1,998

ports. In order to arrive at the amount saved, this result is multiplied by the average cost of imports per ton spent by CONASUPO. The net savings in imports is estimated at 1.9 billion pesos for the 1980–1982 period. Thus the value of the incremental production and import savings attributable to SAM amounts to 75.8 billion pesos, whereas the incremental costs are 105.7 billion pesos. It would appear that the incremental costs attributable to SAM's production activities exceed the value of the benefits with a 1.39:1 cost-benefit ratio. However, there are three additional factors that would tend to reduce the costs and increase the benefits. First, the rapid inflation in 1980–1982 overestimates the costs. Second, many expenses labeled as SAM did not really correspond to SAM, but these cannot be disaggregated. Third, the overvalued peso means that the true value of the import savings would increase using a shadow exchange rate. On the other hand, if local production were valued at international prices rather than the higher support prices, the benefits would decrease.

In addition to the costs associated with SAM's production goals, there were sizable outlays for the postharvest, processing, distribution, and consumer subsidies that related to SAM's consumption goals. Using the same methodology as for the production side, the incremental cost attributable to SAM for these postharvest activities amount to 94.8 billion pesos during the 1980–1982 period. Per capita consumption of maize rose 16% from an annual average of 174 kilograms in 1977–1979 to 202 kilograms in 1980–1982, reflecting the previously mentioned production gains. From an income-distribution perspective, the subsidized consumption undoubtedly increased the effective demand of hundreds of thousands of low-income consumers. However, the generalized nature of some of the subsidies meant that all consumers of the products, for example, tortillas, benefited regardless of income. Assessing nutritional improvement due to SAM is similarly hampered by lack of data, as is discussed in Chapter 12.

Conclusions

SAM succeeded in significantly stimulating production of basic staples. It met its output goals in maize and beans in its second year, reaching self-sufficiency levels. However, self-sufficiency is a moving target and requires ever-increasing efforts. Production shortfalls in SAM's third year in the face of declining incentives once again forced Mexico to increase its grain imports. Still, SAM revealed that its strategy was capable of eliciting a major supply response, primarily through yield enhancement—with peasant producers playing a significant role.

If one nets out the value of the incremental production, the resultant cost to the government attributable to SAM averaged about U.S. $275 million a year, approximately 1.5% of the federal budget. From these figures, it would appear that SAM's fiscal burden was much less than many critics have claimed. Nonetheless, it also appears that production costs, relative to benefits, were higher than expected or desired. The value of SAM's consumption and political benefits will depend on the weights that policy makers attach to them. That is a sociopolitical judgment. Thus, in the end, the evaluation of SAM remains inescapably entrapped in controversy, a characteristic perhaps intrinsic to all countries' food policies.

CHAPTER 12

,

SAM's Influence on Food
Consumption and Nutrition

Jacobo Schatan W.

Unlike many previous food policies in Mexico and other countries, SAM's approach emphasized and made explicit its consumption and nutrition goals rather than just its agricultural production goals. In this chapter we first review the nutritional situation that Mexico addressed and then critically analyze SAM's nutrition objectives and the instruments it used to guide the strategy toward the attainment of those objectives. However, it is difficult to determine—or even to estimate—the effects of the SAM program on food consumption and nutrition among Mexico's diverse population. To begin with, no representative data on nutritional status were collected during the SAM period. Furthermore, the period was very short, while changes in the levels and patterns of food consumption tend to take place at a relatively slow pace. It is true, however, that negative changes, such as those caused by a sudden fall in the purchasing power of a family, a community, or large portions of the population (as is happening in Mexico today) may occur, and can be detected, promptly.

Suppose that a food and nutritional survey similar to the one done by the National Nutrition Institute (INN) in 1979, shortly before SAM began operating, had been carried out soon after the official termination of SAM (the first quarter of 1983). What would it have found? Quite possibly a situation worse than that of 1979, because of the effects on food consumption of the 1983 fall in real income. In other words, advances that might have been recorded in 1981 and 1982 may have been annulled by the present economic crisis.

In spite of the difficulty of measuring SAM's impact on nutritional status, it is clear that the production increases of SAM detailed in Chapter 11 significantly expanded food availability. Apparent per capita consumption of maize rose 38.9% between 1980 and 1982. Given that maize and the other basic grains tend to be more significant (and meat less so) in

the diets of the lower income, the target consumers of SAM, it is probable
that they particularly benefited from these production gains.

Given the lack of further empirical data on consumption and nutritional status, this chapter focuses on additional problems of evaluating
SAM's nutritional impact due to the methodologies used in setting nutritional objectives and measuring effects. It is hoped that such a critical
analysis will be useful in formulating more effective approaches to evaluation of nutritional impact in Mexico and elsewhere (see the selected bibli-
ography on evaluation of nutritional impact at the end of the chapter).

That SAM has disappeared as an institution does not mean that its im-
print has vanished. The strength of its ideas remain. It could not be otherwise, because they pointed in the right direction: solving the food problems of the majority of the Mexican population. But the direction to be
followed is one thing, and the vehicle to be used is another; I believe that
some of the components of the vehicle used by SAM should be modified.

Before we examine some of the methodological issues complicating
SAM's nutritional evaluation, we need to examine Mexico's food and nutritional situation during the second half of the 1970s, the time when the
SAM strategies and their underlying approach were conceived and
nurtured.

Food and Nutrition during the 1970s

Recent surveys and studies show that during the 1970s the food and nutritional situation of Mexico's poorest rural and urban sectors worsened
considerably. While during the period from the mid-1950s to the mid-
1960s consumption of vegetable and animal foodstuffs sustained an
overall growth rate well above that of the population, in the following
decade such progress slowed significantly and, in some cases, worsened in
absolute terms.[1] For example, for maize the annual rate of consumption
growth decreased from 5 to 2.3%; for beans, from 7.7 to less than 1.5%; for
oils and fats, from 8.4 to 3.7%; for rice, from 6 to 3.6%. Only wheat showed
an acceleration of consumption, from 3.4 to 6.6% per year, but this was
mainly due to the sharp increase in animal consumption of this grain as
feed. If this factor is excluded, the rate of increase in human consumption
also decreased from 4.3 to 3.8% from one period to the next.

However, the growth rates of animal foodstuffs (except for eggs) continued to be high, well above the level of demographic growth. This
reflects the change in the average Mexican diet from a predominantly

1. Data are taken from SARH (CESPA), *El desarrollo agropecuario de México: Pasado y Perspectivas,* vol. 5, *La problematica alimentaria,* Table 6, 1982.

plant-based one to one in which animal foodstuffs play an increasingly important role.

Consumer prices rose faster than nominal wages, especially for maize and beans—the two basic foodstuffs of the Mexican diet—which are precisely the items showing the greatest decrease in consumption. This particularly affected the large number of peasants in the country, because the rise in the price of their main consumer good paralleled a decrease in the real sales price of their main marketable product, maize.

If the situation is examined in per capita terms, the results are even more dramatic. For three of the main basic products—maize, beans, and rice—annual growth rates in the two periods were 1.6 and 0.1%, 4.6 and −5.4%, and 2.9 and -0.8%, respectively. Surveys done by the INN in 1979 show that approximately 90% of the rural population suffered some degree of caloric and protein malnutrition, and that half exhibited a severe caloric deficit. These studies also showed significant setbacks in some areas of the country compared to the situation seen in 1968, when another important nutritional study was carried out. For example, in the central and southern areas, the daily intake of calories fell from 1900 to 1750, while in the southeast it fell from 2000 to less than 1600 calories per day.

In short, then, one can see a significant nutritional deterioration, especially in rural areas and among the low-income urban population. At the same time, particularly in urban areas, there was an aggressive penetration of industrialized foods of all kinds, especially the snacks known as "junk food," which are in large part controlled by transnational corporations.

SAM's Nutritional Objectives

In order to confront these problems, SAM set itself the goal of improving the nutritional condition of 35 million Mexicans with different degrees of deficiency, and it called this group the "target population." Within it, those people (almost 20 million) in the worst condition (two-thirds in rural areas and one-third urban) were grouped in a category known as the "priority target population" in order to give them preferential attention and to begin to solve their problems as quickly as possible. Although it is not explicitly stated, we may infer that the 15 or 16 million people remaining in the larger target population were to be taken care of gradually and in the longer term.

The concrete objectives proposed by SAM for both population groups were the same: to reach the INN's recommended levels of 2750 calories and 80 grams of protein per day (though not defining the time frame within which each group was to achieve the goal).[2] Later in this chapter we see the contradictions of such a standard of minimal nutritional requirements. For the time being, we simply state that according to such a standard, practically all of the Mexican population should be considered as underfed, and not just the 35 million people categorized as target population. This minimum is higher than any other recommended national level in the world, including highly developed countries, those with extreme weather conditions, and those in which the main activities of the population require a high intake of energy food.[3]

Regardless of the calorie and protein standard suggested by INN and adopted by SAM, what is worth emphasizing about the SAM strategy is the application of the target-population concept for the purposes of economic policy. Proving that there is a high degree of malnutrition in Mexico was certainly nothing new, but identifying and selecting the less-favored segment of society to get preferential attention can be considered a definite improvement over traditional approaches.[4]

The spatial distribution of the preferential target population allowed for the identification of critical areas in the country,[5] areas in which state actions were to be concentrated to solve the most urgent nutritional problems. A careful examination of such critical areas permitted the demonstration of the close correlation between nutritional deficiencies and deficiencies in other central aspects of welfare, such as housing, education, and health. These critical areas included 688 rural municipalities and the districts of the state of Oaxaca, in addition to the poor shanty towns around large cities SAM was able to define with a good deal of precision the geographic contour of extreme poverty in Mexico.

Let us now return to INN-SAM's targets for calorie and protein consumption. It is worth stressing that, aside from their very high absolute level, it is not clear whether we are dealing with biological requirements of intake (that is, *real nutritional requirements*) or gross food produced. Apparently we are dealing with the latter, which would mean that the true requirements—established by INN itself at around 2080 calories and 62 grams of protein per day—were expanded by approximately 30% to cover the losses between farm and home and in household waste.[6] This means that the 2750 calories and 80 grams of protein correspond, strictly speak-

2. If we assume that the *non*-preferential target population has lower nutritional deficits than the preferential target population, even with more limited actions on the part of SAM such a group should be able to reach the nutritional target long before the preferential group.
3. According to the Fourth World Food Survey of the Food and Agricultural Organization (FAO), the highest caloric requirement in the world is that for Finland, with 2710 calories per day.
4. It is worth stressing that the new Programa Nacional de Alimentación has maintained this concept, and this, I believe, is highly positive.
5. These areas were identified through work done jointly by SAM and COPLAMAR.
6. This is according to verbal information provided to the author by a high official of INN at the time when nutritional goals of SAM were being drafted.

ing, to gross supply, and not to *real intake*. Confusion over the stage at which requirements are measured may have serious consequences, since these figures are used as the basis—at least in theory—to determine production volumes (or, rather, supply) that will be required to meet the *true requirements* of the population. Such a confusion leads to an overestimation of the additional productive effort required because, as has been seen in some recent documents post-SAM (but no doubt reflecting the essence of its propositions), comparing a nutritional target of 2603 calories[7] to the real average intake of less than 2100 calories per day makes a diffrerence of 25%. Such a difference is due exclusively to the comparison of two links of the food chain that are not directly comparable for the purpose indicated.

When we examine some of INN's 1979 data in rural and urban surveys, we may conclude that the coefficients of caloric and protein adequacy are much higher than those implicit in this difference of 25%. Among urban groups with incomes close to the general minimum wage, the coefficient is quite close to 1.0.[8] In Mexico's rural areas, only the poorest, such as Oaxaca, the southwest, and the south Pacific, show indices of caloric deficiency between 0.7 and 0.8; most exhibit coefficients higher than 0.8 and many average higher than 1.0. In protein, indices of adequacy are, in general terms, somewhat lower than those of calories, but in only three of the nineteen nutritional regions into which the country was divided for the purposes of the study was the index of protein adequacy lower than 0.75.

One cannot conclude then, that the average deficit for the whole of the first six deciles' households is higher than 25%. Obviously, there are groups that are well below such requirements and whose deficit is probably over 40%, both in calories and in protein (that is, with a coefficient of adequacy lower than 0.6). However, most likely such groups are concentrated in the first two deciles' households, since, going from one decile to the next, food consumption increases considerably and the deficit begins to decrease until it disappears around the fifth decile.[9]

7. Toward the end, SAM had changed its targets to 2603 calories and 73.9 grams protein per day.

8. Variations in urban areas recorded by the survey were moderate: in calories they ranged from 0.88 for the poorest group to 1.02 for those better off (income ranged between −15 and +15% with respect to the general minimum wage for that year), and in protein they ranged between 1.01 to 1.25.

9. According to the data of ENIGH 1977, monthly expenses for food and beverages of the first decile households were 250 pesos, those of the second, 550 pesos, and the fifth, 1468 pesos, almost six times the average of the first decile. In any case, this does not mean that *all* the members of the households placed over and above the fifth decile enjoyed a normal nutrition level. Urban surveys of INN in 1979 showed that there was malnutrition in preschool children, starting at income levels that were higher than the average of the fifth decile. In other words, malnutrition was likely to occur in children even in the seventh decile. This would indicate, incidentally, that variations between the extremes of one decile may be quite high in terms of their coefficient of nutritional adequacy. This, then, clearly indicates the inadequacy of placing the first six deciles' households in the same category and working with the average of all, as SAM did.

If nutritional problems are examined from an overall perspective, one realizes that Mexico as a whole has sufficient food, but its distribution— due to the very unequal distribution of resources and of income—is most inequitable, as is well known. Therefore, it is not absolutely necessary to increase production—or supply—by percentages as high as those resulting from applying the coefficient of over 25% to meet the real nutritional requirements of the deficiency sectors; rather, action could be taken to redistribute supply in terms of needs and not of economic demand.

I would like to insist on this aspect, not only to shed light on the methodological confusion described above, but to warn against unnecessary pressure on the productive resources of the country, which are, certainly, not unlimited. This becomes more evident if we take into account the concrete possibility of reducing the huge losses of products, not only between the store and the kitchen, but in higher proportions, in the long journey from the countryside to the retail store. Adequate food planning should set itself the goal of drastically reducing losses (for example, with the improvement of transportation and storage services, through the improvement of packing and handling techniques) and the waste that takes place in homes; this alone would increase food intake as well as availability.

We may conclude, then, that the nutritional targets of SAM-INN overestimated the caloric and protein deficit of the *whole* of that 60% of households that constituted the target population, underestimated the deficiencies of the poorest groups that could be called "especially" preferential target population, and set no standards of time frames to solve these. Having said this, however, one must stress that merely identifying the critical areas and developing the concept of target population reflects a specific and welcome concern for less-favored segments. Unfortunately, such a positive attitude and willingness are weakened when there is no quantitative or qualitative definition of the specific and concrete effort needed to solve the problems of those groups.

The Instruments

The Basic Recommended Basket

As was indicated in Chapter 3, the Basic Recommended Basket was one of the main tools created by SAM to tackle the problem of food deficits, and several versions were developed for the different groups of the population. Thus, first a national basic recommended basket was developed, and then one for the target population. Their calorie and protein contents were the same (2750 calories and 80 grams of protein in the initial stage, and 2603 calories and 73.9 grams of protein in the later phase), but with a different composition: the basic basket for the target population had 20% less animals foods (allowing for a cheaper price) than the national basket. This meant that the national basic recommended basket was destined for the 40% of the remaining households. In addition, three regional baskets were prepared, for the north, and the Gulf area, and the southeast, where the main basic product varied.

According to SAM's proposals, the basic recommended basket had the following fundamental requirements:

1. to meet the minimum nutritional standards;
2. to consider production costs of the primary products that affect the final price of foodstuffs included in it;
3. to consider the purchasing power of the population;
4. to consider national and regional consumption habits;
5. to consider the coutnry's potential in terms of the human and natural resources of the agricultural, livestock, and fishing sectors, and those of the food industry (in order to produce it).[10]

Through a process of adjustment, the basic recommended basket resulted from the so-called current consumption basket, which corresponded to the average pattern of consumption of households in the first six deciles (those with lower incomes) according to data of the national survey on income and expenses made by the Ministry of Planning and Budget (SPP) in 1977. This current basket (1979) permitted, on one hand, the registration of changes in the average diet that occurred in the previous twenty years and, on the other, the identification of the most notorious nutritional deficiencies of the whole population. The current basket was also useful as the point of reference to develop "a rough view of the quantitative food needs of the target population and the potential of several combinations of products that could meet nutritional requirements at low costs." This first view, "through a long process of successive approximations within a model of linear programming, led to quantitative and qualitative measures of food consumption that we have called the basic recommended basket. Its structure and composition were the real expression of nutritional needs of the whole population and of the target one, in both cases related to regional consumption patterns and the purchasing power needed to acquire it."[11]

Initially the intention was to apply a 30% subsidy to the cost of the basic basket for the target population, to be applied through the DICONSA system, whose prices were already lower than private trade, especially in rural

10. Oficina de Asesores del C. Presidente, "Primer planteamiento de metas de consumo y estrategia de producción de alimentos básicos para 1980–82" (Mexico City: SAM, March 1, 1980).
11. Ibid., p. 13.

areas.[12] With this, plans were made to increase the number of DICONSA stores in the countryside. Available figures (see Chapter 4) show that DICONSA sales, especially in rural areas and small cities, increased significantly in 1981 and 1982 in real terms, which shows—indirectly—the effectiveness of SAM's efforts to supply a larger number of products at lower prices with the lowest incomes. This basic basket was made up of thirty products combined in variable amounts in order to supply a balanced diet in terms of calories and proteins of vegetable and animal origin (2603 calories and 73.9 grams of protein in the basket's later version).[13] While such an exercise has merit, particularly in its orientation toward nutritional equity, it is nevertheless not very effective as a tool to set standards for planning nor in covering the whole of nutritional requirements.

Let us first take a look at this last aspect. As stated before, it is unclear whether the figures for recommended calories and proteins correspond to effective intake or to availability in the market. If they correspond to intake (which is suggested in one of the headings prepared by SAM on the basic recommended basket—"Net intake, daily grams per capita"),[14] they would represent one of the world's highest levels, which would require an enormous additional effort. On the other hand, there are indications that the figures correspond to availability, since the latest of INN's recommendations reaches 2300 calories per day, which means that the remaining 303 calories correspond to adjustments due to losses. Now if figures given by SAM for current consumption of the target population (2063 calories) correspond to effective intake and not availability, as seems to be the case, then the increase needed to reach the levels of requirements proposed by INN would only be 237 calories per day (that is, 11.5%), instead of the 540 calories (26.2%) presented as the increase needed to reach that goal.

Second, a basket of this sort, which is intended to become a standard, should not try to meet the total nutritional needs whatever their levels, since not all the products consumed by the population are included in it. Besides the thirty products mentioned, there are many other foods that, although consumed in small individual portions, add calories and protein; to these we have to add beverages, especially spirits (soft drinks are included under sugar), which also contribute calories. This is a common mistake in many of the estimates of basic food basket.[15] It is evident that,

12. Ibid., p. 17.
13. The thirty products are maize grains, tortilla and maize dough, flour, sweet rolls, white bread, crackers and pasta, rice, beans, potatoes, tomatoes, chilies, onions, lettuce, carrots, bananas, apples, limes, oranges, beef, pork, poultry and ovicaprine meat, eggs, fresh milk, lard, seafood, vegetable oil, and sugar.
14. Oficina de Asesores del C. Presidente, "Primer," unnumbered table on pages 12 and 13.
15. In a study by ECLA and INCAP related to research on poverty in Central America, there is also a list of the most frequently consumed foodstuffs that together should meet all

were they to be fully implemented, such baskets would inevitably lead to an overfeeding of the target population and therefore to a waste of resources, to say nothing of the damage that such overfeeding would cause to people's health. It would be better for the basket instrument to be designed in such a way as to ensure 80 or 85% of nutritional requirements based on a smaller number of basic foods. The remaining requirements could be covered with other products—recorded or not in current statistics. This would also greatly facilitate the operational side of the basket.[16] Furthermore, the nature of the basket as a standard seems doubtful, unless rationing is contemplated, and this has never appeared among SAM's or INN's proposals. At most, the baskets may constitute a guide, an indicator for planners and managers of the economy about the approximate magnitude of food deficiencies that should be filled, the population groups affected, and their geographic location, thus orienting decisions as to food production and distribution.

Third, it is worth stressing that no reference is made to the treatment of the remaining 40% of households—largely overfed and, as a whole, absorbing almost 60% of food supplies. Although it is easy to understand that SAM concentrated on solving the dramatic problems of the target population, with special emphasis on the preferential target population, it is not easy to understand why there was no emphasis on the overconsuming portion of the population that was responsible for deformation of the diet and waste of resources. It is hard to understand how SAM could have ignored the existence of almost two-thirds of the food supply, a large portion of which was imported, when it was supposed to have been seriously concerned with all matters related to the production of food.

Promotion and Social Communication

SAM actively promoted better eating habits among the Mexican population. It developed several programs of nutritional orientation and communication and undertook educational campaigns, especially about the appropriate combination of foods in meals and about the eating habits of pregnant women and lactating mothers.

The first of these campaigns attempted to highlight the thirty products included in the basic basket, in order to teach the population—and especially mothers—their nutritional value and how to combine them cor-

---

the nutritional requirements of the Central American population; this list ignores the contribution made by a variety of other foodstuffs: ECLA, "Nota sobre las canastas básicas de alimentos en los páises del istmo centroamericano. Caracteristicas y resultados." Documento CEPAL/ME/SEM 4/6 (March 29, 1981).

16. For example, if the list products were limited to twelve—maize, tortillas, white bread, rice, beans, beef, pork, poultry, eggs, milk, vegetable oil, and sugar—88% of the caloric requirement and 87% of the protein requirements would be met (estimated on the basis of data for the basic basket proposed by SAM, unchanged).

rectly for the greatest nutritional benefit.[17] SAM developed a "Table of Food to Improve Nutrition" that was colorful, very simple, and attractive. During the fifteen months of the campaign, the table was published over 100 million times in newspapers, magazines, textbooks, posters, calendars, and in larger format. It was placed in schools, stores, hospitals, offices, and other public locations.[18]

These ideas were grasped by a large percentage of the population, especially those with higher degrees of schooling, as was proven in the evaluations done by SAM throughout its existence. For example, the idea of combining different types of food in meals was understood by 92% of those interviewed during the course of such evaluations. However, it was found that in practice only 15% of the population combined food correctly, while 66% combined only two groups out of four, and the rest consumed, in the same meal, food of only one group. This reveals the frequent gap between understanding a message and actually applying it. Nutritional education needs to be maintained for long periods and should be complemented by all sorts of aids, from the distribution of recipes to constant teaching in school.

The second campaign, designed to promote breastfeeding, was also very intense. Using all the media and with the participation of a famous actress, information on the advantages of breastfeeding was publicized, as was knowledge about adequate diet for pregnant women and lactating mothers. Along with the penetration capacity of DICONSA's 15,000 stores, SAM carried out a complementary campaign to stress the importance of good eating habits—especially for mothers and children—and information on hygiene, food preparation, and preservation.

The response to the campaign on breastfeeding was very high and, surprisingly, almost as high among men was women. This is particularly important because of male influence in the increasing trend to abandon breastfeeding. The short period during which SAM was able to carry out these campaigns does not allow us to verify the degree to which its objectives were attained; it could not have substantially penetrated the countryside. But it would be highly desirable for future food efforts to revive these social-communication campaigns, which acquired considerable impetus under SAM's aegis.

Conclusions

SAM's production gains undoubtedly contributed to increased per capita consumption of basic grains. Additionally, the increased consumer

17. It combined those foodstuffs providing mostly energy with those providing mostly protein, minerals, and vitamins.
18. SAM, *El SAM como estrategia de comunicación social, 1980–1982* (Mexico City: SAM, 1982).

subsidies and the expanded network of rural state stores probably improved the food access of the target group. The consumer education campaigns undoubtedly enhanced the nutritional quality of the dietary practices of many. Thus, although we lack hard data on changes of nutritional status during SAM, the existing indirect evidence points toward a positive impact.

On the conceptual level, SAM constituted an advance for food policy in Mexico by attaching an explicit priority to nutritional improvement of vulnerable groups. The methodologies for setting nutritional targets can be refined, but it is hoped that the strategic nutritional thrust of SAM can be continued in the formulation and implementation of the country's future food policy.

Bibliography

Austin, James E. 1978. "The Perilous Journey of Nutrition Evaluation." *American Journal of Clinical Nutrition* 31(2324): 38.

Austin, James E. 1980. *Confronting Urban Malnutrition.* Baltimore: Johns Hopkins University Press for The World Bank.

Austin, James E., and Marian Zeitlin, eds. 1981. Nutrition Intervention in Developing Countries: An Overview. Cambridge: Oelgeschlager, Gunn & Hain.

Pinstrup-Andersen, Per. 1981. "Nutritional Consequences of Agricultural Projects: Conceptual Relationships and Assessment Approaches." Working paper no. 456. Washington, D.C.: The World Bank.

Sahn, David E., Richard Lockwood, and Neven S. Scrimshaw, eds. 1984. *Methods for the Evaluation of the Impact of Food and Nutritional Programmes.* Tokyo: United Nations University.

CHAPTER 13

SAM, Energy, and Structural Change in the Agricultural Sector

C. Peter Timmer

SAM was promulgated as the sowing of Mexico's petro dollars. One cannot isolate the agricultural sector from the macro policies associated with the surging economic eminence of Mexico's energy sector. In this chapter, I attempt to elucidate those links for Mexico and other countries. I examine data through 1980, the beginning of SAM. The findings suggest that SAM held the potential to serve as an antidote to previous macro policies' erosive effects on agriculture. In Chapter 14 the specific income effects of SAM's crop support prices and input subsidies are modeled in more detail. A major purpose of this chapter is to present a generalizable analytic methodology for understanding how the impact of a national agricultural development strategy can be significantly affected by other macro policies, particularly those emanating from the energy sector. Food policy is inextricably linked to energy policy in Mexico and much of the rest of the Third World. Tracing those links is essential to impact analysis.

The extensive quantitative literature on structural change during the process of economic development contains few, if any, indications that changed petroleum prices play a major role. Chenery and Syrquin (1975) do not mention energy in their empirical analysis of the factors explaining the structure of economies at various levels of development. Similarly, the

This chapter is based on a paper prepared for presentation at the 13th Pacific Trade and Development Conference hosted by the Asian Development Bank and the Philippine Institute for Development Studies, held in Manila on January 24–28, 1983. I would like to thank Carol F. Timmer for substantive and editorial assistance. This chapter is the result of three converging bodies of my work over the past decade: analysis of the interaction of energy and food prices using a simple micro-macro adjustment model (Timmer, 1975); analysis of the role of food price policy in both short-run and long-run efforts to alleviate hunger and poverty (Timmer, 1980; Timmer, 1982a; Timmer, Falcon, and Pearson, 1983); and current research sponsored by the Division of Research of the Harvard Business School on macro food policy and structural change (Timmer, 1982b).

more prescriptive and historical treatment by Johnston and Kilby (1975) of the role of agriculture in structural transformation omits any reference to energy.

Since these studies were done, energy prices have been recognized as having a significant impact on a country's economy. Taylor puts the macro case succinctly:

> Oil shocks affect rich and poor countries alike. The price of a key raw material goes up, and sooner or later the outcome will be an increase in final output prices. If wages are more or less fixed in money terms, the real purchasing power of labour incomes will fall. Aggregate demand and capacity utilization will decline; investment in consequence may be cut back, leading to slower growth. The outcome is . . . stagflation, which standard fiscal and monetary policies will be powerless to offset. (1982: 329)

In a footnote, Taylor notes that "this chain of consequences is now very familiar."

From a sectoral perspective, agriculture might be especially hard hit. Modern agricultural technology, with its dependence on fertilizer and fuel for tractors, trucks, irrigation pumps, and processing, is highly intensive in its use of nonrenewable energy resources. Traditional but low-yield agriculture is intensive in its use of renewable energy resources, such as labor and organic manures (Pimentel et al., 1974). However, its long-run sustainability does not compensate for its inability to increase productivity as rapidly as population growth. Consequently, in the face of higher energy prices the agricultural sector is expected to stagnate. This would be consistent with the pre-SAM performance of the sector in Mexico. SAM, through its subsidies, could potentially avoid these adverse price effects at the micro level.

The consequences of higher energy prices are equally devastating at the micro level, especially for poor consumers who devote a larger share of their budgets to food than do average or upper-income consumers (Timmer, 1981). Soedjatmoko, the rector of the United Nations University, makes the connection to energy prices in the following way:

> Rising fuel prices, boosting transportation and agricultural costs, will inevitably push food prices beyond the reach of hundreds of millions of already hungry people. Rising populations, despite the best efforts to reduce fertility rates, will continue to increase the demand for both food and energy. The developing countries will not be able to solve their food problem without solving their energy problem, and without a satisfactory solution to both, their economic growth will be severely constrained. The centrality of the food and energy nexus calls for a comprehensive policy approach. Only through a clear understanding of this food-energy pivot can the situation be turned around. (1981:7)

Again, as noted in Chapter 12, SAM's consumer subsidies were aimed at holding back food price increases for the poor.

A plausible case can be made that, at the macro, sectoral, and micro levels, higher energy prices for oil-importing countries (and for the agricultural sector and the poor consumers even in oil-exporting countries) have an impact on agricultural growth and on the path of structural change. Actually measuring this impact is complicated because energy prices affect both the rate of growth of total economic output and the structure of that output. Hence the empirical specification of the structural impact must carefully sort out direct from indirect effects, short-run from long-run effects, and static from dynamic effects. No formal, quantitative models exist that are capable of such subtlety of specification, although the work of van Wijnbergen, of Bruno and Sachs, and of Buiter and Purvis is clearly headed in the right direction. None of these models, however, addresses the agricultural sector directly.

What is attempted here is a hybrid, an attempt to model the effects of oil price changes on agriculture heuristically rather than formally, coupled with a direct quantitative test of several "reduced-form" hypotheses that emerge from the heuristic perspective. In several cases, these hypotheses are directly counter to the conventional view outlined above for all three levels—macro, sectoral, and micro. The reason is that the heuristic models include price and substitution effects that are nearly always ignored in macro and sectoral models of structural change.

If one is to test quantitatively the impact of a variable on a system, the variable must vary. When macroeconomic trends and agricultural production trends are all moving more or less smoothly and simultaneously on long-run paths of structural transformation, it is virtually impossible to sort out what is causing what. The oil price shocks of the 1970s offer the analyst an opportunity to follow the impact of a major perturbation as it ripples through the macro economy and into the agricultural sector. No other single macroeconomic event in the recent history of Mexico or elsewhere has so opened windows of opportunity for understanding cause-and-effect relationships between the food and agricultural sector and basic macro variables.

Higher energy prices substantially alter, either directly or indirectly, the basic macro prices in oil-importing and oil-exporting countries. When these macro prices—foreign exchange rates, interest rates, wage rates, and, indirectly, the rural-urban terms of trade (or food prices for short)—change, major shifts in both income distribution and resource allocation occur. Changed relative capital-labor prices alter basic decisions of technique in the economy, strongly affecting the demand for unskilled labor. Pressures on the balance of payments and foreign exchange rates alter basic incentives to export and import. As these mechanisms work themselves through interdependent sectors of an economy, the ultimate impact of higher energy prices has some surprising twists. After nearly a decade of empirical experience with these prices, it is even possible to see some of these surprises in the data for Mexico and other countries.

The Heuristic Model

A sudden rise in energy prices sends shock waves through almost all sectors of an economy, with ramifications from impacts on balance of payments, investment and aggregate growth of changed farming techniques on peasant plots, and altered food-consumption patterns in most households. Some of these shock waves have similar impacts in both oil-importing and oil-exporting countries; others have distinctly opposite effects depending on the direction of oil trade. For the most part, macroeconomic effects of energy-price changes depend on whether the economy is a net importer or exporter of energy, whereas microeconomic effects at the farm and household level are common to all countries (if oil-exporting countries actually permit internal energy prices to rise to reflect international opportunity costs).

The agricultural sector's adjustment to higher energy prices reflects the net, and often conflicting, pressures from the micro and macro adjustment processes. The empirical experience from 1960 to 1980 of Mexico and six other Asia-Pacific developing countries analyzed in this chapter (Indonesia, Republic of Korea, Malaysia, Philippines, Sri Lanka, and Thailand) reflects drastically different agricultural-sector adjustments. The intent here is to develop a heurisitic model and empirical specification that captures the diversity of these sectoral adjustments.[1]

The logic used to develop the model has four discrete steps, and only the first is tested empirically here. The other steps are covered in three other studies complementary to this one. In all cases, the starting point for analysis is the path of agriculture's share of an economy's total domestic product. The basic trajectory of this share during the course of a single country's economic development, or cross-sectionally from poor countries to rich, is the steady decline shown in Figure 13.1. The explanation for this decline—Engel's law, improvements in agricultural technology, and limited opportunities for countries to export agricultural commodities when domestic markets are satisfied—are well known and are only indirectly important to the effects discussed here.

What is important is that any realistic discussion of the impact of energy prices on structural change in agriculture can take place only in the context of the normal structural transition of an economy in the course of economic development. This chapter relies on the Chenery-Syrquin analysis of this transition to provide a starting point. The analysis ex-

---

1. Oil price effects are not symmetric, since macro effects depend on whether a country imports or exports oil, but micro and sectoral effects depend only on oil prices.

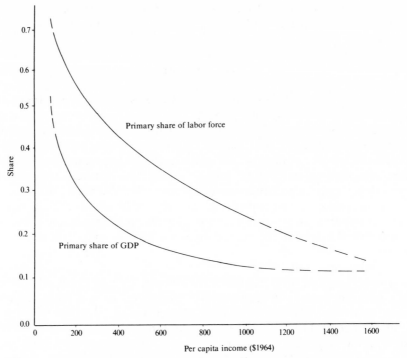

Figure 13.1. Primary sector share of the GDP and labor force

Data from Chenery and Syrquin, 1975.

plores how energy prices might cause the "normal" path of structural adjustment to be altered and what the consequences of different paths might be.

A normative element underlies the analysis as well: an important objective is to find policy interventions that might speed the favorable aspects of structural adjustments while slowing, alleviating, or ameliorating the negative welfare consequences. In particular, these negative welfare consequences might be reflectecd in skewed income distribution and increases in hunger among both the urban and rural poor. The four basic steps in the logic of the model are as follows.

Energy Prices and Structural Change

Energy prices affect structural change in agriculture primarily by altering the relative terms of trade between the agricultural and nonagricultural sectors of an economy. This effect is caused mostly by macroeconomic pressures on the balance of payments and foreign exchange rates in the face of changed energy prices. As Schuh (1976) points out, agriculture tends to produce a relatively large share of tradable goods and services. Consequently, energy-exporting countries like Mexico usually face a deterioration of the rural-urban terms of trade when oil prices rise, as well as a more rapid decline in agriculture's share of the gross domestic product (GDP) than would be expected on the basis of the Chenery-Syrquin analysis, which depends only on per capita incomes and country size. Oil-importing countries, on the other hand, should see improved rural-urban terms of trade and a relatively healthier agriculture sector when oil prices rise. Their agricultural sectors' shares of the GDP should not decline as rapidly as in the Chenery-Syrquin model. The detailed mechanisms by which this effect is mediated by macroeconomic forces are developed in the outline and specification of the empirical model.

Energy Prices and Income Distribution

Larger than expected agricultural shares of the GDP in low-income countries also increase the share of national income received by the poor, for example, the bottom 40% of the income distribution. This is consistent with the finding by Chenery and Syrquin (1975: 61), but the causal mechanisms via higher energy prices need explanation. Labor productivity in the agricultural sector remains far below that of the rest of the economy until very late in the process of economic development (only in the past decade has it reached parity in the United States). The mirror image of this low productivity, however, is low wages in the rural economy, which encourage the creation of jobs when the rural economy is healthy and has rising income in total. For the poorest segment of the population, income security comes with having adequate land to feed and clothe a family or with access to a regular job, even at low wages. For the land-scarce economies of Asia, a healthy and dynamic rural economy is likely to provide rapid job creation for unskilled rural landless workers due to rural household expenditure patterns that favor labor-intensive goods and services. These jobs help improve the distribution of earned incomes as measured by the share received by the bottom 20 or 40% of the income distribution.

It is worth noting that such an improvement in income distribution is not inconsistent with a negative nutritional impact if the higher food prices that create rural dynamism actually reduce the food intake of the poor. Clearly, the impact of higher energy prices on rural income distribution and alleviation of hunger may be favorable in the long run while having extremely serious consequences in the short run. At this point, a broader food policy perspective, one that incorporates both productivity and consumption consequences in the planning process, is essential (Timmer, Falcon, and Pearson, 1983). Further empirical work to measure some of these effects is needed (see Chapter 14).

Agriculture and Economic Growth

A healthy agricultural sector in a poor country contributes materially to more rapid and equitable growth for the rest of the economy as well. Early development theorists saw this connection clearly: "Industrial and agrarian revolutions always go together, and . . . economies in which agriculture is stagnant do not show industrial development" (Lewis, 1963: 433). It is important to distinguish between higher incomes in agriculture, which are the joint outcome of agricultural prices (relative to inputs) and of agricultural production, and rising productivity in agriculture. Higher productivity is consistent with Lewis's view of agriculture's role in development even if agricultural incomes are depressed, a chronic condition for the agricultural sector in developed countries. However, most agricultural economists believe that there is a significant link between growth in agricultural productivity and higher incomes generated by incentives for the agricultural sector in low-income countries (Schultz, 1978; Peterson, 1979; Timmer, Falcon, and Pearson, 1983).

For agriculture to play its essential role in the overall process of economic modernization, appropriate incentives to the rural sector are needed. Higher energy prices force oil-importing countries to provide at least some of these needed incentives. In the intermediate-to-long run, these same higher energy prices will actually cause these economies to grow more rapidly than they would if agricultural incentives continued to be depressed by discriminatory and distorted macro price policy. On the other hand, oil-exporting countries can see their agricultural sectors decimated in the absence of conscious macro policy interventions designed to maintain the health of the rural economy. An "overvalued" exchange rate leads to imports of apparently cheap food and to reduced agricultural exports, although foreign currency reserves are not under pressure because of oil earnings. For countries with relatively small rural populations anyway (Kuwait or Saudi Arabia), this may cause few welfare problems. But countries such as Indonesia, Mexico, and Nigeria are likely to view with alarm the rapid demise of their rural sectors due to "Dutch disease," that is, the relative decline in output and employment in export- and import-competing sectors of an economy whose currency is appreciating due to a boom in natural resource-based exports.

Structural Adjustment, Agricultural Price Policy, and Protection

Energy prices also affect the agricultural sectors of rich countries, perhaps even more strongly because of the greater reliance of industrial agricultures on energy-intensive inputs. Most rich societies have found ways to alleviate the pain of structural transformation from agricultural to industrial economies by protecting their domestic farmers. In effect, industrial societies have used price policy (implemented with agricultural commodity trade barriers) to improve profitability of agriculture. The same policy instruments, when used by poor countries along with other macro policies, usually have effectively discriminated against their agricultural sectors.

Given this pattern of mirror-image policies, how has the changed energy price environment affected the structural transformation process in rich countries? In particular, do the basic macro mechanisms connecting oil prices to agricultural profitability that are so potent in poor countries still operate in the OECD economies? Whatever the answer, the implications are important for agricultural price policy and for efforts to understand the origins and impact of agricultural protection. The tension between agriculture's role in the economic growth process and attempts to protect the welfare of farmers caught in the pressures of the inevitable decline of agriculture during that process is an ongoing challenge to policy makers and researchers.

Specifying the Relationship between Energy Prices and the Agricultural Sector

The mechanisms that cause higher energy prices to improve the profitability of agriculture in oil-importing countries (and vice versa in oil-exporting countries) are convoluted and work primarily through macroeconomic forces. The convolutions cause both logical and econometric problems. Even if the *structure* of causation can be specified, actual estimation may be impossible due to data problems or irreconcilable problems of simultaneously determining all important variables. No claim is made here that all these problems have been solved. However, an effort has been made to address them directly, primarily by attention to specification of structural relationships as a first step, coupled with a second step of estimating procedures designed to overcome the most obvious difficulties created by simultaneity. The logic of the procedures draws heavily on two-stage least squares, although the actual details vary slightly.

As noted previously, the starting point for the analysis is the Chenery-Syrquin specification of sectoral shares of the GDP as a function of per capita incomes and country size. The log quadratic function used by Chenery and Syrquin is used here partly to compare with their work and partly because the fit is good and it is easy to interpret results. Chenery and Syrquin's results refer only to a "primary" sector that includes mining and other natural-resource extraction along with agriculture, while the results here refer specifically to agriculture alone. Consequently, it is necessary to reproduce Chenery and Syrquin's results for the specific

268   *Impact*

country sample and sectoral definition used here. The basic equation used
as a reference throughout the rest of this chapter is as follows:

$$AGSHR_{i,t} = f[A_i, LX1_{i,t}, (LX1_{i,t})^2, \log N_{i,t}, (\log N_{i,t})^2], \quad (1)$$

where

$AGSHR_{i,t}$ = agricultural share of the total GDP for country i and year t, in
current currency units;

A_i = country-specific intercept term;

$LX1_{i,t}$ = log per capita income in U.S. dollars, converted at the annual
average foreign exchange rate, for country i and year t,
deflated by the U.S. GDP deflator;

$N_{i,t}$ = national population for country i and year t;

i = countries in sample, including Mexico, Indonesia, South
Korea, Malaysia, Philippines, Sri Lanka, and Thailand;

t = year of sample data, from 1960 to 1980 (21 observations).

The issue is how the general process of structural change as captured in
Equation (1) is modified by changed energy prices. In the intermediate
run, after the initial shock, but before technology itself has been changed
to fit better a new environment of scarcer energy resources (as argued in
the induced-innovation hypothesis presented by Hayami and Ruttan,
1971), the primary vehicle for changing the performance of the agricultural sector is agricultural prices. In a narrow sense this may mean
prices (and price policy) for specific agricultural commodities, such as
rice, wheat, rubber, or cotton. At the sectoral level, however, the terms of
trade between total agricultural output and output from the rest of the
economy provide the major connection between the impact of energy
prices on macroeconomic variables and the rural economy. Of course,
macro variables are not the only factors that influence the rural-urban
terms of trade. Specific price and trade policies also play major roles.

When physical agricultural output and the level of per capita incomes
are held constant, it is obvious that the rural-urban terms of trade affect
the agricultural share of the GDP via basic national income accounting
identities. However, the issue here is not how agriculture *responds* to
greater financial incentives, but whether oil price increases add to or de-
tract from those incentives. For this purpose it is important to establish
empirically the relationship between agriculture's share of the GDP and
the rural-urban terms of trade and then to establish any role of energy
prices in affecting these terms of trade. In order to eliminate the effects of
simultaneous equations bias, both steps must be done in tendem.

$$AGSHR_{i,t} = f[A_i, (Y \text{ and } N \text{ factors}), TT_{i,d}], \quad (2)$$

where

(Y and N factors) = Chenery-Syrquin structural adjustment variables as in Equation (1);

$TT_{i,t}$ = rural-urban terms of trade, defined as

$$\frac{AgGDP_{i,t}/AgGDPK_{i,t}}{(GDP_{i,t} - AgGDP_{i,t})/(GDPK_{i,t} - AgGDPK_{i,t})},$$

where

$AgGDP_{i,t}$ = agricultural GDP for country i and year t, in current currency units;

$AgGDPK_{i,t}$ = agricultural GDP for country i and year t, in constant currency units;

$GDP_{i,t}$ = total GDP for country i and year t, in current currency units;

$GDPK_{i,t}$ = total GDP for country i and year t, in constant currency units.

Equation (2) represents a naive attempt to capture the impact of the rural-urban terms of trade, as reflected in sectoral income components of the national income accounts, on agriculture's share of the GDP. It is perhaps surprising that the simple correlation between AGSHR and TT is nearly zero. Consequently, more sophisticated attempts to measure the impact of TT turn out to be only a little superior to the naive specification of Equation (2). The more important reason for removing simultaneous equations bias is the structural understanding provided by learning the factors that explain movements in the terms of trade.

$$TT_{i,t} = f[A_i, OILMSH_{i,t}, PPFEX_{i,t}, DCERPR_t, DANC_t], \qquad (3)$$

where

$OILMSH_{i,t}$ = oil imports as a share of GDP for country i in year t (net exports are negative);

$PPFEX_{i,t}$ = purchasing-power parity of the foreign exchange rate for country i and year t, defined as

log(foreign exchange rate to U.S. dollar) - log(GDP deflator for country i/U.S. GDP deflator);

$DCERPR_t$ = World Bank index of international market cereal prices for year t, deflated by U.S. GDP deflator;

$DANC_t$ = index of noncereal agricultural prices on international markets for year t, calculated from World Bank indices as follows:

$$\frac{\text{Agricultural Commodities Index} - 0.1133 \text{ Cereal Price Index}}{0.8867},$$

deflated by U.S. GDP deflator.

The estimated value of TT from Equation (3) (termed TTFIT) can be inserted in Equation (2) in order to eliminate the simultaneity caused by the
accounting definitions. The impact of oil prices is captured in the oil-
import-share variable, which is the joint product of oil import (or oil export volume) and oil prices. In the long run, oil trade volumes are no
doubt partially a function of prices, but within the year-to-year variations
examined in this model, volumes are presumed to be largely independent
of price. Consequently, the impact of changed oil prices are determined by
tracing equivalent movements in OILMSH. Long-run effects will be
magnified in oil-exporting countries if there is a positive supply response
and will be dampened in oil-importing countries if there is a negative demand response.

The actual rural-urban terms of trade in most countries are not simple
translations into the domestic economy of international prices for traded
agricultural commodities as converted at existing exchange rates. However, only these variables are present in Equation (3) (along with
OILMSH). Most countries have active price policies for important agricultural commodities, especially basic cereals, and an added variable for
price policies relative to each other would also be desirable. However, TT
is defined in such a way that all countries have a value of one for 1970, and
so cross-country comparisons of price *levels* do not show up. Only relative
movements in time are captured. Cross-country levels are captured implicitly by separate country intercepts when TTFIT is inserted into Equation (2) and the whole equation is reestimated.

Although world prices for cereals and other agricultural commodities
might legitimately be taken as exogenous, along with energy prices, for a
given country's terms-of-trade formation, the same thing cannot be said of
its foreign exchange rate. Indeed, a major part of the argument is that the
pressures of rising oil prices on the balance of payments causes oil-
importing countries to devalue their currencies, thus creating improved
incentives for exports to pay the import bill. Because international oil export payments must balance oil import payments, Burniaux (1981) calls
this a "global Walras effect," that is, an "adding up" condition that requires the global value of oil exports to equal the value of oil imports,
abstracting from volumes in transit and payment terms. This effect is an
important reason why partial-equilibrium models of oil price impact have
been so misleading. Changed oil prices clearly have general-equilibrium
effects in both importing and exporting countries, and a primary vehicle
for these effects is the foreign exchange rate and mechanisms for financing the balance of trade. Much important work in the past decade in international finance has focused on these relationships (Krause and Sekiguchi, 1980; Branson, 1981; Burniaux, 1981; Cline and Weintraub, 1981;
Sachs, 1981). The purpose here is not to repeat either the theories or empirical findings of this work, but to be reasonably faithful to both while
retaining the heuristic flavor of the argument. Since none of the countries

examined has had floating (market-determined) exchange rates, only a loose relationship with causal variables is expected.

$$PPFEX_{i,t} = f[A_i, OILMSH_{i,t}, DCERPR_t, CABAL_{i,t}, LX1_{i,t-1}], \qquad (4)$$

where

$CABAL_{i,t}$ = current account balance (exports minus imports, as a share of GDP, for country i in year t).

Equation (4) is specified to capture four relatively separate effects on real exchange-rate determination.[2] First, the share of oil imports should be positively associated with pressures to devalue, that is, with higher exchange rates relative to the dollar. International cereal prices may play the same role as energy prices for cereal import-dependent countries, with negative effects for cereal-exporting countries. Since the variable is not weighted by import or export shares, the sign of any possible impact is an empirical issue. It is likely to vary from country to country when separate analyses are performed.

The current account balance as a share of the GDP (CABAL) should be a leading factor influencing exchange-rate formation. Negative current account balances (imports greater than exports) must be financed by foreign borrowing or, temporarily, through reductions in foreign exchange reserves. When foreign confidence and interest in investing in a country is held constant, CABAL should be negative. Since the inclusion of OILMSH controls to some degree for investor confidence, a measure of investor interest is sought. The lagged value of LX1, per capita income, is included for this reason. Other things being equal, countries with higher per capita incomes should attract more foreign investment, thus alleviating pressures to devalue. Consequently, LX1 (lagged) should be negative.

Chenery and Syrquin note (1975:159) that even their basic structural equation, represented in Equation (1), suffers from simultaneously determined dependent and independent variables, as per capita incomes and sector shares mutually influence each other. Chenery and Syrquin use an instrumental-variables technique (using rank of country in per capita income rather than income itself) to show that virtually no changes result in either patterns or parameters.

However, even if simultaneity may be assumed away as empirically unimportant for this model, the influence of energy prices on the determination of per capita income cannot be neglected. Fortunately, careful specification of an equation to explain LX1 in terms of exogenous

2. Note the "real" aspect of PPFEX, composed of two components: (1) LFEX—the "actual" exchange rate, and (2) minus log(relative inflation). The definition of PPFEX provides for *automatic* devaluations in the face of differential inflation rates. Whether actual macro policy is so accommodating is a separate, and interesting, issue.

variables, including energy price-related variables, reduces any simultaneity problems as well.

$$LX1_{i,t} = f[c, LX1_{i,t-1}, PPFEX_{i,t-1}, INV_{i,t-1}, OILMSH_{i,t}], \quad (5)$$

where

$INV_{i,t-1}$ = gross domestic investment as a share of the GDP, for country i and year $t-1$.

In equation (5), per capita income is explained in terms of the previous year's income (with variables in logs, the coefficient should be close to one), the foreign exchange rate in the previous year (a positive coefficient would indicate the stimulating effects on growth of an export-oriented strategy that uses an undervalued exchange rate to provide greater incentives to export and restrict imports), investment in the previous year as a share of the GDP (higher values should increase incomes through a positive incremental capital-output ratio), and the oil import share in the current year. According to the logic quoted from Taylor early in this chapter, the coefficient of OILMSH in Equation (5) should be negative, reflecting the fiscal drag of higher energy prices on oil-importing countries.

One additional element of structural adjustment may also be significantly affected by energy prices. Equation (5) shows that per capita incomes are likely to be influenced by lagged investment shares, but naturally these investment shares themselves may also be partially determined by a country's oil import status and by the foreign exchange and current account balance variables that are intimately tied up with oil import status. Equation (6) represents a fairly simplistic attempt to isolate some of these effects.

$$INV_{i,t} = f[A_i, LX1_{i,t}, OILMSH_{i,t}, PPFEX_{i,t}, CABAL_{i,t}]. \quad (6)$$

It may be necessary to use two-stage least squares techniques to estimate Equation (6) satisfactorily. However, since the INV variable is used only in lagged form in other equations, the structural understanding is more important than removing the simultaneity bias.

Figure 13.2 summarizes the relationships and connections among the variables included in Equations (2) through (6), those addressed specifically to the impact of energy prices on the share of agriculture in GDP. Variables that are considered exogenous to the whole system include population, the oil import share, deflated cereal and noncereal agricultural prices in international markets, as well as the current account balance as a share of the GDP. Sach's (1981) work shows clearly that the current account balance is endogenous in a system such as this, but since the real exchange rate and oil import share variables are included within the analysis, any bias resulting from treating CABAL as exogenous in Equations (4) and (5) is minor. All other variables are endogenous, but

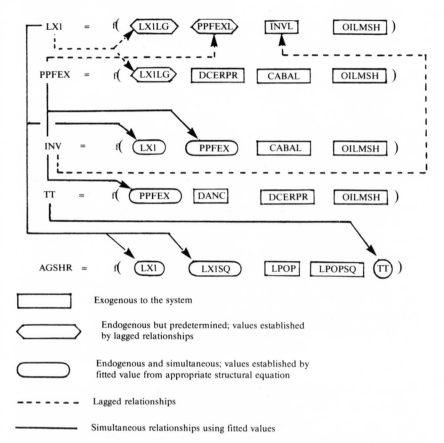

Figure 13.2. Structure of relationships among agricultural share of the GDP, oil import share, and macroeconomic variables

some are predetermined due to their inclusion in lagged form, whereas others are independent; for the latter, fitted values from structural equations are used in the estimation rather than raw values, assuring independence of error terms.

Regression Results

As Figure 13.2 indicates, the structural equations of the heuristic model must be estimated in sequence in order that fitted values of endogenous variables can be used as independent variables in subsequent equations. Because of this use of fitted values to replace observed values for important endogenous variables, a rather generous interpretation of statistical significance was used to decide whether to retain an independent variable or to exclude it. All variables were retained whose coefficients exceeded their standard errors. In fact, most coefficients have t-statistics of well over one, and very few variables were excluded on statistical, as opposed to heuristic, grounds.

Data

Nearly all of the basic economic data for this analysis were drawn from the World Bank data tape "Economic Data Sheet 1—Update 1981," which includes annual information on national accounts, prices, exchange rates, and population for the period 1960–1980, a total of twenty-one observations. When comparisons are made with the Chenery-Syrquin results, it should be noted that the period analyzed here is of similar length, but one decade later, since their analysis covered 1950 to 1970.

International agricultural commodity prices were taken from the series maintained by the Commodities Projection Department of the World Bank. Oil import and export data were obtained from the IMF current data tapes except that for Sri Lanka, where it was necessary to resort to national statistical sources. Occasional missing values for particular years for other variables were filled in through similar resources.

Per Capita Income

Regression 1 in Table 13.1 shows that the log of per capita "real" income is determined primarily by the lagged value of the variable. In addition and as expected, the lagged foreign exchange rate, lagged investment share, and oil import share also significantly (but sometimes marginally) influence per capita incomes. The investment share coefficient is easiest to interpret. At the mean values for the variables in this sample (shown in Appendix Table 1), the coefficient of 0.703 attached to INVL reflects an incremental capital-output ratio of 1.42. This is obviously too low for a "pure" ICOR and no doubt reflects the fact that INVL is picking up the contribution of a whole previous series of investments and not just the previous year's investment share. This "excluded-variables bias" creates no special problems when the equation is used for predictive purposes.

The contribution of the lagged foreign exchange rate is small but positive. Holding all other things constant, including domestic inflation relative to inflation in the United States (taken to represent external trading partners), a 10% devaluation of the exchange rate will contribute an increase of about 0.1% in the next year's per capita income. The cumulative effects of such a devaluation, again assuming no differential inflation exists, is substantially larger. After five years, the initial devaluation leads to approximately 5% higher per capita incomes, due to the importance of the foreign exchange rate in other structural equations.

Table 13.1. Estimated parameters in structural equations (t-statistics in parentheses)

| Dependent variable | Regression number | R^2 | Constant | Independent variable | | | | | | | | | | |
| | | | | Log per capita income | | | Purchasing power exchange rate | | | INVL | OILMSH | CABAL | DCERPR | DANC |
				LXI	LXI1LG	LXIFIT	PPFEX	PPFEXL	PPFEXFIT					
LXI	1	.971	-0.0704 (0.7)	0.989 (47.0)				0.00927 (1.6)		0.703 (2.9)	-0.417 (1.7)			
PPFEX	2	.996	A_i	-0.379 (11.1)								1.240 (3.2)	-0.612 (1.7)	-0.00120 (1.7)	
	3	.992	A_i								0.930 (1.7)	-1.909 (4.0)	-0.00236 (2.4)	
INV	4	.732	A_i			0.162 (3.4)			0.177 (1.4)		-0.354 (2.1)	-0.440 (3.7)		
	5	.869	A_i	0.169 (22.1)			0.171 (11.7)				-0.337 (5.0)	-0.480 (8.1)		
TT	6	.574	A_i						0.119 (1.7)		2.229 (7.6)		0.000814 (1.3)	0.00298 (2.6)
	7	.576	A_i				0.0857 (2.0)				2.226 (8.1)		0.000932 (1.5)	0.00248 (2.5)

The oil import share has similar short-run and long-run impact, such as Taylor outlined in the argument earlier. In the short run, an increase in OILMSH of 1% causes per capita income to fall by 0.4%, at mean values. (The longer-run impact is discussed in the context of full simulation runs for various oil import and export circumstances.) As noted earlier, changes in OILMSH in the short run are meant to reflect changes in international oil prices. For example, for a country expending 2% of its GDP to import oil, a 1% increase (to 3%) would reflect a 50% increase in oil import prices with import quantities held constant.

Foreign Exchange Rate

The factors influencing the (log of) purchasing-power foreign exchange rate (PPFEX) are reflected in Regression 2 in Table 13.1. Deflated world cereal prices have a marginally negative impact on PPFEX, either because they reflect export opportunities or perhaps because they are correlated with inflation rates in the United States. It should be noted, however, that deflated noncereal agricultural prices, as measured by DANC, are not as significant as cereal prices, and the inclusion of DCERPR improves the significance of the other variables in the equation.

The current account balance (CABAL) also has the expected negative effect on PPFEX when other variables are held constant. Similarly, for a given current account balance, larger oil imports lead to a devaluation of the real exchange rate. This impact is both quite significant and large in quantitative terms. As the heuristic model argues, because the foreign exchange rate is an important explanatory variable in all the other structural equations, this significant effect of oil import share in determining the equilibrium exchange rate is crucial to understanding the indirect role of trade and macro variables in explaining structural change in agriculture.

Lagged per capita income is also important in explaining the formation of the exchange rate. The highly significant coefficient of −0.379 presumably reflects the role of a healthy economy in attracting foreign investment, both short run and long run, and thus supporting a relatively lower exchange rate (measured by units of domestic currency per U.S. dollar) than would otherwise be possible for a given oil import bill and current account balance.

The per capita income term is also important for isolating the specific effects of the other independent variables. Regression 3 in Table 13.1 shows what happens when LX1LG is deleted from the equation. The coefficient of DCERPR doubles and rises in significance, and the CABAL coefficient rises threefold in magnitude and greatly in significance, while the OILMSH coefficient drops slightly in magnitude and considerably in significance. It is clear that LX1LG is important for understanding the structural role of oil and the other variables in exchange rate formation.

Investment Share

There is no econometric reason why the investment share of the GDP needs to be estimated. It is present only as a lagged variable in the per capita income equation. But for dynamic simulations, the factors that influence investment and hence income growth must be understood. Regressions 4 and 5 in Table 13.1 show that current incomes and exchange rates as well as oil import share and current account balance all contribute significantly to short-run investment share.

Regression 4 uses fitted values of per capita income (from Regression 1) and of foreign exchange rate (from Regression 2) as independent variables, in addition to OILMSH and CABAL, which are treated as exogenous in this analysis. Regression 5 uses actual current values of the endogenous variables. Apart from the very sharp increase in the significance of the coefficients for all four independent variables, very little is changed. The estimated parameters themselves are nearly the same, thus lending considerable confidence in their robustness in the face of specification changes.

Investment is financed from both domestic and international savings, and Regressions 4 and 5 reflect this dual composition. The per capita income variable captures the marginal domestic savings rate directly, while the OILMSH variable captures the switch in expenditure patterns when oil prices change. At least in the short run, higher oil prices lead to lower savings and investment domestically. At the same time, a widened current account balance must be financed by at least short-run capital inflows (or lower foreign exchange reserves, not included in this model), which raises measured investment in the national income accounts.

It is necessary for both OILMSH and CABAL to be in the regression for this dual role to show up. When CABAL is excluded, the coefficient attached to OILMSH is positive and significant. Since CABAL is influenced, but is not solely determined, by OILMSH, it may be desirable to add an equation explaining CABAL as a function of OILMSH, PPFEX, and other potential variables, as in Sach's analysis. For the time being, the full simulation runs discussed below assume that CABAL is determined exogenously or is held approximately constant by government intervention.

Rural-Urban Terms of Trade

The ultimate purpose of the structural model is to be able to use fitted values of per capita incomes and the terms of trade in the final model of structural change. Regressions 6 and 7 carry out the final step of the structural model that permits such a treatment. The rural-urban terms of trade depend in quite direct fashion on the foreign exchange rate, the oil import share, and measures of prices in international markets for cereals and noncereal agricultural products.

The positive role of the international market prices for agricultural commodities is the most obvious and direct, and yet it is perhaps the most surprising. All of the countries in the sample have varying degrees of border price interventions designed to separate internal agricultural prices from external agricultural prices. The significant coefficients attached to DCERPR and DANC show that these interventions are not entirely successful and that domestic agricultural prices do tend to follow their international market trends.

However, a 10% movement in DANC has twice as large an impact on the terms of trade as does a similar 10% movement in DCERPR (measured from mean values), despite the large share of agricultural GDP made up by cereal production in the countries in this sample. Consequently, the relatively small impact of international market cereal prices in determining the domestic rural-urban terms of trade reflects the short-run role of price policy implemented via trade interventions in structuring rural incentives. In the longer run, fiscal pressures and problems of balance of payment tend to keep internal commodity prices roughly in line with their international opportunity costs.

The important variables influencing the terms of trade from the perspective of the model here are the exchange rate and the oil import share, both of which are separately significant (especially in Regression 7, where actual rather than fitted values of PPFEX are used). The positive effect of devaluations on the rural-urban terms of trade is confirmation that rural goods and services tend to be more "tradable" than urban goods and services. Consequently, the strong tendency of developing countries to maintain overvalued exchange rates is not just an impediment to efficient resource allocation and rapid growth; it also significantly biases income distribution against the rural sector.

Oil imports force countries to remove some of that bias. As pressures build to create incentives to export in order to pay for the oil imports, the rural sector receives improved terms of trade since it produces many of those exportable goods. According to Regressions 6 and 7, a 1% increase in the oil import bill increases the terms of trade by about 2%. It is important to note that this effect of oil prices is symmetric for oil exports. As oil prices rise in exporting countries, the terms of trade deteriorate for their agricultural sectors. Since the oil share is also a significant factor in determining exchange rate, the ultimate impact of oil prices is even larger. The full effects are explored in the simulation model.

Agriculture's Share of the GDP

Table 13.2 presents the basic regression results relating income, population, and terms of trade to the agricultural share of the current GDP

Table 13.2. Estimated parameters in agricultural share in GDP regressions (*t*-statistics in parentheses)

Dependent variable				Independent variable							
				Per capita income				Population		Terms of trade	
Regression number	R^2	Constant	LXI	LXIFIT	LXISQ	LXIFTSQ	LPOP	LPOPSQ	TT	TTFIT
AGSHR										
8	.934	A_i		-0.425 (5.4)		0.0299 (4.3)	-1.767 (3.0)	0.0470 (2.7)		0.220 (5.3)
9	.954	A_i		-0.312 (4.6)		0.0206 (3.5)	-0.489 (1.1)	0.0097 (0.8)	0.185 (9.6)	
10	.919	A_i		-0.578 (7.1)		0.0437 (6.1)	-0.142 (0.3)	0.00089 (0.1)		
11	.942	A_i	-0.482 (7.0)		0.0348 (5.7)		-2.326 (4.8)	0.0630 (4.5)		0.239 (6.1)
12	.956	A_i	-0.382 (6.2)		0.0267 (4.9)		-0.659 (1.8)	0.0148 (1.4)	0.167 (9.5)	
13	.926	A_i	-0.618 (8.5)		0.0471 (7.3)		-0.714 (1.6)	0.0174 (1.3)		

(referred to here as AGSHR). Regressions 8 and 12 show the fully specified model, first using fitted values from earlier structural equations for all endogenous variables, then using the raw data from the sample. Just as Chenery and Syrquin found, the bias caused by using endogenous variables as independent variables is relatively small. Regression 12, using raw data, has highly significant coefficients in all cases, except those of the population variables, but the size of the parameters for income and the terms of trade are little changed. Again, the real value of the structural model is not its ability to solve specification problems but the genuine insights it brings to the role of the important causal variables.

Regressions 9 and 12 have the same structure as Regressions 8 and 11 with a substitution of the fitted value of the terms of trade for the actual values, and vice versa. Again, the results are little changed, indicating that the explanatory variables used in the terms-of-trade equation are the important forces influencing the role of the terms of trade in AGSHR (despite the relatively low R^2 of about .57 in Regression 6). The results do become significantly different when the terms of trade are dropped from the AGSHR regressions altogether.

Regression 10 shows the results for fitted values of per capita income—the population variables lose all significance—while Regression 13 reproduces the basic Chenery-Syrquin model, with somewhat divergent results. The pattern of signs and relative magnitudes for the two income variables are similar, allowing for difference in sectoral definition (Chenery and Syrquin use total primary production; the model here treats agriculture alone). The population coefficients, however, have inverted signs and substantially different magnitudes due, perhaps, to the inclusion of country-specific intercepts in all of the AGSHR regressions reported here (see Appendix Table 2 for values of the country-specific intercept terms). The terms-of-trade variable is an important determinant of the share of agriculture in the country's GDP. Raising the terms of trade by 10% (also a 10% increase from the mean value of 1.00) raises the share of agriculture in the GDP by roughly 2%.

As Figure 13.2 shows systematically, and Regressions 1, 2, 4, 6, and 8 show quantitatively, a complicated web of external prices, foreign trade variables, and macroeconomic forces all influence the terms of trade. The next task is to understand how all of these factors, and in particular how oil prices, interact to influence per capita incomes and the rural-urban terms of trade. These variables, when population is held constant, determine the pace of structural change in a growing economy. The next section simulates this process of structural change.

Simulating the Impact of Oil Price Changes on Structural Change

With some confidence in the robustness of the specification and empirical estimates of the structural model, it is now possible to ask several important questions about the impact of energy prices on the agricultural sector. First, and perhaps most important, the empirical results amply confirm the earlier heuristic perspective that argued that oil prices would significantly influence the basic terms of trade between rural and urban areas, but that the influence would be through a variety of connected and often offsetting mechanisms. The terms of trade depend on OILMSH directly, with an immediate and direct increase of over 2% for every 1% increase in OILMSH, and on PPFEX, which also depends on OILMSH as well as lagged income, which is influenced by OILMSH directly and indirectly through investment shares and, again, PPFEX.

Second, the agriculture's share of the GDP, which is the measure used here to capture impact on structural change, is quite sensitive to the rural-urban terms of trade when included in a full structural model, although the simple correlation between AGSHR and TT is nearly zero. If incomes and population are held constant when oil prices change (obviously only a hypothetical idea), the marginal change in AGSHR for a unit in OILMSH is 0.52; that is, one direct and immediate effect of an oil price increase that raises a country's oil import share by 1% is an *increase* in the agricultural share of the GDP by about half that amount. This is a symmetric effect, and higher oil prices *reduce* the share of agriculture in oil-exporting countries by a similar proportion.

Third, the real value of the model developed here is not its ability to capture immediate and direct effects of oil prices on rural-urban terms of trade and on the agricultural share of the GDP, although these are important and this ability is a necessary first step. The interesting question has to do with the dynamics of change. What is the impact of oil prices over time as all the various trade and macro adjustment mechanisms work themselves out? Here, of course, the empirical results must be used with considerable caution, for the model is fairly simple, and economic dynamics are inevitably complicated and subject to unforeseen forces and shocks. In particular, long-run dynamics are frequently quite different from the sum of repeated short-run effects, which are the primary effects captured here. With that proviso noted and understood, the results of the dynamic model are extremely interesting.

The Simulation Model

When Regressions 1, 2, 4, 6, and 8 are substituted into the network of connections in Figure 13.2, a dynamic simulation model results. From initial values for the exogenous variables, a time path is traced out for all the endogenous variables, including incomes, foreign exchange rates, terms of trade, and ultimately, the agricultural share of the GDP. As the value of OILMSH is varied exogenously, different paths are generated for these variables. They capture, to the extent possible within the structure of the model, the full dynamic effect of changes in oil prices. An infinity of assumptions are possible as starting points, and diverse questions can be asked for each one.

The results discussed here were generated using mean values of exogenous variables—no single country as an example, but rather a "representative" set of intercept terms that permit both oil-importer and oil-exporter experiences to be simulated—and a constant term in the per capita income equation that causes real per capita incomes to grow by 2.8% per year when oil import share remains at zero throughout the ten-year simulation period. The base rate of growth is important because economies growing more rapidly for "natural" or exogenous reasons are less affected by a given change in oil prices. The model is easily modified to generate alternative growth paths; comparisons with a high growth rate are made in the discussion below.

If the only interest is in the impact of OILMSH on AGSHR, the model can be solved for these two variables. The results are shown in Equation (7).

$$AGSHR_t = 0.308 + 0.550(OILMSH) - (t - 1)[0.0026 - 0.00589 \\ (OILMSH) + 0.1453(OILMSH)^2]. \tag{7}$$

When t = 1, the first year in the simulation model after the initial conditions are inserted, the marginal impact of a 1% change in OILMSH on AGSHR is 0.55, quite close to the short-run, direct effect reported above of 0.52. When t = 10, the end year for the model results reported here, the marginal impact is a function of OILMSH, as is shown in Equation (8).

$$\frac{\delta AGSHR}{\delta OILMSH} = 1.08 + 2.62(OILMSH) \quad \text{(for t = 10)}$$

$$= 1.34 \quad \text{(for OILMSH = 0.10)}$$

$$= 1.08 \quad \text{(for OILMSH = 0.00)} \tag{8}$$

$$= 0.82 \quad \text{(for OILMSH = -0.10)}.$$

Equation (8) shows that the full dynamic effects of oil prices on agriculture's share of the GDP are not symmetric for oil importers and oil exporters, although they are symmetric in the immediate period of adjustment. At the extremes of oil-import and oil-export dependency (OILMSH = 0.10 and −0.10, respectively), the dynamic adjustment coefficient is more than half again as large for oil importers as for oil exporters. Despite all the problems of being an oil exporter due to the macroeconomic and trade biases introduced relative to agriculture, the resources provided by oil exports clearly dampen some of the necessity for adjustment. Oil importers have no such cushion and must adjust more rapidly and completely.

If the basic economy is growing at 7.0% per capita per year instead of the 2.8% assumed in Equations (7) and (8), the dynamic marginal adjustment coefficients are significantly smaller, 1.16 for OILMSH = 0.10 and 0.64 for OILMSH = −0.10. Since oil exports contribute to higher growth, the adjustment process is doubly eased. Even for oil exporters maintaining the purchasing-power parity of foreign exchange rates, however, the adjustments of the agricultural sector are quite significant.

The reference to maintaining this parity is a reminder that a number of assumptions and internal dynamics drive the basic model. For example, when OILMSH = 0.00 for the entire period, the terms of trade gradually decline (by 0.1% per year), AGSHR declines from 0.308 to 0.285, mostly because of income growth, and the purchasing-power parity of exchange rates declines about 0.7% per year as per capita incomes grow (that is, the real exchange rate is gradually revalued relative to the U.S. dollar). Naturally, oil importers or oil exporters will have different internal dynamics. In particular, oil exporters experience a more significant decline in their purchasing-power parity of exchange rates (that is, relative to the U.S. dollar) as oil exports increase foreign exchange earnings relative to all other aspects of the economy. When OILMSH = −0.02, a representative value for an oil-exporting country before the OPEC-led rise in oil prices, PPFEX falls 1.4% per year relative to the U.S. dollar. This is a more significant revaluation of the domestic currency of an oil exporter against the U.S. dollar than when OILMSH is zero. Oil importers of similar dependence (OILMSH = 0.02) maintain their exchange rates relative to the U.S. dollar almost exactly. Since the United States was an oil importer throughout this period as well, such a close correspondence lends some confidence to the results being generated by the model.

The Impact of Changed Oil Prices on AGSHR

Oil prices and oil import or export status have two different effects on the path of agricultural share of the GDP over time. Figure 13.3 shows both effects. In the first instance, oil import or export status determines *which* basic path a country is following, with oil exporters consistently having smaller agricultural shares of the GDP than oil importers, holding all other aspects of the economy constant. Naturally, the differences in paths are magnified the higher the oil price, that is, the larger OILMSH is in absolute value for any given quantity of oil trade. Figure 13.3 shows three basic paths of AGSHR over time, for OILMSH = 0.02, 0.00, and −0.02, representative of oil importers, self-sufficiency, and oil exporters before the rise in oil prices. Examples would be Thailand and the Philippines as importers, Malaysia as nearly self-sufficient, and Indonesia and Mexico as exporters.

It is apparent from Figure 13.3 that AGSHR declines less rapidly for oil

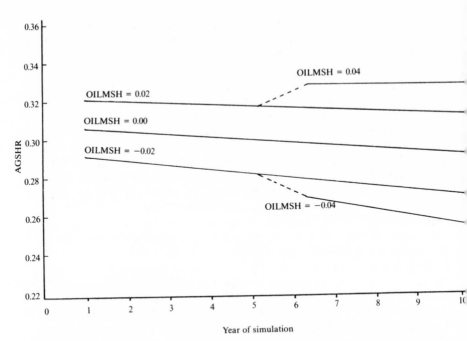

Figure 13.3. Basic simulation: Relationships between AGSHR and OILMSH over time

importers than for oil exporters. Roughly half the difference in rate of decline is due to differences in per capita income growth rates (4.2% per year for the oil exporter versus 1.4% per year for the oil importer), while the rest is attributable to relatively improved terms of trade for agriculture in oil-importing countries.

Figure 13.3 also demonstrates the impact of doubling oil prices at the start of year 6 in the ten-year simulation run. This is accomplished by exogenously changing OILMSH from 0.02 and −0.02 for oil importers and oil exporters, respectively, to 0.04 and −0.04. The results are quite dramatic. For the oil importer there is an immediate increase in AGSHR in year 6 from 0.313 to 0.324, and its dynamic path from then on is nearly constant; that is, the doubling in oil prices *stops* the structural transformation of the economy, and the agricultural share of the GDP stabilizes. Much of this effect is through sharply reduced growth in per capita incomes, but the effect on terms of trade is also quite significant.

The picture is sharply different for oil exporters, where AGSHR drops immediately from 0.278 to 0.268. From there, the rate of decline in AGSHR year by year is accelerated. By year 10, the agricultural share of the GDP is 0.249 for the newly enriched oil exporter, while it remains as high as 0.323 in the impoverished oil importer.

It should be recognized that the agricultural sector of the oil importer is only *relatively* healthier than that of the oil exporter, which has higher levels of output and rates of growth. However, for a given level of per capita income, this different *relative* performance of the two sectors can have important implications for income distribution and the degree of poverty as well as long-run effects on the base for overall economic growth. Destroying their agricultural productivity base is not likely to be good long-run development strategy for the heavily populated oil-exporting countries.

The Impact of Changed Oil Prices on Agricultural GDP

Although the relative effects of oil prices on the agricultural share of the GDP are important in and of themselves, it remains true that overall agricultural performance is also an important test of the impact of oil prices. Unfortunately, the model specified and estimated here has several important handicaps for addressing the issue of overall agricultural growth. First, the ultimate dependent variable in the model is the agricultural share of the GDP, not agricultural GDP itself. Total real GDP per capita (in U.S. dollars) is estimated by Regression 1, and per capita agricultural GDP can be calculated implicitly by multiplying AGSHR by the estimated per capita income. This, of course, is a rather indirect approach to the issue.

Second, the model does not incorporate either a short- or long-run agricultural supply response to the rural-urban terms of trade. Although some quantity response is implicit in the overall estimated coefficient for the terms of trade in the AGSHR regression, much of its significance is no doubt due to the direct price effect on rural incomes. Within the time horizon of the model presented here, the "intermediate run" to be purposefully vague, the magnitude of any additional supply response to that already captured in Regression 8 is a subject of considerable controversy. An experiment with an additional short-run elasticity of 0.1 and a long-run supply elasticity of 0.4 is reported later. These values are larger than those reported by Griliches (1960) for the United States and smaller than those reported by Peterson (1979) for a large and diverse sample of countries.

Third, the model implicitly assumes a sophistication of macroeconomic management that matches the performance of the United States over the 1960–1980 period in terms of determination of inflation and exchange rates. Where differential inflation exists, the model automatically devalues, or revalues, the currency relative to the U.S. dollar to maintain the initial purchasing-power parity, adjusted for "real" changes due to income changes and oil importer or exporter status. In fact, of course, many countries have not managed either zero differential inflation or automatic devaluation when differential inflation occurred.

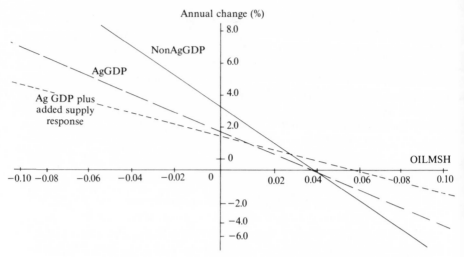

Figure 13.4. Basic simulation: Relationship between OILMSH and average annual change in AgGDP and NonAgGDP between year 1 and year 10

The importance of the first two problems is illustrated in Figure 13.4, which shows annual average growth rates for agricultural GDP (AgGDP) and for nonagricultural GDP (NonAgGDP) as a function of OILMSH. The apparently paradoxical result relative to the earlier results, with respect to AGSHR as a function of OILMSH, is that the rate of growth of agricultural GDP *falls* as a country moves progressively from oil exporter (OILMSH = -0.1) to oil importer (OILMSH = 0.1) status. The paradox is resolved, however, by noticing that the slope of the AgGDP growth rates is less steep than for NonAgGDP. In fact, for OILMSH greater than 0.04, AgGDP growth rates are higher than NonAgGDP growth rates, but both are *negative* from that point on.

Oil export revenues determine the overall growth rate of the economy, with higher oil prices speeding growth for exporters and impeding it for importers. At the same time, the *structure* of what growth is occurring is determined by the terms of trade, which are also heavily influenced by OILMSH. Indeed, if a positive supply response is added to the agricultural GDP growth rates previously calculated, as noted above, the structural effects of OILMSH are even more striking. Figure 13.4 shows that AgGDP growth rates have an even flatter response to OILMSH when the supply response is incorporated. For all OILMSH levels greater than 0.01, AgGDP grows faster than NonAgGDP, and this is now true for positive rates of growth as well. When OILMSH = 0.04, the assumed supply response permits AgGDP to grow about 0.5% per year although NonAgGDP is completely stagnant.

The Role of Policy in Determining Agricultural GDP

In one sense the model developed here is intimately connected to issues of basic development policy, for it incorporates core macroeconomic variables that all governments attempt to influence: investment, foreign exchange rates, rural-urban terms of trade, and eventually per capita incomes. In another sense, the model has little policy relevance, for all these variables are determined exogenously; there are no obvious policy levers to pull.

As noted earlier, the policy levers are implicit in the workings of the model itself. The base for evaluation of the performance of the macroeconomic variables is performance of U.S. macroeconomic variables, in particular the degree of inflation and the value of the U.S. dollar relative to its major trading partners. No attempt has been made here to model independent paths for the U.S. variables relative to those for the representative example simulated in this chapter. To do so would involve modeling the current account balance, domestic money supplies, and the factors leading to differential inflation. Oil prices no doubt affect all of these in nonsymmetric ways for oil importers and exporters. Such a modeling effort is simply beyond the scope of this discussion, and somewhat cruder mechanisms for simulating macro interventions are used here.

A second obvious role for government policy is in determining the rural-urban terms of trade that so strongly condition production incentives in the countryside and the relative costs of food for consumers. The present model determines the terms of trade on the basis of the foreign exchange rate and on exogenous prices. Most governments also influence this basic macro price through a variety of trade and subsidy policies, some commodity-specific (e.g., a floor price policy for grain), some sector-specific (e.g., a subsidy for fertilizer), and some quite general (e.g., a uniform tariff barrier to protect domestic industry). None of these potential policy interventions are modeled specifically. Indeed, their differential impact from one country to another in the sample used for empirical analysis is captured only indirectly, through country-specific intercept terms in the AGSHR equation. Since these intercept terms capture a host of other country-specific factors as well, especially agricultural resource endowment relative to population and the level of biological, chemical, and mechanical technology used in agriculture, their estimated values are not very revealing about different levels of agricultural incentives across countries. Understanding these differentials requires much more country-specific or commodity-specific analysis (see, for example, Timmer and Falcon, 1975).

The impact of the exchange rate policy warrants particular scrutiny. All of the simulations of oil price impact on agriculture have held current account balances constant as a share of the GDP (at the average level for the entire sample of roughly a 2% deficit), while simultaneously allowing foreign exchange rates to be determined entirely by real variables. Any differential inflation relative to the United States is immediately translated into exchange rate changes. While this may be a reasonably accurate description of exchange rate formation in the intermediate-to-long run (after all, Regression 2 for PPFEX has an R^2 of .996), it clearly misses the political decision-making element in the timing and degree of currency devaluations. In addition, the shock of major dollar flows to oil exporters from oil importers has tended to mask underlying equilibrium trends, thus complicating the task of determining the appropriate exchange rate in the first place. A good example is the overvaluation of the Mexican peso and the resultant delays in devaluing.

Two exchange rate interventions are simulated here. The first is for an oil exporter who finds domestic inflation running faster than U.S. inflation at the same time that foreign exchange reserves are rising and the current account balance is positive—for example, Indonesia from 1974 to 1978. Here the experiment is to leave the nominal exchange rate intact, thus leading to a significant *revaluation* of the domestic currency in real terms relative to the U.S. dollar. The second simulation treats the mirror image case in which an oil importer decides to devalue in the face of a current account deficit, even though differential inflation is negligible. This example is representative of a country that consciously uses its foreign exchange rate for export promotion and to generate faster economic growth through international trade (see Corden, 1974; Bhagwati, 1978; Krueger, 1978; Dornbusch, 1980). An example is South Korea in the mid-to-late 1970s.

Figure 13.5 shows the growth in agricultural GDP over time for an oil-exporting country. Before the increase in oil prices, oil exports made 2% of the GDP (OILMSH = -0.02). As an oil exporter, its initial agricultural GDP is small, but total growth is rapid due to the contribution of oil revenues. In the absence of oil price changes and with careful macro policy management (either with respect to domestic inflation or to the foreign exchange rate), agricultural GDP grows from $79.50 per capita in year 1 to $102.50 in year 10, a growth rate of 2.9% per year.

When oil prices double at the start of year 6, the immediate effect is a *fall* in agricultural GDP, but then a more rapid rate of growth, with the year-10 value nearly the same as without an oil price increase. However, substantial agricultural output is lost due to the higher oil price, a loss shown by the cross-hatched area in Figure 13.5.

Even this loss is predicated on *good* macro policy management. What if the management is not so good? For example, assume a 20% differential in inflation that is not corrected by a devaluation. (In Indonesia by 1978, the cumulative differential had reached about 50%). This is accomplished in the model by reducing the intercept term in the foreign exchange equation before year 6, at the same time that oil prices double. The effect on

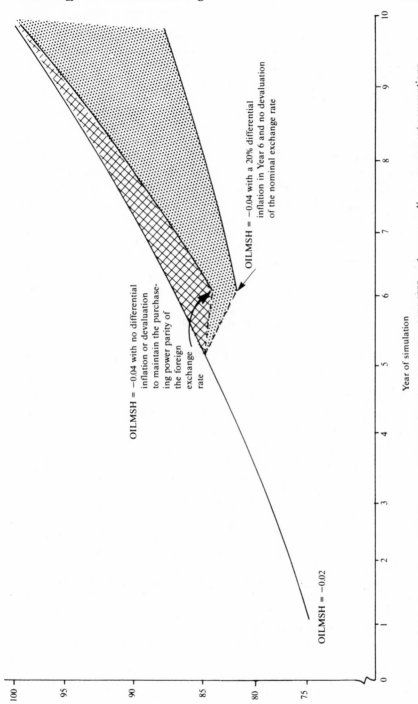

Figure 13.5. AgGDP in an oil-exporting country over time for alternative OILMSH and macro policy management assumptions

agricultural GDP is startling and dramatic. Instead of the small drop that occurred earlier and then more rapid growth than in the baseline path, the failure to devalue in the face of significant differential inflation causes a sharp fall in agricultural GDP, from about $87.60 to $85.40 per capita in year 6, and then a much slower rate of growth in agricultural GDP (which is part of a much slower growth in total GDP as well). Instead of a growth rate of 4.1% per year after the oil price rise when macro policy is managed carefully, the growth rate is only 1.6% in the face of a progressively overvalued exchange rate. The total losses in agricultural GDP, above the losses incurred even with good macro management, are shown by the dotted area in Figure 13.5. These additional losses are roughly two and a half times as large as the original loss and widen every year due to slower growth rate.

By contrast, Figure 13.6 shows an oil-importing country under a similar set of circumstances. With OILMSH = 0.02, agricultural GDP starts at a higher level than in the oil-exporting country, but it grows more slowly, only 1.0% per year. When the oil price shock comes before year 6, agricultural GDP rises, but then the growth rate is even slower, with year-10 output slightly less than in the baseline path. The cross-hatched area shows the increase in agricultural GDP caused by the oil price increase.

Suppose now that a deterioration in the current account balance due to oil imports or other forces, or a specific policy decision, leads to a devaluation of the domestic currency by 25%.[3] Again, the impact on agricultural GDP is startling and dramatic. The immediate increase in per capita agricultural GDP is nearly twice as large as before, jumping to $91.80 from the baseline path of $87.75. More important, the growth rate is now sharply higher at 2.7% per year.

This transformation in the growth rate is especially interesting in comparison with the rapid deceleration of growth in the agricultural sector of the oil-exporting country when differential inflation is not corrected by exchange rate devaluations. Now the differential paths in relative structure of the economy are reinforced by larger output and more rapid growth in total for the agricultural sector of the oil-importing country. Its agricultural GDP is growing at 2.7% per year while that of the oil exporter is only 1.6% per year, under this combined set of circumstances.

This combined picture begins to sound familiar, for the rapid demise of the agricultural sectors of such oil exporters as Mexico, Nigeria, and Iran are common knowledge. Some of this demise is revealed by the model as it is presently constructed, and more could certainly be built in if current account balances (and possibly inflation) were made endogenous to the

---

3. The intercept term in the equation for foreign exchange was modified by the same absolute value for the oil importer as for the oil exporter, but the antilog yields different percentage changes from the base foreign exchange rate.

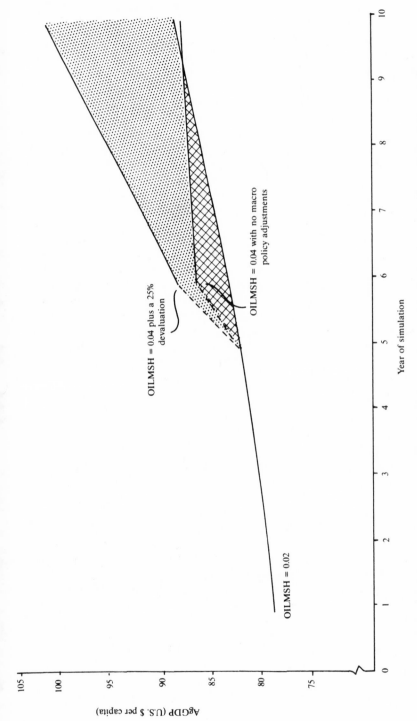

Figure 13.6. AgGDP in an oil-importing country over time for alternative OILMSH and macro policy management assumptions

model and dependent on OILMSH. But clearly there are policy choices at work here as well, as the Indonesian devaluation in November 1978 reveals. The policy debate leading up to that devaluation, and the actual experience since then, are strongly reflected in the logic of the model developed here. It is exciting to see the model capture these experiences and show the general message as well, which is the importance of sound macro policy management to the health of the agricultural sector, whether a country is an oil importer or an oil exporter. For Mexico, the failure to devalue sooner may have been a critical deficiency for SAM.

There is little point in running simulations of alternative levels of the terms of trade as a reflection of an agricultural price policy intervention. The marginal coefficient of 0.220 for the terms of trade in the AGSHR regression reflects both the immediate and the ultimate impact of any price policy changes, so long as the model has no feedback effects from agricultural share of the GDP to overall growth rates, exchange rates, or investment. All of these feedback effects are likely to exist, however, and are the subject of both modeling efforts and empirical estimation in future work. For the time being we must be content with the obvious conclusion that price policy is extremely important to the agricultural sector even in direct and short-run terms. Its long-run and dynamic role is likely to be even larger and more important. Some of the more dynamic effects are discussed analytically and intuitively in Timmer, Falcon, and Pearson (1983), but very little empirical evidence is presently available.

Summary and Conclusions

Although the data were not available to apply the simulation model to the 1980–1982 SAM period, the model's linkages enable us to posit some additional explanations of SAM's performance. Without policy interventions such as in SAM, the model would have predicted rising energy input costs in agriculture due to oil price rises and increasing cheap food imports due to an overvalued exchange rate depressing agricultural prices. These negative incentives would result in declining production. Lower production and higher costs would lead to rising food prices unless food imports increased sharply. In fact, this was the pre-SAM pattern revealed by the model's analysis of the pre1980 data. SAM broke this pattern.

Production and consumption rose and imports decreased during the initial two years of SAM. The explanation lies with the policy measures in SAM's self-sufficiency strategy that created higher incentives for the rural sector. The rising costs were offset by SAM's subsidies for seeds, fertilizer, and credit. Crop prices were kept high through SAM's price support policy. Consumer demand was preserved by SAM's subsidized food prices. In effect, SAM protected the rural sector's terms of trade, thereby offsetting the negative effects of the previous macro policies.

It is important to note, however, that in the third year of SAM these subsidies were reduced, the exchange rate had become significantly overvalued, the fiscal deficit soared, and inflation was rampant. Farmers' costs soared, and their support prices, in real terms, deteriorated. The rural-urban terms of trade reversed and production incentives eroded. It is not surprising that 1982 production fell from its previous highs. SAM's performance in its final year was significantly affected by the changing macro-policy environment, especially the budgetary costs of offsetting a negative macro price environment.

This chapter represents an evolution of work and thought over several years, and yet it is clearly only the very first step into an empirical understanding of the impact of energy prices on structural adjustment in the agricultural sector in Mexico and elsewhere. If nothing else, however, it establishes some of the perspectives that will need to be drawn on to build that understanding. First, the major point of this discussion is that the question cannot be approached in any meaningful fashion without addressing basic macroeconomic trends and forces in the economy as well as the efficacy of macroeconomic policy.

Second, there are tantalizing hints in the work here and from other literature in the field to suggest that some very important redistributions of income between rural and urban areas are occurring as oil prices change. These redistributions are probably positive in the sense that higher oil prices raise rural incomes relative to urban incomes, but there may be net food-consumption changes for the landless poor in both urban and rural areas that would significantly alter this conclusion. Both the direct income-distribution effects and the disaggregated food-consumption consequences need to be evaluated within the consistent framework of a much more detailed and carefully specified macro model (see Chapter 14).

Third, the model developed here is a rough and preliminary attempt to capture the general-equilibrium consequences of oil price changes on a particular domestic economy. It is not, however, a global general-equilibrium model of those consequences. Not all countries can maintain an undervalued exchange rate to spur exports and growth, nor are the growth paths generated necessarily consistent with global trade balance or flow of oil and other resources. A preliminary world agricultural model in a general-equilibrium framework is being investigated by Jean-Marc Burniaux and colleagues, especially Jean Waelbroeck at the World Bank. The results obtained with the country-specific model developed here are at least roughly consistent with initial results achieved by Burniaux, but further consideration of global equilibrium effects is in order.

As a final point, the complexity of this issue is obvious. At the same time, however, a fairly simple macro model with realistic parameters estimated from readily available data generated some complex insights that seem to mirror the varied experiences of Asian-Pacific countries as energy prices changed. This approach is as relevant to post-SAM food policy analysis as it was to SAM and pre-SAM policy. Complexity is not a cause for despair, but rather a catalyst for the development of intuition and insight. The analytic framework of economics can help build such intution and insight when it is used to address the real problems of economic development.

Appendix Table 1. Statistical characteristics of basic variables

			Range	
Variable	Mean	Standard deviation	Minimum	Maximum
LX1	5.5625	0.698205	4.200	7.128
LPOP	17.2689	0.778683	15.898	18.801
TT	1.0008	0.141132	0.709	1.410
INV	0.2049	0.059560	0.045	0.357
AGSHR	0.2916	0.103404	0.098	0.588
PPFEX	3.2478	2.064620	0.765	6.956
CABAL	−0.0216	0.049966	−0.219	0.128
OILMSH	0.0006	0.044536	−0.202	0.095
DCERPR	63.4585	15.745800	44.719	117.694
DANC	47.2390	10.123000	35.062	67.808

Appendix Table 2. Values of country-specific intercepts in regressions (*t*-statistics in parentheses)

Regression number	Constant[a]	South Korea	Malaysia	Mexico	Philippines	Sri Lanka	Thailand
1	−0.070 (0.7)			not significant			
2	8.017 (48.8)	0.673 (9.7)	−4.658 (64.3)	−2.923 (36.7)	−4.153 (66.1)	−4.161 (74.7)	−2.998 (51.4)
3	6.325 (75.3)	0.208 (2.6)	−5.159 (63.6)	−3.637 (53.9)	−4.533 (59.5)	−4.404 (60.4)	−3.259 (43.5)
4	−1.724 (1.7)	−0.141 (1.6)	0.796 (1.3)	0.424 (1.1)	0.751 (1.4)	0.720 (1.3)	0.563 (1.5)
5	−1.718 (14.4)	−0.154 (10.5)	0.748 (10.8)	0.385 (8.6)	0.714 (11.5)	0.688 (11.1)	0.535 (11.9)
6	0.287 (0.6)	−0.411 (9.1)	0.432 (1.2)	0.321 (1.2)	0.193 (0.6)	0.327 (1.1)	0.077 (0.3)
7	0.499 (1.7)	−0.393 (9.6)	0.280 (1.2)	0.208 (1.3)	0.051 (0.3)	0.196 (1.0)	−0.017 (0.1)
8	18.142 (3.5)	−0.113 (2.2)	−0.304 (4.3)	−0.204 (4.4)	−0.115 (2.4)	−0.378 (6.2)	−0.103 (2.2)

Appendix Table 2. (Continued)

Regression number	Constant[a]	South Korea	Malaysia	Mexico	Philippines	Sri Lanka	Thailand
9	6.989 (1.9)	−0.212 (4.9)	−0.366 (6.1)	−0.284 (7.3)	−0.212 (5.3)	−0.444 (8.5)	−0.195 (5.0)
10	4.503 (0.9)	−0.167 (2.9)	−0.239 (3.1)	−0.233 (4.6)	−0.163 (3.1)	−0.326 (4.9)	−0.146 (2.9)
11	23.141 (5.5)	−0.081 (1.9)	−0.304 (5.2)	−0.183 (4.7)	−0.089 (2.2)	0.374 (7.5)	−0.077 (2.0)
12	8.591 (2.7)	−0.185 (5.0)	−0.340 (6.6)	−0.267 (7.8)	−0.192 (5.5)	−0.417 (9.4)	0.174 (5.2)
13	9.546 (2.4)	−0.134 (2.8)	−0.226 (3.5)	−0.207 (4.7)	−0.137 (3.1)	−0.312 (5.6)	−0.118 (2.7)

[a]Indonesia is represented by the constant term. All other coefficients are relative to this constant term and, hence, to Indonesia.

Bibliography

Behrman, Jere. 1982. "Review Article on Hollis B. Chenery, *Structural Change and Development Policy," Journal of Development Economics* 10(3): 313–324.

Bhagwati, Jagdish. 1978. *Anatomy and Consequences of Foreign Exchange Control Regimes.* vol. 11 of *Foreign Trade Regimes and Economic Development,* for the National Bureau of Economic Research. Cambridge: Ballinger Press.

Branson, William H. 1981. "Macroeconomic Determinants of Real Exchange Rates." Working paper No. 801. Cambridge: National Bureau of Economic Research.

Bruno, Michael, and Jeffrey Sachs. 1979. "Macro-Economic Adjustment with Import Price Shocks: Real and Monetary Aspects." Working paper No. 342. Cambridge: National Bureau of Economic Research.

Buiter, William H., and Douglas D. Purvis. 1980. "Oil, Disinflation, and Export Competitiveness: A Model of the Dutch Disease." Working paper No. 592. Cambridge: National Bureau of Economic Research.

Burniaux, Jean-Marc. 1981. "First Experiments with a World Agricultural Model in a General Equilibrium Framework." Discussion paper No. 8205, Université Libre de Bruxelles: Centre d'Economie Mathematique et d'Econometrie.

Chenery, Hollis, and Moises Syrquin. 1975. *Patterns of Development, 1950–1970.* London: Oxford University Press.

Clark, Kim B., and Richard B. Freeman. 1980. "How Elastic Is the Demand for Labor?" *Review of Economics and Statistics* 62(4): 509–520.

Cline, William R., and Sidney Weintraub. eds. 1981. *Economic Stabilization in Developing Countries.* Washington: The Brookings Institution.

Corden, W. Max. 1974. *Trade Policy and Economic Welfare.* Oxford, England: Oxford University Press.

Dornbusch, Rudiger. 1980. *Open Economy Macroeconomics.* New York: Basic Books.

Griliches, Zvi. 1960. "Estimates of the Aggregate U.S. Farm Supply Function." *Journal of Farm Economics* 42: 282–293.

Hayami, Yujiro, and Vernon Ruttan. 1971. *Agricultural Development: An International Perspective.* Baltimore: Johns Hopkins University Press.

Johnston, Bruce F., and Peter Kilby. 1975. *Agriculture and Structural Transformation.* New York: Oxford University Press.

Krause, Lawrence B., and Sueo Sekiguchi, eds. 1980. *Economic Interaction in the Pacific Basin.* Washington: The Brookings Institution.

Kreuger, Anne O. 1978. *Liberalization Attempts and Consequences.* Vol. 10 of *Foreign Trade Regimes and Economic Development,* for the National Bureau of Economic Research. Cambridge: Ballinger Press.

Lewis, W. Arthur. 1963. "Development with Unlimited Supplies of Labor." In A. N. Agarwala and S. P. Singh, eds., *The Economics of Underdevelopment.* New York: Oxford University Press.

Peterson, Willis. 1979. "International Farm Prices and the Social Cost of Cheap Food Policies." *American Journal of Agricultural Economics* 61(1): 12–21.

Pimentel, David, et al. 1974. "Workshop on Research Methodologies for Studies of Energy, Food, Man, and Environment: Phase 1." Ithaca, N.Y.: Cornell University Center for Environmental Quality Management.

Sachs, Jeffrey D. 1981. "The Current Account and Macroeconomic Adjustment in the 1970s." *Brookings Papers on Economic Activity* No. 1, 201:282.

Schuh, G. Edward. 1976. "The New Macroeconomics of Agriculture." *American Journal of Agricultural Economics* 58(5): 802–811.

Schultz, Theodore W., ed. 1978. *Distortions of Agricultural Incentives.* Bloomington: University of Indiana Press.

Soedjatmoko. 1981. "UNCNRSE: First Step towards a Broader Vision?" *ASSET* 3:7. Tokyo: United Nations University.

Taylor, Lance. 1982. "Back to Basics: Theory for the Rhetoric in the North-South Round." *World Development* 10(4): 327–335.

Timmer, C. Peter. 1975. "Interaction of Energy and Food Prices in LDC's." *American Journal of Agricultural Economics* 57(2): 219–224.

_____. 1980. "Food Prices and Economic Development in LDC'S." In Ray A. Goldberg, ed., *Research in Domestic and International Agribusiness Management,* vol. 1, pp.53–67. Greenwich, Conn.: JAI Press.

_____. 1981. "Is There 'Curvature' in the Slutsky Matrix?" *Review of Economics and Statistics.* LXII(3): 395–402.

_____. 1982a. "Developing a Food Strategy." In H. Carter, ed., *Food Security in a Hungry World: A Proceedings,* pp. 120–129. Davis: University of California.

_____. 1982b. "The Financial Aspects of Macro Food Policy." Paper presented at conference on Food and Finance, Minneapolis, Minn., September.

Timmer, C. Peter, and Walter P. Falcon. 1975. "The Political Economy of Rice Production and Trade in Asia." In Lloyd Reynolds, ed., *Agriculture in Development Theory,* pp. 373–408. New Haven: Yale University Press.

Timmer, C. Peter, Walter P. Falcon, and Scott R. Pearson. 1983. *Food Policy Analysis.* Baltimore: Johns Hopkins University Press.

van Wijnbergen, Sweder. 1980. "Oil Price Shocks and the Current Account: An Analysis of Short Run Adjustment Measures." Development Research Center. Washington: World Bank.

Waelbroeck, J., et al. 1981. "General Equilibrium Modeling of Global Adjustment." World Bank Staff Working Paper.

CHAPTER 14

SAM's Impact on Income Distribution

Bill Gibson, Nora Lustig, and Lance Taylor

A primary goal of SAM was to combat the crushing poverty and stagnation characteristic of traditional agriculture.[1] According to a household income-expenditure survey done by the Ministry of Planning and Budget in 1977, 59.5% of agricultural households earned less than the minimum wage, and 36.8% earned less than *half* the minimum wage. Those earning above the minimum wage accounted for only 8.6% of the total (Lustig, 1980). In Mexico, the rural poor consist largely of agricultural workers and peasants who produce maize and beans (Shejtman, 1981).

Malnutrition is a gnawing and persistent aspect of rural poverty. According to recent estimates (see Chapter 12) about 35 million Mexicans do not consume widely recommended minimum levels of calories and proteins. Of these, 19 million are critically undernourished and the overwhelming majority, some 12 million, live in the countryside. As part of a program to stave off increasing rural hunger, SAM policy makers proposed a series of measures intended to raise rural incomes of the poorest peasant producers. As indicated in previous chapters, these measures included an increase in the support price for maize (and other grains) and increases in subsidies for improved seeds, fertilizers, and credit.[2]

In the year following the implementation of the SAM policies, maize and bean production showed a significant increase (see Chapter 11). How growth in real income was distributed among rural producers, however, is a much more elusive question. Reliable data on the distribution of income

1. The tendency toward stagnation in traditional agriculture is evident in the following data: the 1970–1977 average annual rate of growth in physical production of maize was only 1.38%, while beans actually contracted on average by 2.44%. For more details, see Luiselli (1980) and Martin del Campo (1980); for more on the objectives of SAM, see Chapter 3 of the present volume.

2. In 1960, the maize support price was 800 pesos per ton, but by 1979 it had fallen to 606 pesos per ton (1960 pesos). For a discussion of Mexican agricultural policy, see Hewitt de Alcantara (1978).

Table 14.1. Social classes and productive sectors in the Mexican CGE

Social classes		Productive sectors	
1.	Peasants	1.	Maize and beans
2.	Agricultural workers	2.	Other agriculture and livestock
3.	Agricultural capitalists	3.	Petroleum
4.	Urban workers	4.	Fertilizer
5.	Urban capitalists	5.	Food processing
6.	Merchant capitalists	6.	Industry and construction
7.	Urban marginals	7.	Services
		8.	Commerce

in developing countries is not easily obtained and, moreover, the effects of SAM policies are intermixed with other short-term fluctuations as well as government policy initiatives. In the absence of concrete survey data, the approach used by SAM policy makers was to mathematically simulate probable effects of alternative policies and programs.

This chapter provides some evidence on the distributional consequences of SAM-styled policies using a short-run economy-wide simulation model. A multisectoral computable general-equilibrium model (CGE) built for the purpose of analyzing SAM policies is discussed, and simulations of SAM programs are presented.[3] The model shows that SAM policies essentially redistribute income from the urban to the rural sectors without necessarily improving the overall distribution of income. In particular, the rural working class may actually suffer a deterioration in real income if SAM policies are not augmented by additional expansionary measures.

The Model

The CGE used in this study accounts for seven social classes[4] and eight productive sectors, as shown in Table 14.1.[5] The agricultural sectors are assumed to have given supplies. "Flex-prices" balance given supply with demand that is determined by income-expenditure relations of the social classes plus exogenous elements of final demand. In the remaining "fix-price," nonagricultural sectors, supply curves are horizontal at prices

3. See Taylor (1979, 1983), Dervis, de Melo, and Robinson (1982), and Scarf and Shoven (1983) for a general discussion of CGEs. For a more neoclassical CGE for the Mexican economy, see Serra-Puche (1983). For a detailed discussion of the present model, see Gibson et al. (1984, 1986). Taylor and Lysy (1979) discuss the effects of various "closure rules" in CGEs.

4. See Lustig (1980) for a discussion of the data base from which the class categories were drawn.

5. See the appendix, footnote 10, for data sources.

determined by costs and fixed markup. Output is determined in fix-price sectors by the level of effective demand. Excess capacity in these sectors allows output to rise with no corequisite increase in price (see the appendix to this chapter for the algebraic statement of these assumptions). The formal model does not include factor markets. The money wage is given exogenously and there is no "capital" explicitly accounted for apart from intermediate goods. Neither is there an explicit investment function. Investment is taken as given in real terms and macroeconomic equilibrium is achieved through Keynesian quantity adjustment and Kaldorian "forced savings."[6]

To see how the model basically works, consider an exogenous increase in investment demand for output in the nonagricultural sectors. As nonagricultural output rises to meet the new demand, gross profits increase with the level of economic activity. With fixed savings propensities, higher profits generate higher savings, but not at a rate that will entirely finance the new level of investment. The shortfall is made up by an improvement in agricultural terms of trade that, with fixed agricultural supply, must rise as demand from the urban sector increases. The inflation in food prices, together with the assumption of fixed money wages and markups, implies that real income will shift from workers to capitalists in both rural and urban sectors. Since workers make little contribution to aggregate savings, the process of "forced saving" increases the pool available to finance the initial rise in investment. Although urban employment rises in the new equilibrium, per capita consumption of workers falls. Higher fix-price levels of output cause noncompetitive intermediate imports to rise, and with fixed exports, foreign savings must increase. Government savings also go up since indirect tax receipts rise with inflation and economic activity and direct taxes increase with income. In the new equilibrium, all three components of savings contribute to the higher level of investment.

Results of the Simulations

This section describes the results of numerical simulations of SAM policies with emphasis on the distribution of income among the social classes listed above. Three sets of simulations are discussed, labeled 1 through 3 in the following tables and text:

Scenario 1. the basic SAM policy package, subsidies of 30% for fertilizers and 75% for improved seed varieties used in the maize and bean sector and an increase in the maize-bean support price of 15%. Supply in the maize-bean sector is assumed to increase by 2%.

Scenario 2. scenario 1, but with a subsidy of 18% on the final price of food processing. The subsidy is assumed to be applicable only to the pop-

6. For a simplified version of the present model, see Cichilnisky and Taylor (1980).

Table 14.2. Distribution of real income (% of total)

Scenario	Peasants	Ag. wkrs.	Ag. caps.	Urb. wkrs.	Urb. caps.	Merch. caps.	Urb. margs.	Total
Base	4.05	3.08	4.91	37.41	30.74	14.2	5.64	100
1a	4.63	3.00	5.80	36.65	30.41	13.93	5.57	100
1b	4.41	3.09	5.06	37.12	30.53	14.14	5.65	100
2a	4.79	2.88	6.66	36.09	30.36	13.76	5.47	100
2b	4.55	2.97	5.88	36.58	30.50	13.97	5.54	100
3a	4.43	3.02	5.73	36.96	30.28	13.99	5.58	100
3b	4.21	3.12	5.00	37.43	30.40	14.19	5.65	100

ular classes: peasants, urban and rural workers, and urban marginals.

Scenario 3. scenario 1, but with a 15% increase in nominal wages for both urban and rural workers.

Each of these scenarios is run under two assumptions about supply response in sector 2, other agriculture. Scenarios 1a, 2a, and 3a assume rigid supply and consequently show the greatest redistributive effect through the process of forced savings. Since supply response in the second sector is so important in determining the distribution of real income, scenarios 1b, 2b, and 3b allow for a 2% increase in output in sector 2, the same rate of expansion as in the maize-bean sector.

Distribution of Income

Table 14.2 shows how various policy initiatives change distribution of real income relative to the base solution of the model.[7] The first row of the table gives the proportions of total income earned by each social class in the base state to which comparisons of three scenarios are then made. The results indicate that, as a whole, the redistributive effect of SAM-like interventions is quantitatively small. If the objective is a fundamental restructuring of the distribution of income, terms-of-trade policies, subsidies, and so on will probably be insufficient.

All scenarios essentially redistribute income from the urban to the rural sectors. Indeed, the sum of income proportions of the first three classes rises in every scenario relative to the base state. Note that the urban-rural redistributive effect is the greatest when SAM measures are combined with a consumer subsidy on processed foods and there is no supply response in sector 2 (scenario 2a). When accompanied by an increase in nominal wages, the rural-urban redistributive effect is less pronounced. In every case, the effect is less with supply response in sector 2.

7. If there is no change in any parameter, the model reproduces the base social accounting matrix for 1975. See the appendix.

Table 14.3 Real income (% change from base)

Scenario	Peasants	Ag. wkrs.	Ag. caps.	Urb. wkrs.	Urb. caps.	Merch. caps.	Urb. margs.
1a	17.04	−0.26	20.89	0.20	1.20	0.53	1.01
1b	11.87	3.13	5.92	1.86	1.99	2.38	2.74
2a	21.72	−3.52	39.54	−0.75	1.65	−0.13	−0.35
2b	16.12	−0.22	23.78	0.97	2.47	1.79	1.36
3a	12.43	0.88	19.90	1.37	1.08	1.26	1.39
3b	7.21	4.32	4.96	3.05	1.88	3.14	3.15

While all simulations show a shift in income from the urban to the rural sectors, there are important differences between the scenarios when classes are considered individually. From Table 14.2, it can be concluded that SAM subsidies and support prices tend to redistribute income to peasants and agricultural capitalists at the expense of the remaining segments of society. In particular, agricultural workers suffer a decline in their share in all but the last simulation. Table 14.3 shows the change in real income by social class. The table confirms that SAM policies reduce real incomes of the rural proletariat in *absolute* as well as relative terms. Without supply response in sector 2, rural workers suffer a 0.3% *decline* in real income (scenario 1a). Consumer subsidies do little to recoup their position: Tables 14.2 and 14.3 show that rural workers lose under scenario 2a and 2b, both relatively and absolutely.

From the point of view of the rural proletariat, the success of SAM policies depends largely on supply response in sector 2. Indeed, without an increase in sector 2 output, real income falls absolutely in the first two simulations and falls relatively in all.[8] Since one of the main objectives of SAM programs was to enhance the real incomes of the rural poor, the simulation results seem to reveal a fundamental design flaw. Moreover, the largest gains from SAM policies are captured by agricultural capitalists. In scenario 1a, their real incomes increase by 20.9% and their relative position by almost one percentage point. Supply response substantially damps the improvement, but the incomes of agricultural capitalists rise in every case.

In the transfer of income from the urban to the rural sector, workers contribute the most. Urban workers show the smallest percentage gains in real income and loss of relative position in all but the last scenario. Urban marginals as well as urban and merchant capital consistently lose relative ground (except in 3b and 1b for urban marginals). Table 14.3 shows that

8. Given the assumption of fixed labor coefficients (see the appendix for details), employment rises in scenarios 1b, 2b, and 3b by the same amount as output: 2%. To compute the change in real per capita income of rural workers, one must subtract 2%.

Table 14.4. Real consumption: Scenario 1a (% change from base)

	Peasants	Ag. wkrs.	Ag. caps.	Urb. wkrs.	Urb. caps.	Merch. caps.	Urb. margs.
Maize and beans	6.8	2.5	−0.0	2.7	−0.0	−0.0	2.4
Other agriculture	9.7	−2.6	5.4	−3.1	−2.6	−2.7	−1.7
Petroleum	19.7	0.5	18.2	0.8	1.6	1.1	2.1
Processed food	16.4	−1.2	11.7	−1.1	−0.8	−1.3	0.3
Industry	21.2	0.4	37.9	1.2	2.8	2.1	2.1
Services	21.4	0.6	22.7	1.0	2.0	1.4	2.1
Commerce	17.7	−0.2	21.7	0.3	1.3	0.7	1.2

all popular classes with the exception of peasants actually lose real income in the SAM cum subsidy scenario 2a. Without an increase in real output, the fiscal stimulus causes sufficient inflation to effect a reduction in real income.

By summing the percentages of the income received by workers, peasants, and urban marginals, it is clear that whether SAM policies improve the overall distribution of income depends upon how agricultural supply responds to the terms-of-trade intervention. Without supply response, income of the popular classes falls as a percentage of the total. With consumer subsidies, income falls whether there is supply response or not. The most progressive intervention is without question the third. But again, even with an increase in nominal wages, income of the popular classes falls as a proportion of the total without supply response. Table 14.2 shows that the only scenario that *simultaneously* improves the percentages of all popular classes is the third, and only if supply response accompanies the wage increase.

Consumption

Table 14.4 shows that the basic SAM package increases peasant consumption of all goods. Observe, however, that agricultural workers, urban workers, and urban marginals suffer a deterioration in their sector 2 real consumption. The table confirms that SAM policies tend to shift purchasing power from the urban to rural sectors, with a consequent decline in food intake of the urban popular classes. In contrast, agricultural capitalists benefit enormously from the favorable shift in the terms of trade. The results are more favorable when supply in sector 2 is allowed to adjust.[9]

9. Simulation results (not shown in the tables) indicate that with the exception of capitalists' consumption of maize and beans, consumption of all classes rises.

Table 14.5. Real consumption: Scenario 2a (% change from base)

	Peasants	Ag. wkrs.	Ag. caps.	Urb. wkrs.	Urb. caps.	Merch. caps.	Urb. margs.
Maize and beans	6.6	1.0	−0.0	1.4	−0.0	−0.0	1.5
Other agriculture	11.1	−6.0	8.9	−6.2	−5.4	−5.7	−3.9
Petroleum	29.0	−0.4	35.5	1.7	3.0	1.7	3.7
Processed food	27.1	−0.6	20.5	1.5	−2.9	−4.2	3.3
Industry	30.9	−0.7	73.7	2.4	5.0	2.8	3.7
Services	31.3	−0.5	44.3	2.0	3.6	2.1	3.7
Commerce	26.6	−0.8	41.8	1.6	2.1	0.3	3.0

Table 14.5 indicates that an attempt to repair the impact of SAM policies on the urban poor through direct consumer subsidies is not entirely successful. Real consumption of processed foods falls for agricultural workers by 0.6%. On the other hand, what urban workers lose in real processed food consumption under SAM policies (1.1%), is more than regained with the subsidy (1.5%). Note that urban marginals' consumption of processed foods increases as a result of both SAM policies and direct consumer subsidies.

The consumption effects of a nominal wage increase, aimed at counteracting the regressive tendencies of SAM, are considered in Table 14.6. Again, the big gains are had by peasants and agricultural capitalists at the expense of the urban sectors. Observe that even when the nominal wages are increased, real food consumption of the urban working class falls (with the exception of maize and beans). Agricultural workers are better off with an increase in nominal wages than with a food subsidy, although in both cases food intake falls (again with the exception of maize and beans). Note also that urban marginals benefit less from the wage increase than the food subsidy. Table 14.7 shows that with supply response in other agriculture, an increase in nominal wages brings positive consumption increments for all social classes. Maize and beans is shown to be an inferior good.

Table 14.6. Real consumption: Scenario 3a (% change from base)

	Peasants	Ag. wkrs.	Ag. caps.	Urb. wkrs.	Urb. caps.	Merch. caps.	Urb. margs.
Maize and beans	6.0	3.0	−0.0	3.1	−0.0	−0.0	2.6
Other agriculture	6.1	−2.0	4.6	−2.7	−2.9	−2.7	−1.7
Petroleum	15.2	2.1	18.3	2.4	2.4	2.5	3.0
Processed food	11.6	−0.0	11.0	−0.2	−0.9	−0.7	0.6
Industry	15.9	1.7	37.0	3.4	3.4	4.2	2.8
Services	15.3	1.4	20.9	1.7	1.3	1.5	2.1
Commerce	12.8	0.9	20.5	1.4	1.1	1.3	1.4

Table 14.7. Real consumption: Scenario 3b (% change from base)

	Peasants	Ag. wkrs.	Ag. caps.	Urb. wkrs.	Urb. caps.	Merch. caps.	Urb. margs.
Maize and beans	3.4	2.6	−0.0	2.3	−0.0	−0.0	2.0
Other agriculture	4.9	2.6	1.7	1.3	0.4	0.8	1.9
Petroleum	8.2	4.7	4.8	3.0	2.3	3.2	3.8
Processed food	7.4	4.7	3.4	2.6	1.5	2.6	3.2
Industry	8.5	4.4	9.2	4.7	3.6	6.0	3.7
Services	7.8	4.2	4.3	2.5	1.2	2.4	2.9
Commerce	7.3	4.2	4.8	2.9	1.7	3.0	3.0

As a whole the simulations show that SAM policies accompanied by an increase in nominal wages are less effective in redistributing income to peasants but more progressive in the overall distribution of income. On the other hand, the consumer subsidy program accelerates the redistribution of income from city to country but is not powerful enough to maintain real purchasing power of the poorer classes. SAM policies accompanied by an increase in nominal wages are far more balanced in effect on the real distribution of income and, as is seen below, bear other favorable macroeconomic consequences.

Macroeconomic Consequences of SAM Policies

Table 14.8 shows disaggregated employment growth data for the fix-price sectors of the model. From this table it can be concluded that all policy scenarios are expansionary in terms of aggregate real output, since employment is determined by fixed labor coefficients and rises for each simulation. SAM measures accompanied by an increase in nominal wages is the most expansionary policy package, followed by the subsidy scenario.

Table 14.9 provides data on the distribution of aggregate savings. As ex-

Table 14.8. Employment growth rates (% change from base)

Scenario	Petroleum	Fertilizer	Processed foods	Industry	Services	Commerce
1a	1.7	0.4	0.7	1.6	2.1	1.9
1b	1.7	1.7	2.5	1.5	1.9	2.1
2a	2.0	0.4	0.8	1.9	2.1	2.0
2b	1.9	1.7	2.7	1.8	1.9	2.3
3a	3.1	0.5	2.1	2.8	3.8	3.4
3b	3.0	1.8	4.1	2.7	3.5	3.7

Table 14.9. Savings composition (% of total)

Scenario	Private	Foreign	Government	Total
Base	82.4	11.5	6.1	100
1a	85.0	11.7	3.2	100
1b	84.3	11.8	3.9	100
2a	86.9	10.6	2.5	100
2b	86.2	10.7	3.1	100
3a	87.0	11.9	1.1	100
3b	86.2	12.0	1.8	100

pected, the SAM subsidies cause government savings to fall in all scenarios. Since investment remains constant in real terms, the loss in government savings must be made up through some combination of private and foreign savings. The table shows that compared to the base-state solution, all simulations bring about an increase in the share of private savings. This implies that, on balance, SAM policies shift real income from classes with low savings propensities (workers, peasants, and urban marginals) to classes with higher propensities (capitalists and merchants). The Kaldorian process of forced savings is accompanied by a Keynesian adjustment in output in all three scenarios. Since intermediate imports rise with output and exports are fixed, this puts pressure on foreign savings to rise. While the percentage change in private savings is the same in the second and third scenarios, the more expansionary character of the third causes foreign savings to play a more prominent role.

The data on the composition of savings provides additional support for raising nominal wages when SAM-like policies are introduced. In addition to the favorable distributive effects seen in the third scenario, there is little additional pressure on the foreign account. The data of Table 14.9 suggest that the reduction in government savings might be accelerated in the third scenario, but in fact the data on absolute changes show that government savings falls by 82% in the second scenario, compared to 54% in the third. Compared to the first scenario, the additional "cost" in terms of foreign savings in the third scenario is not great. In absolute terms, foreign savings rise in the last scenario by 2.8%, compared to 2.4% in the first.

The most obvious objection to the third scenario is the possibility of igniting a wage-price inflationary spiral. The equations of the appendix show that the model has no "general price level," but a GDP deflator may be constructed from the individual sectoral price levels. SAM policies alone cause the GDP deflator to increase by less than 1%, while SAM cum subsidies raise the deflator by almost 3%. It is not surpising that the largest increase, almost 15%, occurs when nominal wages are boosted by the same amount.

Conclusions

The simulations discussed in this chapter show that policies recommended by SAM generate higher incomes for the peasant sector, both relatively and absolutely. In order for an improvement in the standard of living of rural and urban workers to accompany these measures, SAM-like policies should be complemented with nominal wage increases. Otherwise, the measures simply shift income from urban to rural sectors and may fail to improve the overall distribution of income. Of the simulations considered here, only SAM policies coupled with an increase in nominal wages and supply response in other agriculture and livestock improve the percentage of total income accruing to all popular classes. SAM policies by themselves increase the percentage of total income captured by rural, urban, and merchant capitalists.

SAM policies augmented by a system of consumer subsidies are even less attractive in terms of output and income distribution. The results of the simulations show that a coalition of peasants and agricultural capitalists would favor this alternative, however, especially when supply response in other agriculture is limited. Because of their political popularity, these measures would not necessarily be resisted by an opposing coalition of urban classes and rural workers even though their real incomes would suffer as a result.

Finally, it is to be emphasized that the static nature of the model disallows claims about how the economy would adjust over time and therefore provides only a limited picture of SAM-like policies. The simulations considered here are simple comparative static exercises that do not adequately account for dynamic adjustment of a host of important variables. In particular, the rate of investment is held fixed, and there is no monetary-financial feedback onto the real sector of the model and no role for expectations whatsoever. Due to these shortcomings, the model amounts to only a first step in understanding the macroeconomic implications of SAM programs.

Appendix: Model Specification[10]

Variables			Exogenous parameters	
p	price	π mark-up	t	indirect tax rate
x	output	w wage	T	direct tax rate
y	income	I investment	m	noncompetitive intermediate imports
c	consumption	l labor coefficient	w_g	government wages
ρ	retail price	E exports (net competitive)	q	commercial margin
		G government expenditure	p_m	intermediate import price
		μ consumption propensity	s	savings propensity
		θ subsistence consumption	σ	consumer subsidy
		m_c noncompetitive imports	z	proportion of value added
		a input-output coefficient		of peasants and urban
		τ input subsidy		marginals

The material balances for the productive sectors are given by

$$x_i = \sum_{j=1}^{8} a_{ij}x_j + \sum_{j=1}^{7} c_{ij} + I_i + G_i + E_i,$$

$$i = 1, 2, \ldots, 8.$$

Consumption of the ith good by the jth social class is given by a linear expenditure system (Lluch, Powell, and Williams, 1977; Taylor 1979). Consumption is a linear function of total expenditure in excess of subsistence expenditure:

$$c_{ij} = \theta_{ij} + \frac{\mu_{ij}}{\rho_i}\left[(1 - s_j)(1 - T_j)y_j - m_{cj} - \sum_{i=1}^{7}\rho_i\theta_{ij}\right],$$

$$i = 1, 2, \ldots, 7,$$
$$j = 1, 2, \ldots, 7.$$

10. The model data is summarized in an eight-sector, seven-class social accounting matrix for 1975, available from authors. The social accounting matrix is based on an aggregation of a seventy-two sector input-output study in SPP (1981a). The disaggregation of agriculture into maize and beans and other agriculture is taken from the well-known linear programming model for Mexico, CHAC, named after the rain god. Maria Bassoco of the Division of Macroeconomic Analysis of SAM prepared the estimates. Horacio Santamaria of the Coordinación del Sistema Nacional de Información assisted in the dissaggregation. The consumption functions were estimated using a linear expenditure system with data from a 1977 budget study conducted by SPP (1981b). The authors had access to the original computer tapes of this study, from which the class structure was determined. Direct tax rates were taken from Reyes Heroles (1980), as were the proportions of value-added accruing to urban marginals. A more detailed description of sources and methods can be found in Lustig (1980).

The exception is sector 8, commerce, which is determined by real consumption levels in the first seven sectors,

$$c_{8j} = \sum_{i=1}^{7} q_i c_{ij}, \qquad j = 1, 2, \ldots, 7,$$

and fixed "physical" commercial margins q_i. Retail prices are then given by

$$\rho_i = p_i + p_8 q_i - \sigma_i, \qquad i = 1, 2, \ldots, 7.$$

Peasant incomes are determined by a fixed fraction z of value-added in the two agricultural sectors,

$$y_1 = \sum_{i=1}^{2} \left[p_i - \sum_{j=1}^{8} p_j a_{ji}(1 - \tau_j) - p_m m_i \right] z_i x_i,$$

while agricultural workers' income depends on direct labor coefficients and the level of wages:

$$y_2 = w(l_1 X_1 + l_2 X_2).$$

The remaining value-added in the agricultural sectors accrues to agricultural capitalists:

$$y_3 = \sum_{i=1}^{2} \left\{ \left[p_i - \sum_{j=1}^{8} p_j a_{ji}(1 - \tau_j) - p_m m_i \right] (1 - z_i) - w l_i \right\} x_i.$$

Urban workers' income is determined in the same way as rural workers:

$$y_4 = w \sum_{i=3}^{8} l_i X_i + w_g.$$

Urban capitalists earn profits in sectors 5 through 7 (fertilizer and petroleum profits accrue to the state as revenues):

$$y_5 = \sum_{i=5}^{7} \left[(p_i - \sum_{j=1}^{8} p_j a_{ji} - p_m m_i)(1 - z_i) - w l_i \right] x_i,$$

where a share of valued-added is captured by urban marginals. Similarly, merchant capitalist income is

$$y_6 = \left[(p_8 - \sum_{j=1}^{8} p_j a_{ji} - p_m m_8)(1 - z_8) - w l_8 \right] x_8,$$

and urban marginal income is a fixed proportion of value-added in sectors 5 through 8:

$$y_7 = \sum_{i=5}^{8} \left(p_i - \sum_{j=1}^{8} p_j a_{ji} - p_m m_i \right) z_i x_i.$$

The nonagricultural fix prices are then given by costs marked up at a given rate:

$$p_i = (1 + \pi_i)(1 + t_i)\left(\sum_{j=1}^{8} p_j a_{ji} + wl_i + p_m m_i\right), \qquad i = 3, 4, \ldots, 8.$$

The model thus amounts to a system of eighty-four equations in eighty-four unknowns, six output levels, two flex prices, six fix prices, seven retail prices, fifty-six consumption levels, and seven incomes. Flexible prices equilibrate given supply and general equilibrium demand in the first two agricultural sectors. In the nonagricultural sectors, outputs adjust to the levels of demand.

Bibliography

Chichilnisky, G., and L. Taylor. 1980. "Agriculture and the Rest of the Economy: Macro Connections and Policy Restraints." *American Journal of Agricultural Economics* No. 62.

Dervis, K., J. de Melo, and S. Robinson. 1982. *Planning Models and Development Policy.* London: Cambridge University Press.

Gibson, B., N. Lustig, and L. Taylor. 1986. "Terms of Trade and Class Conflict in a Computable General Equilibrium Model for Mexico." *Journal of Development Studies,* forthcoming.

Gibson, B., L. Taylor, and N. Lustig. 1984. "Ventajas comparativas y autosuficiencia alimentaria: Una comparación un model de equilibrio general con dos especificaciones de precios." *Investigación Económica* 168 (April–June).

Hewitt de Alcantara, C. 1978. *La modernizición de la agricultura mexicana: 1940–1970.* Mexico City: Siglo XXI Editores.

Lluch, C., A. Powell, and R. Williams. 1977. *Patterns of Household Demand and Saving.* New York: Oxford University Press.

Luiselli, C. 1980. "Agricultura y alimentación: premisas para una neuva estrategia." In N. Lustig, ed., *Panorama y perspectivas de la economía mexicana.* Mexico City: El Colegio de Mexico.

Lustig, N. 1980. "Politicas de consumo y distribución del ingreso." Sistema Alimentario Mexicano Subproyecto No. 10. *Demografía y Economía* 50.

Martin del Campo, A. 1980. "Transformación agraria y neuvas opciones para el desarrollo." In N. Lustig, ed., *Panorama y perspectivas de la economia mexicana.* Mexico City: El Colegio de Mexico.

Reyes Heroles, J. 1980. "Welfare Effects of Short-Run Macroeconomic Policy in a Dual Economy: The Case of Mexico." Ph.D. diss., Massachusetts Institute of Technology.

Scarf H., and J. Shoven, eds. 1983. *Applied General Equilibrium Modelling.* Cambridge: Cambridge University Press.

Serra-Puche, J. 1983. "A General Equilibrium Model of the Mexican Economy." In H. Scarf and J. Shoven, eds., *Applied General Equilibrium Modelling.* Cambridge: Cambridge University Press.

Shejtman, A. 1981. *Economía campesina y agricultura empresarial: Tipología de productores del agro mexicano.* Mexico City: CEPAL.

SPP. 1981a. *Sistema de cuentas nacionales.* Vols. 1, 7. Mexico City: Coordinación General del Sistema Nacional de Información.

_____.1981b. *Encuesta nacional de ingresos y gastos de los hogares: 1977.* Mexico City: Coordinación General del Sistema Nacional de Información.

Taylor, L. 1979. *Macromodels for Developing Countries.* New York: McGraw-Hill.

_____. 1983. *Structuralist Macroeconomics.* New York: Basic Books.

Taylor, L., and F. Lysy. 1979. "Vanishing Income Redistributions: Keynesian Clues about Model Surprises in the Short Run." *Journal of Development Economics* No. 6.

PART V

THE FUTURE

CHAPTER 15

SAM's Successor: PRONAL

Although SAM ended with the administration that had set it in motion, the Mexican government's interest in food policy remained strong. The present administration very early developed the new National Food Program, PRONAL, which was designed to grapple with the same problems of agricultural production, distribution, and consumption that SAM had faced.

In this three-part chapter we present the thinking of current Mexican policy makers on their country's food problems, from perspectives of both strategy and implementation. The first part presents the official summary of Mexico's new food program, with an introduction by President Miguel de la Madrid. The second part, written by Maria de los Angeles Moreno, is an analysis of the background against which the new policy was formulated and a discussion of its key strategic thrust. In the third section, Clara Jusidman de Bialotovsky examines some of the obstacles to even defining, and then actually implementing, a food policy in Mexico, and how PRONAL is addressing those problems.

The National Food Program, 1983–1988

Introduction, by Miguel de la Madrid Hurtado, Constitutional President of the Mexican United States

The National Development Plan for 1983–1988 sets forth as a major task for Mexican society the satisfaction of the basic needs of the popula-

Translation of official government document: Introduction by President Miguel de la Madrid, Summary Description of PRONAL by the executive branch, October 17, 1983; minor format adjustments were made in the translated version for the sake of editorial consistency.

315

tion and the permanent enhancement of welfare. Food unquestionably plays a significant role in this task. It is the starting point for satisfying basic needs.

Reaching the food objectives is not an easy task. It demands profound changes in production, distribution, and consumption patterns. It requires the application of an orderly and systematic effort, recognizing clear priorities and paying attention to precise strategies. The National Food Program, whose formulation is an explicit commitment of the National Development Plan, responds to the requirements.

By making specific the features incorporated in the Plan's social policy, the National Food Program ensures the achievement of food-sovereignty objectives and the attainment of food and nutrition conditions that allow the full development of the capacity and potential of every Mexican.

Under the concept of food sovereignty, the nation retrieves and retains exclusive control of decisions regarding the satisfaction of the population's basic food needs. The exercise of food self-determination refers to consumption as well as to production and distribution standards and includes the required technologies for achieving the goal.

Within the framework of the national aim of maintaining and enhancing our national independence, the achievement of an independent food chain, almost invulnerable to risks and critical situations, both internal and external, is a fundamental requirement for the construction of a society that guarantees, under the principles of a state under law, individual and collective liberties in a full democratic system and under conditions of social justice.

Furthermore, the path to an egalitarian society goes through a stage of overcoming poverty conditions and, particularly, the malnutrition still affecting important groups of our society.

If in situations of rapid economic growth it is necessary to ensure the most efficient use of our resources, which are always scarce compared to the size of our needs, this efficiency becomes indispensable in crisis circumstances, such as the current conditions.

The Program's policies meet these demands by clearly setting priorities for the groups and areas to be served. Toward the same end, the producers and foodstuffs that will receive the preferential support of the state are precisely defined.

The operation of the Program must ensure that the various actions of the public sector contained in it will be directed with greater certainty and consistency toward the established priorities.

The National Food Program is a first step in the programmatic integration of public actions connected with the food chain, for the purpose of introducing more order and better utilization of resources.

In the framework of the mandatory aspect of the National Planning System, and in light of the proposition of rearranging the economic and social policy instruments, the joining and balancing of the efforts of the participating institutions is pursued, without superimposing new structures.

The creation of the National Food Commission covers the need for a permanent forum of communication and coordination of the institutions directly responsible for the implementation of policies and the application of actions.

The formulation of the program is the result of a cooperative effort made by an intersecretarial group that integrated and joined the proposals of the administrative sectors connected with the so-called food chain.

The concerns and suggestions of several social, political, and professional groups during the campaign for the election of the president were also incorporated. Likewise, an effort was made to capture the ample and varied experience available in the country regarding the analysis of nutrition problems and the formulation of food strategies and policies.

Because of its strategic character and its contents, implying involvement of various sectors in its implementation, the National Food Program is defined as a Special Program, the guidelines for which shall be mandatory for the different levels of the federal public sector. In turn, the consistent application of the policies provided will stimulate the participation of the social and private sectors in the direction required for reaching the previously defined objectives. Coordination with state governments and harmonization with the different sectors will ensure the necessary commitment to this great undertaking and will guarantee its impact throughout the nation.

Nine chapters are included in the National Food Program. They present, as a whole, an all-embracing view of the problems affecting the food chain, and they propose solutions in order to improve its performance.

The first chapter includes the objectives and broad lines of economic and social strategy that, as stated in the National Development Plan, support the propositions of the Program.

The second chapter describes the characteristics of the nutrition and food problems, their dynamics and evolution. The domestic and international context is also analyzed, as well as its limitations and the interaction existing between the various phases of the food system.

The third chapter presents the objectives and strategy guidelines as a qualitative expression of the changes pursued in the short and medium term, under a rational approach in the use of resources mobilized by the entire society.

The fourth chapter presents the goals of the National Food Program by defining a set of priority foodstuffs and the required production to overcome critical nutrition situations. The Program also determines the target and priority population groups who will be attended in a special and hierarchical fashion in the short and medium run.

The fifth chapter sets forth specific conditions for each phase of the domestic food chain: production, processing, marketing, consumption, and nutrition; strategic guidelines are described for directing actions aimed at modifying the current situation.

The sixth chapter deals with economic and social policies as they affect the different phases of the food chain and influence the success of the defined strategies. References are made to policies on financing, taxing, pricing and subsidies, professional training and preparation, social organization, technology, and social communication.

The seventh chapter discusses product programs and reviews the progress achieved in this field. Emphasis is also placed on such matters as the structure and development of programs for each one of the goods constituting the basic list of priority foods.

The eighth chapter refers to the organization and operation of the National Food Commission and to food programming within the National Planning System, defining the coordination levels required to put the Program into practice.

The ninth chapter draws together and integrates, by strategic element and by action in the four phases, public expenditure programs included in the Federal Expenditure Budget, as well as ongoing actions. This first effort toward the programmatic integration of the public activities around the defined food strategies will enable the National Food Commission to review the consistency of actions in order to evaluate them and direct them toward a more efficient fulfillment of Program priorities.

The large challenges faced by the nation in the food area demand national unity and the organized participation of all sectors in order to overcome them.

Within the framework of the National Development Plan, the National Food Program will integrate and give force to the efforts of the social body for reaching one of the priority objectives of development: the achievement of sufficient food fairly distributed to all Mexicans.

Official Summary Description of PRONAL

The Importance of Food in the National Development Strategy

Mexico currently faces the most serious economic crisis in its modern history, within the framework of a world situation of transition, insecurity, and uncertainty.

The international recession is reaching its most acute manifestations in widespread unemployment, the reduced flow of goods between countries, the high levels of foreign debt, the drastic decrease and, in some nations, reversal in the evolution of domestic production, the persistence of acute inflationary processes, and, as a consequence of all the foregoing circumstances, the rise of serious social conflicts.

The impact of the international situation, together with the vulnerability of the domestic economic system because of structural limitations, the persistence of social and economic imbalances and inequalities, the lack of integration of productive processes, and the scarcity of resources for financing growth, explain the difficult set of circumstances that the nation must face.

An acknowledgment of the size of problems has been the starting point for the current administration. The fundamental guidelines for overcoming them consitute an essential part of the National Development Plan 1983–1988.

According to the Plan, the current and future challenge is how to recover the growth capability of the country on a basis of greater social equality, legal security, consistency in development advances, and efficiency in the use of resources, while at the same time qualitative changes are introduced in the production and distribution network in order to secure firmly the independence of the nation as well as its political, social, and cultural democratization.

The strategies for the reorientation and modernization of the production and the distribution networks attempt to promote an industrial sector internally integrated and competitive abroad as well as an agriculture and livestock sector with improved standards of living and social participation in rural areas, which ensures basic staples for the population and provides an increasingly modern and functional service sector adequate to the needs of production and consumption.

The conditions and levels of nutrition are closely linked with the levels of family income, which, in turn, are conditioned by the access to production resources or the possibilites of obtaining productive and well-paid employment.

The goal is to generate employment at an annual rate between 3.5% and 4% starting in 1985. While this goal is being reached, the state is committed to maintaining special programs for the protection and direct creation of work positions in order to protect the income and consumption levels of the most unprotected strata of our population.

In the same direction, the Plan attempts to increase the relative income participation of the lower-income families that are a majority in the rural environment, as well as to favor an improved distribution of income by giving special attention to the urban working population.

The purposes of income redistribution mentioned above are a fundamental premise of the National Food Program. It is not sufficient, obviously, to reach an adequate production of food if at the same time the purchasing power of people is not supported.

Due to the multisectoral nature of the food question and the impact on it of the various economic and social policies, the elements related to the satisfaction of the nutrition and food needs of the population constitute one of the guiding threads of the National Development Plan.

An Analysis of the Food Situation

The food phenomenon is highly complex, as it involves both primary production and processing activities, distribution, marketing, and consumption. It also constitutes a dynamic and interdependent process in its phases, and it has different spatial expressions. A large number of production units, with heterogeneous characteristics, participate in the production and distribution of food. Likewise, food habits and requirements are different according to the ethnic, cultural, sex, age, and income particularities of the population.

The International Framework. The food situation of the world has become a central topic for discussion and analysis, the formulation of initiatives, as well as confrontations derived from conflicting interests. The call for international cooperation in order to eliminate hunger and malnutrition has been a fundamental concern of all international bodies during the current decade. Results, however, have been minimal and the situation is tending toward deterioration. In practical terms, we see the imposition of criteria derived from an international order that by its very nature is obviously unable to satisfy, at a minimum, the vital needs of the large majority of the world population.

Food deficiencies range from simple caloric deficit to critical levels of malnutrition resulting in their extreme in death. Ten percent of the world population, approximately 500 million people, barely survive, and 1.36 billion, 27% of the total, do not meet their nutrition needs. The number of starving persons in the world has been calculated to have grown by 50 million between 1977 and 1980.

The current international economic order favors big economic powers, and consequently the problem of starvation is increasingly concentrated in poor countries and, still more alarming, in the poorest area of the latter, namely the rural areas.

It is possible to state, in general terms, that the so-called food crisis does not represent a transitory set of circumstances, but rather a structural phenomenon made manifest by situations of false scarcity, strong variability in international prices, and a clear control of food price quotations by countries having large reserves.

The growing external food dependency in several developing countries, both for foodstuffs and inputs, means that, in critical situations of foreign-exchange scarcity such as the present one, the availability of foodstuffs for marginal urban and rural populations deteriorates still further. Another contributing factor is the strong restriction on financing for the food-production sectors.

Domestic Consumption of Food. The determinants of the nation's nutritional situation are the unequal income distribution, the reduction of growth in the national production of basic staples, the deviation toward nonfood purposes, and deficient internal distribution. We cannot also help but recognize the impact of demographic, cultural, and educational factors on the behavior of consumption. Although malnutrition affects large groups of society, it should be stressed that

it is twice as frequent in rural areas on the average;

its prevalence and acuteness is greater in the south, center, and southeast of our country;

it especially affects children, pregnant women, nursing mothers, and old people.

Of the 2 million babies born every year in this country, 100,00 die in their early years of life due to factors related to malnutrition, and 1 million babies survive with physical or mental handicaps due to food deficiencies.

According to the results of surveys made in 1979 by the National Nutrition Institute, 19 million people suffered serious deficits in their calorie and protein intakes; of these, 13 million lived in rural areas and 6 million in urban centers. In the former, approximately 6 million were less than 14 years old, and in the latter, 2.7 million.

As a percentage of total population, 18% are fed varied, rich, sufficient, and almost always balanced diets, though some engage in excessive and imbalanced diet fads.

In recent years, a diet emphasizing the consumption of animal protein has spread, together with the accelerated growth of livestock breeding. Similarly, the influence of certain advertisements has favored the substitution of processed products with scarce or zero value for traditional foods with high nutritional contents.

Although the average per capita consumption of calories and protein is not far from recommended levels, the poor distribution of food among various population strata is the reason why approximately 40% do not cover their minimum necessary requirements.

The recent tendency in food prices has contributed to a deterioration of consumption levels. In current circumstances, with high inflation rates and decrease of paid jobs, the impaired purchasing power of low-income groups brings about a degradation of their food consumption.

One fact that is important in the food perspective of the nation is that in 1988 the population of Mexico will reach 83 million people, which is an 8 million increase with respect to 1983.

The Availability of Food. After 1940, rapid urbanization and income distribution and growth influenced the evolution and orientation of the production network in regard to food consumption.

Regional consumption habits were merged into a more homogeneous pattern, while a food-processing industry was created that, in turn, enhanced and disseminated a new scheme of consumption patterns, strongly influenced by models developed abroad.

Preferential attention and speed in the development of certain industrial production activities meant a relative retardation in other areas of production activities and a growing gap between producers and consumers.

The intervals between primary producers, industrial producers, and consumers were covered, in most cases, by a large number of middlemen. The result was the rise of a disjointed, insufficient food chain, squandering resources and production, dependent on inputs and equipment, and essentially oriented to the satisfaction of an urban consumption pattern.

The Primary Production of Food. For many years government actions were focused on areas with high production potential, leading to the promotion of irrigated farming. Products designed for export markets, for livestock breeding, and for industrial processing were favored, displacing the production of maize and beans to rainfall-dependent areas.

This consolidated the rise of a modern farming sector, to the detriment of a large number of peasants made up of small landowners, *ejido* members, tenants, and settlers, as well as a growing group of landless salaried workers.

Up to the middle of the 1960s, the production growth of staples amply covered effective domestic demand and permitted the export of surpluses. The significant decrease that was, however, felt in public investment in rural areas from the late 1950s to the early 1960s, the unequal exchange relationships between the rural and the urban industrial sectors, and the influence of agrofood transnational corporations in favor of intensive livestock production led to a drop in the growth rate of food production, mostly since 1965.

Regulation of the foodstuff markets and a policy of price controls have kept the farmers' revenues below the growth rate in the remaining areas of the economy. Such a policy has clearly fostered a resource transfer from rural to urban areas.

Moreover, rural populations have been disadvantaged in the supply of manufactured goods, inputs, and equipment as well as consumption goods and services.

In this context, harvested areas, crop mix, and the average yield per acre underwent important changes, all detrimental to food production.

The latter phemonenon was enhanced by the influence of agroindustrial corporations, most of them transnationals. Through financing, input supply, technical assistance, and agile procurement mechanisms, maize and wheat were shifted to fodder uses, and sorghum and soybean production, among others, were expanded and linked to intensive production of hogs, poultry, and dairy cattle.

Livestock production, as any other primary sector, has witnessed the decrease of its participation in the gross domestic product from 5.3% in 1960 to only 0.2% in 1980. Its growth, however, was the highest in the whole sector. This situation may be explained as a consequence of technological progress and the growing demand for animal protein from certain population strata.

There is a noticeable insufficiency and lag in the generation of technological innovations, of equipment and inputs adequate to the needs of small landowners and dryland producers. As a consequence, a vast sector of primary food production is seen as suffering serious technological retardation, which causes a low productivity level.

In regard to fisheries, although the last decade was characterized by rapid growth, the exploitation of fishing species was more oriented to industrial uses. Capture for human consumption does not encompass many varieties and is basically restricted to high-priced species designed for export and high-income urban markets.

The Food Processing Industry. As to the composition and evolution of the food-processing industry, we may state that the industry is characterized by a high level of heterogeneity in regard to ownership, size, and productivity of facilities, jobs generated, and types and levels of production. There are many family-type operations, equal to 70% of the total but contributing only 2% of the production of the industry, while a small number of large corporations, 1%, generates 65% of the total production.

Generally speaking, the high dynamism observed up to 1970 in the output of the food-processing industry should be recognized. But its evolutionary pattern favored the displacement of small facilities, reduced the capability for the creation of new jobs, and fostered a high degree of concentration.

In recent years, the rise of a small group of large corporations, controlling certain production lines, has subjected primary producers to its requirements and altered the composition of demand.

Dependence from Abroad. As a result of the decrease in the dynamism of domestic food production, the loss of control of the process, and the disjointedness among its phases, as well as the changes in demand composition, growing amounts have been imported in recent years, not only of grains, oilseeds, and dairy products, but also of inputs and machinery.

This commonly observed trend is alarming, as the volume of imports has grown at an accelerated pace. While in the 1965–1969 five-year period 283,000 tons of basic grains, oilseeds, and sorghum were imported, in the 1980–1982 three-year period over 20 million tons of the above products were imported.

On the other hand, milk and egg imports in the period 1972–1975 were, respectively, 0.3 billion liters and 170 tons. The estimated imports for the two-year period 1980–1981 are 7.5 billion liters of milk and 18.4 million tons of eggs.

There is also an enormous dependence regarding inputs, equipment, and, more seriously, seeds and genetic lines.

The Commercial and Distribution Apparatus. The commercial and distribution apparatus of foodstuffs is characterized by the existence of a network of middlemen and transporters who take a significant proportion of the final value of the products handled by them, depressing prices obtained by the primary producers and increasing them for the consumers.

Moreover, deficiencies in the facilities and services for adequate reception, conditioning, warehousing, transportation, distribution, and marketing of food are currently the cause of wastage amounting, as an average, to 10% of grain and cereal crops, 30% of fruit and vegetables, and 50% of fish and seafood.

The official distribution system currently represents approximately 24% of the domestic total, although its share is sometimes larger in certain areas and products, particularly grains and marginal areas.

Facing the drop in domestic production and in order to ensure food supplies, on one hand, and raw material supplies for agroindustry on the other hand, the domestic storage actions of the state have been progressively replaced by import purchases and distribution.

Limitations faced in regard to warehouses are their scarce number and poor distribution throughout the national territory, the lack of specialized terminals for the rapid and efficient handling of growing volumes of grain imports, and technical limitations in several available facilities.

Beyond the physical loss of product volumes brought about by such causes, significant wastage is generated in the nutritional quality of these, due to degradation in the original nutrient content.

Concerning the transportation of supplies, transporters of food and livestock products are highly fragmented. Unorganized participation amounts to 23,000 truckers. Their inefficient operation causes a 10 to 1 ratio between the cost of transportation of food supplies and the cost of local transportation.

The dispersion of the population over a large number of villages, together with its excessive concentration in large urban centers, gives rise to great difficulties for food distribution systems. In recent years this situation, together with the increase in grain imports, contributed to serious bottlenecks in the national transportation network.

Besides, more than 7 million people living in remote villages are not served by any permanent form of land communication.

Food distribution to the consumer markets is also characterized by an excessive number of middlemen and high profit margins for merchants.

Modernization and increased efficiency in trade activities has mostly taken place in the urban environment through a system of large stores and commercial chains against which small retailers can hardly compete. These large commercial chains cater to the higher-income groups and charge the lowest prices in the market because of their operational volume and efficiency level.

Small retailers still play, however, a significant role in rural areas, in small towns, and in marginal areas of the large urban centers. But their dependence on wholesale traders, the low volume of their transactions, and the high cost of money considerably increase their overhead so that, paradoxically, trade serving the lower-income population turns out to be the most expensive.

Support Policies for the Food Chain. Although public contribution to the development of the food chain has been significant, one cannot help but acknowledge that some of the problems currently faced were enhanced by the scattered and sometimes limited scope of the respective public actions. The sectoral functioning of offices and agencies empowered with handling the policies connected with the food sector, as opposed to the intersectoral character of the food issue, has been the main cause of the lack of an integrated handling of public actions.

We should add the lack of complementarity between the various public policies, although in recent years great efforts were made to operate as integrated packages some of the government support measures aimed at agricultural, livestock, and industrial producers.

The lack of integration and complementarity in public actions affecting the food chain becomes more evident in regional operations, reflecting, in addition, the enormous gap between decisions made at the highest level and the actual implementation of the instruments.

The Objectives and Strategy Features

The objectives of the National Food Plan constitute the qualitative expression of the changes or transformations in food and nutrition matters to be achieved in the 1983–1988 period. The attainment of such objectives will be made possible only through actions following the strategic guidelines presented in the National Development Plan and in the specific features of the Program.

The National Food Plan is aimed at the following general objectives:
to strive for food sovereignty;
to reach food and nutrition conditions allowing the full development of each Mexican's skills and capabilities.

Food sovereignty as a fundamental objective of the National Food Plan finds its place in the revolutionary tradition of preserving and safeguarding for the nation decisions regarding every significant aspect of the welfare, freedom, and security of Mexicans.

Food self-determination, a fundamental aspect within the concept of national sovereignty, must be pursued in consumption standards as well as in production and distribution standards. It includes the free and sovereign election of the components of this pattern. It implies, similarly, an autonomous capacity to guarantee the satisfaction of the minimum actual requirements of the whole population with our own resources. It also supposes technological self-determination.

Food sovereignty involves ensuring access of the whole population to foodstuffs enabling them to reach full development, both physical and mental. It is not enough to produce sufficient amounts of staples; it is also necessary to intervene in the handling and distribution processes, removing current deviations and wastes.

The National Food Plan is based on fostering the domestic production of food, especially in rainfall-dependent areas, without neglecting irrigation areas, as well as the productivity increase of the food-processing industry and its efficiency in marketing.

An element of paramount importance in the concept of food sovereignty is the possibility of deciding and the obligation to define in what foodstuffs we wish and must be self-sufficient, without concessions, and in what other terms it is impossible to be self-sufficient within a certain time horizon.

The specific objectives of the Program reflect the commitments assumed by the current administration in matters of food.

In this sense, the Program shall be preferentially devoted to improve the nutritional levels of low-income populations, as well as those in particularly affected areas and social groups. It will also be directed to favoring a more equitable distribution of foods among economic strata, social groups, and areas.

For the immediate future the purpose is to secure the current level of food and nutrition for the majority and, facing the present crisis, to protect the income of our population.

For the medium term, the objectives of the Program stress attention to the most nutritionally vulnerable groups. Regarding this purpose, the emphasis is placed on poor families, preschoolers, pregnant and nursing mothers, as well as the residents in the south and southeast areas of the territory.

To decrease the waste and the exaggerated consumption of certain population strata and to distribute better the available foodstuffs are the basic purposes of any food strategy.

The objectives of the Program point toward a change in the functional and economic relationship between farming, livestock, and fishing activities and all other sectors and aim to close the gap between the welfare and participation levels of the rural population and the urban population.

Complementary objectives of the Program consist of linking the agroindustrial productive structure to the demands of social consumption among the majority sectors and multiplying productive and well-paid work positions, particularly in the rural environments.

The development of an autonomous technological pattern, in keeping with our domestic needs, must contribute to the attainment of all these goals.

The consistent integration of such actions requires the definition of strategic channels in order to achieve better and fairer nutritional situations. In this sense, the strategy guidelines of the program are as follows:

The various phases of the food process—production, processing, marketing, and consumption—shall be considered in an integrated and dynamic fashion, in order to enhance efficiency all along the chain, preventing and solving imbalances in its operation, while at the same time encouraging participation by all primary producers, in keeping with the strategy of integrated rural development.

The policies and actions of the Program shall be operated with selective and specific criteria adequate to the characteristics and needs of the target population.

The definition of the policies and actions of the Program shall be basically oriented toward the care and promotion of small- and medium-sized productive units involved in the various phases of the process; it must take into account the producer-consumer binomial, which is typical of all the participants in the food chain.

The territorial deconcentration of the productive agrofood apparatus and particularly of the industrial and commercial network shall be supported, in the process of more efficient local integration of productive capabilities in order to satisfy regional needs.

Regional food-production systems will be strengthened, seeking local self-sufficiency in basic products and a more equitable space and social distribution of foodstuffs.

The conservation and adequate use of natural resources will be secured, avoiding the achievement of extensive quick results at the cost of their exhaustion.

Public expenditure will be rationalized, while the programs and budget resources designed for activities of the food process shall be clearly identified and submitted to a strict order of priorities.

Consistency and coherence shall be brought to all policy instruments involved in the food process in order to guarantee that their effects are compatible and correspond to the general and specific objectives of the Program.

Short- and Medium-Term Requirements

At the present time a great productive effort is indispensable for the protection of the levels of food and nutrition that have already been achieved by important groups of the population. At the same time, the significant problems of marginality and malnutrition require structural changes in general patterns of consumption that will bring about greater balance between resource utilization and food needs in addition to a more equitable distribution of food. This will compensate for the effects of the crisis and open superior possibilities for future social development.

The quantities of the principal foods set out below that will be required by Mexican society as a whole from 1983 to 1988 were determined in accordance with the economic conditions estimated as prevailing during that period.

Because the foreseeable evolution of effective food demand relative to food requirements will not satisfy the minimum nutritional needs of a broad sector of the population, quantification was made of the additional production efforts required to achieve the objective of assuring sufficient and accessible food for the entire population and, particularly, for the Target Population and the Preferential Population, as defined by the Program.

On the basis of analysis and interpretation of the composition levels of total demand, the National Food Program proposes orientation of said requirements in accordance with the needs of the population in order to fit a socially more adequate consumption pattern to Mexico's productive possibilities.

Human Consumption Requirements. As a result of the contraction of economic activities, effective demand for the majority of food products could experience a quite limited, or negative, rate of growth for the short term. This possible general reduction in demand would result from a decrease in the consumption of broad sectors of the population whose income has been most affected by the present situation.

Reactivation of the whole of economic activity is foreseen for 1985 and beyond, and, therefore, it is possible that development of effective demand for food destined for human consumption will become positive again and, likewise, that some products during certain years will achieve rates of growth superior to the rate of growth of the population. Nevertheless, taking the entire 1982 to 1988 period as a reference, the expected growth is small (see Table 15.1).

With regard to the energy contribution of food, the minimum recommended by the National Nutrition Institute is 2300 calories per person per day. This figure represents an average through which it is sought to cover the energy requirements of the entire population, which varies according to age, weight, and activity. With respect to protein, the minimum recommended is 60 grams per person per day.

The estimated energy consumption of high- and medium-income strata of the population is much greater than the recommended mean requirements.

Table 15.1. Expected effective demand for food for human consumption in 1984 and 1988
(thousands of tons)

Product	1984		1988	
	Maximum	Minimum	Maximum	Minimum
Maize	9,080	8,744	9,415	9,217
Beans	1,135	1,130	1,230	1,205
Wheat	3,065	2,722	3,532	3,295
Rice	660	594	735	692
Sugar	3,319	3,102	3,729	3,529
Vegetable oil	665	598	759	717
Whole milk[a]	8,810	7,869	10,395	9,511
Eggs	795	692	936	858
Beef	1,040	941	1,227	1,128
Pork	710	656	855	777
Poultry	490	437	585	533
Fish	982	835	1,520	1,290

[a]Millions of liters.

The fundamental problem is presented by the population located in the low-income strata, in which repercussions of the effects of the economic crisis could result in a substantial reduction in the caloric intake for the short term. This average reduction could reach almost 18% in the population stratum dependent on agricultural activities and 10% among the population dependent on nonagricultural activities between 1982 and 1984. Forecasts indicate that this intake will thereafter gradually increase until in 1988 it reaches a level similar to, but still less than, the 1982 level.

With regard to protein consumption, the figures indicate that the low-income population consumes barely more than one-half of the intake of medium- and high-income population in the agricultural as well as in the nonagricultural environment.

According to the projection, the decline in protein consumption could become quite serious between 1982 and 1984, changing after 1984 until it reaches a level similar to 1982's average consumption, in 1988.

On the average, the drop in protein intake would be 15%; however, the drop in the consumption of animal-origin protein would be greater than 50%, still further aggravating the alimentary situation of the low-income population, because the drop in protein consumption could become significant not only quantitatively but also qualitatively.

The failure to take adequate emergency measures could result in the severe deterioration of the food situation of this population. Prompt attention to this situation is a necessary condition for attaining better levels of well-being and making more balanced and sustained social and economic development possible in the future.

The Target Population and Its Requirements. The group defined as the *Target Population* in the National Food Program is the low-income population representing 40% of the entire population. Its magnitude will be on the order of 30 million persons in 1984 and 33 million by 1988.

Preference will be given to the most vulnerable groups within the Target Population, which are made up of preschool children and pregnant and nursing women. It is estimated that this population will consist of 6.7 million persons in 1984 and of 6.3 million by 1988. It constitutes the *Preferential Population* of the National Food Program.

The advisability of centering greatest attention on a reduced number of *Priority Foods* and basically defining the quality and quantity of the basic intake, particularly of low-income families, is also considered by the Program.

For the purposes of this Program the following products and their derivatives are considered Priority Foods: maize, wheat, beans, rice, sugar, vegetable oils and fats, poultry, eggs, milk, and fish. Within this framework the state will concentrate its effort by product line for the benefit of the Target Population and the Preferential Population, taking into consideration regions, states, and municipalities requiring special attention. The most affected municipalities are concentrated in the states of Oaxaca, Chiapas, Guerrero, Hidalgo, Puebla, Tlaxcala, San Luis Potosí, Querétaro, Tabasco, Yucatán, México, Guanajuato, Michoacán, Zacatecas, Morelos, and Jalisco.

The actions of the state will be specifically directed in accordance with these definitions; agreements will be reached with the private and social sectors in order to guarantee the timely and sufficient supply of these Priority Foods.

Necessary Foods are other types of meat, vegetables, tubers, and fruits to which attention will be paid by the National Food Program for supervision of present and future supplies within a plan reorienting the population's consumption patterns in the short term toward assuring efficient use of resources and suitable levels of nutrition within a framework of food.

Priority and Necessary Foods have been defined in the terms of natural products from which, in certain cases, a series of processed foods are generated. The Basic Popular Consumption Package includes the range of these processed foods, whose production and distribution are the objects of the public support and incentives established by the Immediate Economic Reordering Program for the protection of the consumption levels of the medium-income sectors.

Additional Requirements. In order to protect the Target Population's consumption, an estimate has been made of the quantity of each of the Priority Foods that must be brought within its reach to avoid the drastic reduction in the intake that would be forced by the drop in its purchasing power.

Table 15.2. Per capita Priority Food requirements for the target population 1984 and 1988 (kilograms per person per year)

Product	Effective demand			With additional contribution	
	1982	1984	1988	1984	1988
Maize	124.4	134.8	124.5	136.8	130.1
Wheat	26.6	16.5	25.1	36.1	35.6
Beans	15.3	14.9	15.1	16.1	16.3
Rice	6.0	4.1	5.7	6.1	—
Sugar	29.3	22.1	28.1	27.1	29.6
Vegetable oil and grease	5.5	3.4	5.1	6.8	—
Whole milk	58.7	27.3	53.3	57.7	66.6
Eggs	5.8	2.8	5.3	6.0	8.0
Poultry	2.4	0.7	2.2	2.5	—
Fish	2.9	1.3	2.7	4.3	5.7

An estimate of additional requirements has been made, including the minimum necessary quantities of those foods for which consumption by the Target Population was deficient even prior to the beginning of this final stage of the crisis. If the quantities of the additional requirements are indeed very high for 1984, the number of products involved decreases in 1988 in accordance with the substantial growth of effective demand resulting from the recovery of the domestic economy and the structural changes foreseen for the period 1985–1988 (see Table 15.2).

It must be emphasized that, if a part of the state's efforts in favor of the Target Population does indeed reside in facilitating its access to these conditional requirements of each product (a matter that basically corresponds to the spheres of distribution and consumption), from the production point of view it is necessary to guarantee sufficient supply of Priority Food destined to the satisfaction of not only the expected effective demand but also the additional requirements.

The high-income population—the source of powerful effective demand and, therefore, of free market movement—will have the capability to induce the diversion of natural resources for the production of food to satisfy its needs; rationalization of this phenomenon is required to prevent the satisfaction of the excessive consumption of a few being translated into scarcities in the consumption of the many.

Food Requirement Implications. Although Food Program goals solely refer to requirements originated by human consumption needs, an estimate was made of the total foreseeable domestic demand for the 1983–1988 period to support the projections of the total availability of food products required.

In order to calculate total domestic demand, one must add to the estimated volumes of effective demand and additional consumption proposed for purposes of human consumption and for animal and industrial consumption requirements, handling losses, seed needs, and the requirements of a technical reserve to provide security for the population's consumption. In thousands of tons, the estimated figures of the total domestic demand for maize, beans, wheat, and rice for 1988 are, respectively, 16230, 1380, 4625, and 780.

Therefore, the strategy of the National Food Program will guide a gradual restructuring of the present consumption pattern not only to achieve food adequacy and diversity but also to gradually reverse the trends of the productive structure to bring about a superior and more efficient use of the country's resources.

Phases of Strategic Outline

The Production Phase. The solution of the nutritional problems affecting the Program's Target Populations necessitates that priority be given to the domestic production of grain and agricultural products for human food. Therefore, the Program proposes that greater attention be paid to a Food Priority Table that will take into consideration nutritional balances in agreement with food habits, the population's income levels, product purchase prices, and Mexico's potential resources, seeking complementarity among the various activities.

It is proposed by the Program to give priority to the population's nutritional requirements by the domestic production of food with Mexico's own resources and technology, by emphasizing the production of foods of vegetable origin, particularly grain, by assignment of concentrated support to rainfall zones, and by reorientation of production in irrigated areas.

The medium-term objective is to increase the supply of primary products, which form a part of the Priority Food Table, within the framework of improving the producers' standard of living and greater social and economic participation, thereby strengthening the domestic food system.

The short-term objective is protection of the productive capacity of and employment within the primary sector, in order to maintain the levels of food production that have been reached and the income of farm, livestock, and fishing producers.

The strategy features are

reordering of production incentives and support in order to increase their effectiveness, particularly for the production of the food included within the Priority Food Table;

an increase in farm, livestock, and fishing productive capacity;

an increase in the productivity of the resources in farm, livestock, and fishing production, by their utilization according to their occupational, domestic, and social needs;

the grant of legal security to landholders;

strengthening producers' organizations;

integration of consumption, transformation, and marketing phases.

The Transformation Phase. Through conservation systems, the industry can assure the availability of seasonal products during the entire year, facilitate their movement to consumption centers, make possible the utilization of certain products for food purposes, and facilitate the introduction of new products into diets and the addition of enrichers into natural foods.

In addition to action in the mandatory flow of the National Planning System, the National Food Program considers of supreme importance the handling of regulatory and development policies in the induced flow of private planning, the reconciliation of the social and private sectors, and coordination with state governments. All of this will be accomplished through the corresponding sectoral programs.

Medium-term objectives comprise a selective increase in the production of processed foods derived from the Priority Food Table for the purpose of making the consumption of primary foods more stable and accessible and to overcome the seasonal characteristics of production processes; making agroindustrial productive structure conform to the social-consumption demands of majority sectors; utilizing agroindustry as a rational and efficient axis of integration of each product in the food chain; achieving greater fluidity of goods; and reducing the intermediary links between primary producers and the consumer—all of which will have a consequent effect on food availability and prices.

The specific objective is to raise the income levels and improve the working conditions of the population engaged in agroindustrial activities.

For the short term the purpose is to maintain, at least, the production of the Basic Popular Consumption Package, which is defined by the Immediate Economic Reordering Program, as a framework for coordination of the actions of the private and social sectors.

The strategic central outline of this phase consists in making the food chain operate more efficiently from the primary producer to the final consumer, through organization of agroindustrial activities in such a manner as to integrate the links in the chain of the operational plan and through coordination of the backup actions of the public agencies involved.

The main strategic thrusts for this phase are to be found on two planes: economic reordering and structural change.

The Program's short-term strategy, a product of the Immediate Economic Reordering Program, consists in developing the production, the supply, and the control of the Basic Popular Consumption Package.

The National Food Program establishes the following strategic elements of structural change in the agro-alimentary industry:

reorganization of state participation in the food industry in accordance with social priorities;

recovery and preservation of domestic investment in the basic and strategic food industry;

reorientation of the food industry toward the production of priority foods;

the development and strengthening of integrated food agroindustry through the participation of primary producers;

domestic integration of the food industry and a reduction of its dependence on foreign technology, inputs, and equipment;

promotion of the nutritional enrichment of staples and better utilization of foods having a high nutritional value.

Marketing Phase. In the food process, marketing is the link among the consumption, production, and transformation phases. It encompasses the estimation of input requirements for primary production, the assembling of output, that is, collection, storage, and transportation, which brings it to industry and the consumer.

The objectives for the medium term are to assure timely supply of the products included in the Priority and Necessary Food Table at prices within the reach of the majority of the population, and to increase their availability and quality through a substantial reduction in losses and the retention of their nutritional characteristics. It is sought, in addition, to foster development and modernization of the system for the collection and supply of food to such a degree that it will be organically linked to the other phases of the food chain and will provide clear guidelines for primary production.

In the short run it seeks to assure timely supply of the Priority Foods at suitable prices and quality, especially of the Program's Target Population and the regions most affected by the loss of income due to a drop in employment levels.

The following are the principal strategic elements for achieving the marketing objectives of the National Food Program:

modernization of the basic food market collection system and the protection of producers' income;

integration of the transportation infrastructure and its service in accordance with the dispersion of the producers and the location and size of markets;

modernization of the final distribution system to guarantee the availability of basic products to the consumer at just prices within their reach;

the supply of priority foods in the zones most affected by nutritional deficiencies;

an overhauling of product subsidies so that they are increasingly given to the final product;

orientation of technological research toward the origination of new procedures for food conservation and handling.

Consumption and Nutritional Phase. Consumption is the phase in the chain permitting the conversion of farm, livestock, and industrial products into food that provides the nutrients required for living a healthy and normal life.

Consumption is conditional on three main factors: the actual availability of food in time and space, prevailing food patterns, and the possibility of effective access to the products, whether through purchasing power or production for self-consumption. These factors lead the National Food Program to stress the necessity of raising income levels and reducing inequalities in distribution.

One of the Program's strategies points out the need for consideration of the producer-consumer binomial in the definition of policies, to such an extent that their participation in the production and distribution of food generates sufficient income for improvement of their food-consumption levels and of their lives in general.

The quantitative and qualitative improvement of the diet of population groups affected by nutritional deficiencies is the principal objective of the consumption policies of the National Food Program. It is also sought to redirect and diversify the food habits of the population toward suitable social and nutritional patterns.

A short-term objective of the Program, supported by food marketing and distribution actions within the framework of the Immediate Economic Reordering Program, is the protection of the adequacy of caloric and protein intake by the poorer levels of the population whose income has been affected by the present economic situation.

The nature of the consumption phenomenon, which belongs to the private sphere of individual and family-group decisions, is a determinant in the location of the state's strategies, policies, and actions as they relate to behavioral inducement.

In its assistance to groups and regions facing critical malnutrition situations, the state is actively participating in the direct or subsidized distribution of food rations.

In accordance with the foregoing, the National Food Program sets out the following strategy features:

an increase in the quantity and quality of consumer information and knowledge for making decisions regarding food consumption;

support of a more balanced and just relationship between the consumer and the other agents in the food chain and between various strata of consumers and regions;

strengthening nutritional research and technological development;

support of food and nutrition monitoring systems;

the promotion and support of nutritional food consumption by vulnerable population groups.

Guidelines for Implementation of Economic and Social Policies Supporting Food Chains

To a great degree, the possibility of achievement of the objectives and goals of the National Food Program depends on the efficient coordination of the supporting policy instruments lying along the food chain within the strategic course introduced in each one of its phases. Such efficiency will require an orderly and integrated application of the supports, which will assure proper and rational use of resources, avoid distortion of the food processes, and bring about the structural changes in the food and nutritional field that society is insisting on.

In accordance with the priorities and guidelines of the National Development Plan and, where applicable, the special sectoral and regional programs, the specific programs for coordinated implementation of each one of the support policies under consideration are being grouped within the framework of the National Food Commission.

The objectives of the financing, taxation, price and subsidy, input, equipment, science, technology, organization, professional formation and training, and social-communication policies for the purposes of national food strategy are set out below.

Financing Policy. The medium-term objective of the financing policy is to assure the harmonious integration and operation of the phases of the food chain, the supply of timely, proper, and sufficient financing to the phases and adaptation of the financing operations to the actual and characteristic operating conditions of the participants.

It seeks in the short term to keep farm, livestock, and fishing operations, and the food, input, equipment, and transportation systems in operation by the issuance of sound credit, preferentially to small- and medium-sized units, in accordance with established priorities.

Taxation Policy. The medium-term aims of the taxation policy are to support an efficient interrelation of the food chain by fiscal incentives, to discourage the production and consumption of foods with low nutritional values, and to adapt instruments to producers' characteristics.

For the short term, it seeks to stimulate the production of Priority Foods and protect and support the income and purchasing power of the Program's objective population.

Price and Subsidy Policy. The medium-term objectives are to balance the factors of production and the agents taking part in the food-chain stages to encourage the establishment of more equitable exchange terms between the farming and livestock sector and the industry and service sectors; to bring about technological changes in the phases in accordance with the characteristics of the factors involved; and to rationalize subsidies in order to favor the primary producer and the final consumer.

The short-term goals are protection of the consumption of priority products in marginal rural and urban zones, discouragement of speculation and price increases, adequate compensation of primary producers, and protection of consumers' purchasing power.

Input and Equipment Policy. For the medium term, the objectives are to guarantee the timely and adequate access of the producers of *Priority Foods* to necessary inputs, machinery, and equipment, to increase production capacity in these items, and to support the development and adoption of technology in the production of equipment, intermediate goods, and machinery.

For the short term, the policy seeks to keep installed plant capacity in operation, to improve the distribution of inputs to producers, and to support the production and incorporation of domestic technological advances, which will permit the present production problems to be solved.

Science and Technology Policy. The medium-term endeavors are the development of Mexico's own scientific and technological capacity, support of the domestic economic integration of the various phases of the chain, improvement of knowledge of the potential of natural resources and their use and development for food purposes, and the application of technology for their exploitation and transformation.

For the short term, it seeks to improve technical assistance in the activities of the food chain, in order to increase production and avoid waste and to make certain that the transfer of technology is made through an evaluation, selection, and control process in accordance with domestic priorities, terms, and needs.

Organization Policy. For the medium term, it seeks to encourage organizational forms involving various activities and those of a vertical type between producers and consumers, eliminating intermediaries and strengthening the food chain, and to secure greater participation of farm and livestock or fishing producers in food transportation, marketing, and industrialization.

In the short term, the organization of farm, livestock, and fishing producers will be strengthened, the organizations of small merchants and carriers will be fostered, and consumers' organizations will be promoted in order to confront speculation and the cornering of the market in priority foods.

Professional Formation and Training Policy. The regulation and rationalization of the formation of technicians and professionals taking part in food research, production, and distribution is the medium-term goal; so also is the development of the knowledge, abilities, and skills of the participants in the chain, in order to make their participation in planning and organization and an increase in the training coverage of rural wage earners, small and medium food producers, merchants, carriers, and industrialists possible.

It seeks in the short term to encourage greater utilization of the institutional infrastructure available for professional formation and training, to adapt present programs to the characteristics and needs of their users, and to speed up and enlarge the regional coverage of the programs.

Social Communication Policy. The medium-term objectives are to foster a closer relationship between the participants in the food chain in order to reduce intermediaries, to make the role and advances of the food program known, and to inform the participants in the chain of the support and incentives offered by the state and of the way in which their participation in the process can be made more efficient.

It seeks in the short term to create an awareness in the domestic population of the importance of food priorities, to raise consumption and nutritional levels, and to make better use and conservation of society's food resources. In addition, an effort is being made to disseminate the objectives, strategies, and actions of the National Food Program, and to call on the domestic community to participate in carrying it out.

Product Programs

In order to contribute to more effective integration of the food chain, the integral process of each product in accordance with the Priority and Necessary Food Table must be considered.

' Thus, food strategy will be concentrated in programs by products, which consider completely a product from its primary production, storage, conditioning, and transformation up to its marketing and consumption. A diagnosis of the process will bring about recognition of the elements and their relationships that in each phase hinder the fluidity of the process, so that it will be possible to focus the instruments of the state and the actions of the social and private sectors on the specific resolution of the critical points by product within the framework of the general phase and support programs.

The Specific Development Program is one of the mechanisms adopted by the federal government in order to give integral treatment to the problem by products. The main instruments of industrial promotion are oriented toward protection of the purchasing power of wages for the principal items of family expenditure. The program will make an integral analysis that will include a diagnosis of the recent evolution of the product or product lines involved, the balance of supply and demand for the 1983 to 1988 period, the objectives sought, a definition and quantification of supports and incentives, the obligations of the production and distribution sectors, and mechanisms for coordination of programs and evaluation of the performance relative to obligations.

The strategy of each development program is reflected in specific actions on the distinct gradients established by the Planning Law; that is, the aspects of agreements with state governments and municipal authorities are specified for each product or product line, the desired and suitable level of direct and obligatory participation of the federal government is established, and, finally, the production and distribution needs of the social and private sectors are defined.

Because of their impact on family expenditures, some of the priority food programs are in the process of preparation.

As a part of this effort, the new Milk Production, Supply, and Control Program (1983–1988) was published in the *Federal Official Journal* on April 5, 1983.

The Institutional and Programmatic Framework

The 1983–1988 National Development Plan foresees the accomplishment of this National Food Program, the nature of which is defined by Article 26 of the Planning Law, which provides that special programs must refer to the integral development priorities of Mexico established by the Plan or to activities related to two or more coordinating agencies of the sector.

In accordance with the law, the agencies and states shall prepare annual programs that shall include the corresponding administrative and political and social economic aspects. In addition, the special program shall remain in force for a period that shall not exceed the constitutional term of the governmental administration in which it is approved. Actions subject to coordination shall be specified in agreements with state governments and in agreements with interested social groups.

In order to assure the general coherence, consistency, and coordination of the Program, the National Food Commission was established. The president of the republic will head it and the members will be the secretaries of the ministries of the Treasury and Public Credit, Planning and Budget, Commerce and Industrial Development, Agriculture and Water Resources, Health and Welfare, Agrarian Reform, and Fishing, the head of the Department of the Federal District, and the general directors of the National Popular Subsistence Company, the National System for Integral Family Development, and the National Nutrition Institute.

The Commission shall have a coordinator appointed by the president of the republic. In addition, a technical committee is established that will have representatives from each of the ministries that make up the Commission. There will also be subcommissions to deal with specific aspects.

The technical committee secretariat will be the responsibility of the Ministry of Planning and Budget.

Implicit in the creation of the Commission is the operation of new administrative agencies. It will seek to make better use of the resources, capabilities, and projects now existing in the food and nutrition fields. In addition, it shall stay in close communication with other coordinating agencies in fields related to its object, such as economic, farm and livestock, and health.

The Commission shall establish three permanent coordinating agencies in order to make the plan more flexible. Besides said general coordination, a second level will be provided for each phase of the food chain, and the third level of coordination will be related to each one of the support policies.

The work of the coordinating agencies of the second and third levels shall be performed by subcommissions.

In the following list, phase and support policy coordination (left) shall be the responsibility of the entities on the right.

Production	The Ministry of Agriculture and Water Resources
	The Ministry of Fishing
Transformation	The Ministry of Commerce and Industrial Development
Distribution and marketing	The Ministry of Commerce and Industrial Development
Consumption and nutrition	The Ministry of Health and Welfare
Financing	The Ministry of the Treasury and Public Credit
Fiscal Policy	The Ministry of the Treasury and Public Credit
Organization	The Ministry of Agriculture and Water Resources
	The Ministry of Labor and Social Welfare
Training and professional formation	The Ministry of Public Education
Input and equipment	The Ministry of Commerce and Industrial Development
Science and technology	The National Council of Science and Technology
Social Communication	The Ministry of Government

Within the Sole Development Agreement [Convenio Unico de Desarrollo], the federal executive shall propose the establishment of state food commissions in the states of the republic and within the state planning and development committees. These state food commissions shall prepare and supervise the performance of state food programs within the framework of state development plans in accordance with the outlines of the National Development Plan and the National Food Program.

The technical secretariat of the state commission shall be responsible for representation of the Ministry of Planning and Budget in the state.

On the rural scene, the farm and livestock-development districts shall be considered as the geographic coordinating units, because they make possible the establishment of organizational coordinating structures for handling federal resources at the level of programs, which carry out not only productive tasks, but also tasks related to all phases of the food chain.

Within the urban sphere, the coordinating agency can be established at the municipal or delegational level by the creation of specific mechanisms or by taking advantage of similar mechanisms already existing for similar purposes.

At the state, district, and, possibly, at the municipal or delegational level, actions will be coordinated with the private and social sectors within the gradient of coordination with National Development Plan.

Because the execution and control of the National Food Program needs the cooperation of many public and private institutions, agencies, entities, and of the general population, it is necessary to be able to rely on an integral and orderly legal system that will permit orientation and revision of the legislation that is now in force, define principles, regulate food activities and areas of participation, and avoid duplication of, and defaults, in public duties.

Accomplishment of the objectives of the National Food Program will require the establishment of legal standards that will prevent monopolies and monopsonies; that will assure a policy of guaranteed prices of, and timely and suitable credit for, the production of food and inputs; that will regulate the production and distribution of inputs; that will foster the organization and training of producers; that will guarantee the proper distribution of income among the participants in the chain; that will promote the continuation and coordination of public action; that will avoid duplication, divergence, or contradictions in the performance of the public sector related to the production, distribution, and marketing of food; and that will prevent extravagance and waste in the use of resources.

The legislative objective will be promoted by the National Food Commission among related chambers, which are basically responsible for the ordering and systematization of the judicial framework governing food activities.

Programmatic Integration by Phases for 1983

In accordance with the National Development Plan, the participation of the social welfare, communications, transportation, and rural-development sectors in total public disbursements will increase from the average 46.4% observed during the 1977–1982 period to approximately 60% between 1984 and 1988. In conformity with this structural modification of

disbursements, it is also indicated by the Plan that, because of the importance of employment, food production, and social welfare, the average investment participation of the farm and livestock sector will also be
increased.

Public disbursements are assigned in accordance with the strategies
and phases of the food chain in conformity with specific programs and
lines of action. For such purposes, annual efforts towards program integration will be required. The first effort in this respect for 1983 is set out
in the original document of the National Food Program. Included in said
document are the program, subprograms, and actions for 1983 that are
being carried out in almost fifty agencies and entities of the federal
public sector.

This integration of programs prepared in accordance with the Federal
Disbursement Budget and programs that are being carried out is a reflection of public actions in conformity with the basic principles and
strategies of the National Food Program.

Between 1984 and 1988, in accordance with the attributes of the
National Food Commission and the operational mechanisms of the
National Food Program, greater congruence will be sought between the
actions and programs of public agencies and entities and the objectives,
strategies, priorities, and phases of the National Food Program, and the
allotments of public expenditures.

In this respect the Program will furnish a clear framework of strategic
objectives and guidelines in order to channel the use of resources and the
activities of the various institutions involved. It is sought to make more
and more efficient use of the public resources and efforts that bear on the
food chain at the present time.

Strategic Thrusts of the New Policy,
by Maria de los Angeles Moreno

In this section I briefly describe the context in which Mexico's new food
policy was formulated and highlight some of its key dimensions.

Development strategy in Mexico has traditionally stressed food policy,
especially during the past few years. However, the country's present difficult economic situation has pushed it, along with other countries with a
similar degree of development, to reexamine such strategy. The crisis has
exacerbated basic social problems through higher unemployment and
decreasing purchasing power of the population. In the case of food, the
fall in purchasing power has caused a return to consumption patterns that
had long been transcended.

The food system is a complex phenomenon, cutting across all phases of the economic process, from primary production to processing to distribution and commercialization to consumption and nutrition. Furthermore, it is inevitably tied to income distribution. Since food policy is related to and directly affects all those areas, it should be placed within the framework of the overall development strategy.

For years, the agricultural production strategy—implicit within the development strategy—centered around the extension of the agricultural frontier, the incorporation of larger areas under irrigation, and the adoption of technologies to increase yield per hectare, especially in irrigated areas. Land distribution played an important role in this policy of increasing the area of cultivated rain-fed land. For a long period, these actions contributed to the fundamental role of the primary sector, which transferred a significant part of its resources to the promotion of industrial growth. However, close to twenty years ago this dynamism slowed, agricultural production began to grow less rapidly than the population, and the volumes of imports of basic grains and oilseeds increased.

In spite of indications that the agricultural policy was exhausted, inertia blocked the changes that were obviously required. Indeed, public institutions created for the countryside, the lines of technological research and development, training programs, the orientation of investment programs and public expenses and price, subsidies and financial policies for agricultural and livestock had all maintained the same basic pattern for over forty-eight years, with a few temporary changes.

There were two concerns underlying the traditional agricultural model: achievement of a level of agricultural and livestock production sufficient to supply the urban market, and a permanent search for higher economic profitability from the countryside's resources. These concerns are legitimate, but they are also incomplete if the central objective does not include the improvement of the living conditions in the countryside as well as the modification of the terms of trade with urban sectors. Moreover, the concern to increase production and profitability led to the development and use of technologies that were often inappropriate for the soil and weather conditions, the sizes of the plots, and the cultural patterns of Mexico's rural environment.

The current situation of scarce resources and foreign exchange severely restricts the country's capacity to maintain a policy that implies increasing investments per irrigated hectare, a high dependence on seeds, genetic lines, machinery, spare parts, and components from abroad, and whose results necessitate increased volumes of imported foodstuffs.

The heterogeneity of Mexican society, the coexistence of very diverse cultures, and the unequal growth and inequitable distribution of income and resources lead to a range of different—and at times, contradictory—conceptions of the best development alternatives for the country. One of the most notable characteristics of Mexican society has been its capacity to utilize fully its various tendencies for the benefit of the country.

Within this context, there are several currents of thought regarding the alternatives for food policy. Among these, the model of "comparative advantage" stands against the call for food self-suficiency; those proposing free exchange markets oppose those favoring greater state control and regulation; some prefer private property while others favor social ownership of the means of production; one school of thought proposes the nationalization of the food industry while another encourages foreign investment. There is also the thesis that proposes a model of subordination for peasant organization against the one in favor of self-management; one striving for the creation of an entrepreneurial class in the countryside versus one concerned with the protection and preservation of the peasant forms of production. Finally, there is the model defending the conservation of ecosystems—in their natural forms—against one proposing the maximum utilization of all possible land for agricultural and livestock exploitation. These alternatives combine in different ways and are variously adhered to by different people.

The relevance of these factors within the economic and social context of the country is recognized in the National Development Plan for 1983–1988 and in the corresponding National Food Program (PRONAL), which clearly define the conceptualization of the problems and the strategy to achieve truly integral rural development and authentic national food sovereignty.

The National Development Plan points out, as the main objective of the strategy of integral rural development, "the improvement of the levels of welfare of the rural population, on the basis of its organized participation and the full utilization of natural and financial resources, following social criteria for productive efficiency, permanence and equity and strengthening their integration with the rest of the nation." Similarly, it proposes to foster the establishment of more just terms of trade between the agricultural sector and industry and services. It also aims to extend and strengthen the production of basic foodstuffs for popular consumption so as to guarantee national food sovereignty and to improve the nutritional conditions of the majority of the population, and to coordinate more efficiently agricultural, livestock, and forestry activities with each other and with industry, commerce, and other services.

In terms of food policy, the plan's general objectives are to "strive for food sovereignty and to reach the food and nutrition conditions that will allow for the full development of the capacities and potential of every Mexican." It states, moreover, that the National Food Program is to be fundamentally oriented to improving the nutritional levels of the low-income population, especially those in areas and in groups that are particularly affected by food deficiencies. Furthermore, it aims to foster a more equitable distribution of food.

The plan defines its strategies clearly: to encourage domestic production of food, especially in rain-fed areas; to strengthen basic and rural productive organizations that encourage broader and better participation, considered as the essence of development; to give priority attention to the basic needs of the rural population; to reorient the productive structure so as to set the economic base that will guarantee rural social development and, at the same time, respond to the needs for food and raw materials at the national level; and to change the relations of subordination of the countryside with respect to urban industrial development.

More specifically, the guidelines of the National Food Program establish the need to consider, in a comprehensive and dynamic fashion, the phases of the food process, in order to increase efficiency along the chain by preventing and solving the disequilibrium in its operation. Parallel to this, it proposes to encourage the participation of primary producers in all phases in a form consistent with the strategy of integral rural development.

Along the same lines, it recognizes the heterogeneity that characterizes productive activity and social conditions, especially in the rural context, and the need to understand and to deal with the countryside as an integrated reality. For this reason, it considers primary food producers in their double role as producers and consumers and proposes to promote the diversification and harmonious and balanced development of the activities that foster the optimal utilization of existing resources.

As far as production is concerned, the strategy proposes to encourage primarily the production of maize, beans, wheat, and rice to meet internal demands and to decrease imports. In a complementary fashion, it proposes to prevent the expansion of export agriculture from diverting the resources destined to meet the need for basic foods and strategic products. These constitute the fundamental guidelines for channeling public-sector support to the countryside.

One central aspect of the agricultural and livestock strategy is the priority assigned to the development of rain-fed areas. This is where one can generate very significant social effects from employment, income, and welfare, and where there are greater possibilities for yield and production increases, especially for basic grains. Therefore, the commitment is made to channel state support in a preferential and harmonious fashion to these areas, and especially for the production of basic products, in order to foster technological change and to encourage the increase of cultivated area and of yield.

The policy also recognizes the importance of improving the integration of livestock and agricultural exploitation at the farm level, so that, through their complementarity, overall economic production is increased, food restrictions are reduced, and rural employment is increased.

In the area of the food industry, the strategy is to reorient production toward popular consumption goods with higher nutritional value and to strengthen national control of the industry. The purpose of achieving a greater participation of peasant producers in other phases of the process implies the encouragement of the development of agroindustries with peasant participation and the support of small- and medium-sized food-processing industries.

Within this context, the organized integration of primary producers will also be encouraged, in fishing as well as in agriculture and livestock, in the phases of commercialization and distribution of foodstuffs. As to the role of the state in the commercialization of basic products, the plan orders the reinforcement of the capacity to regulate the supply of these products and their prices. Likewise, it proposes to strengthen its coverage and to broaden the supply of the system of state stores and warehouses for the distribution of basic products. In this phase, a commitment is made to the modernization and development of the commercial and transport infra-structure, through cooperation with the states and the private sector. The promotion of self-sufficiency at a regional level to reduce the unnecessary transportation of foodstuffs constitutes an important alternative for more efficient regional supply.

Finally, national sovereignty is put forth as the fundamental element and guiding thread of the whole strategy, for the determination of patterns of consumption, and for technological models to be adopted in the different phases of the chain.

The alternatives chosen are clear. They imply substantive changes, and they respond to a nationalist view of the future development of the country. Nevertheless, actually producing the necessary structural changes will depend on overcoming inertia and on fulfilling the commitments made by the different public entities according to the Law on Planning, and within the framework of the National Food Commission, which was created as the mechanism to guarantee a more effective and timely execution of the tasks proposed by the National Food Program.

Its success will depend equally upon the active and permanent participation of civil society, of its support for the alternatives chosen, and of the capacity of the Mexican system to respond in an increasingly flexible, well-aimed, and dynamic manner to the challenge of national food development.

Problems of Defining and Operating the New Food Policy in Mexico, *by Clara Jusidman de Bialotovsky*

During the past decade, the growing deficit in domestic food supply and the resulting increased dependence on imports made the solution of food problems a top priority. SAM was a first attempt to give coherence to policy makers' understanding of the problem, as well as to systematize the policies and actions aimed at nutritional deficiencies. Though the country had a long tradition of research on nutrition problems and food and basic agricultural technology, and though important government actions in food marketing and distribution were being carried out, the government had no integral approach or even a definition of a comprehensive and clear policy to tackle the problem.

SAM significantly advanced the conceptualization of the nature and characteristics of the problem. It counted on the most advanced tools of economic and information sciences, secured the support of outstanding specialists in various disciplines, and attracted broad interest at an international level. However, most important was SAM's firm political support, the internal expectations it generated, and the changes and innovations it made in some policies toward small- and medium-sized agricultural producers.

Nevertheless, SAM was unable to attain the committed participation of the main actors in the process—producers and consumers—and it overlapped with existing administrative structures. It also fostered all sorts of public programs, regardless of their relationship to food problems. Consequently, sky-high expenses that really corresponded to standard programs of various ministries and state enterprises, some many years old, were attributed to SAM. To get a sense of this, it is worth examining closely the relevant data in the annexes VI of the sixth Presidential Report to the Nation for 1982 (also refer to the analysis in Chapter 11).

The experiences of the Mexican Food System—both positive and negative—and SAM's body of knowledge and data accumulated over the years are currently being utilized to define and operate a national food policy. One early result of this process is the appearance of the National Food Program 1983–1988 (PRONAL) described in the previous two sections.

Let us now examine some issues that might be fundamental in the successful definition and operation of such a food policy in Mexico. They can be divided into three groups: those pertaining to the nature of the food problem, those that refer to the characteristics of social actors who take part in the process, and those related to the structure and operation of the public apparatus involved in the formulation and implementation of food policy.

It is worth stressing the complexity and extent of the food problem. On one hand, it is a multisectoral issue involving agricultural production, livestock and fisheries storage, collection, transport, industrialization, commercialization, and consumption. It involves, therefore, the daily action of a great number of individuals taking part in the different phases of the chain and also affects the whole of society as consumers. In addition, the food problem not only refers to productive and economic behavior, strictly speaking, but is related to consumption habits and traditions of the population as a whole. It is closely related to prevailing levels of employment and income distribution.

The complexity of the problem poses the risk of scattering the policy actions and the resources allocated to it, thus making it impossible to solve in the short term. For this reason, it is necessary to trace clearly the most important knots in the food process, the concentration of actions and resources to eliminate them, the permanence and maintenance of a consistent food policy, and the necessary participation of society as a whole. It cannot and should not be a policy defined and operated only by the government; it must be understood and taken on as a social responsibility, since only to the extent that this happens will it be guaranteed success and permanence.

The characteristics of the agents taking part in the food chain are essential for the formulation and execution of policy. Mexico is a heterogeneous country, with deep inequalities. From the standpoint of food production and distribution, as in other areas, it is possible to find a large spectrum of situations: from family units producing for their own consumption to large parastatal and transnational agroindustries oriented to production for the market. The sizes, technologies, operation levels, degrees of input use, products, and degrees of organization vary widely, not only along the food chain but also by region and commodity. There are community systems that are vertically integrated, operating with a clearly capitalistic logic, and others where on-farm consumption or marketing of limited surpluses in precarious local markets predominate. This creates asymmetries in the control of resources and in the distribution of benefits and contributes to increased inequality.

In the sphere of consumption, the wide differences in the levels of family income and in consumption patterns also result in heterogeneity (see Chapter 12). An additional characteristic of the actors participating in the food chain is their imbalanced geographic distribution. The situation is extreme: on one hand, a wide geographic dispersion of rural producers and consumers and, on the other, a large concentration of urban consumer markets (see Chapter 9).

Finally, a third characteristic of the actors in this process is the degree of organization of their different categories. It is worth highlighting the almost nonexistent organization of consumers and the precarious or merely formal organization of important sectors of rural producers and of small traders and transporters in the face of powerful organizations of large intermediaries and agroindustries. This results in inequalities in dealing with the state and in the presence of pressure groups that often do not represent the interests of the majority (see Chapters 7 and 8).

The heterogeneity, the unequal geographic distribution, and the lack of organization among producers, traders, and consumers impose basic constraints on the processes of designing and carrying out food policy, since they affect the nature of the programs and mechanisms designed to foster systematic participation of actors in the food chain.

Considering the above, the National Food Program planned the creation of urban food committees in municipalities or urban *delegaciones.* Such committees, made up of local authorities and community representatives, will discuss and formulate actions related to the problems of food supply within their areas. They are to agree on short-term measures at the local level and make policy suggestions to the state food commissions.

In the rural sector, the creation of rural food committees was proposed at the level of agricultural development districts. These districts are gradually becoming the basic units for the programming of agricultural and livestock production throughout the country. The creation of such committees implies the participation of federal, state, and local authorities involved with food production, as well as that of producers within the region. They are to deal with matters related to production scheduling and assuring a sufficient supply of food for the local population.

Food policy can be made more effective by implementing it in various forms that take into account the population and producers to which they are oriented. Generalized subsidies, standard criteria for credit, equal standards for the purchase of products, and many other measures are based on assumptions of social homogeneity and strictly market-oriented production units and consumption units. These assumptions have repeatedly been inadequate. They will have to be radically changed and adapted to the priorities set by the National Food Program for priority products as well as for targeted groups of producers and consumers.

Finally, let us examine how the structure and operation of the Mexican public sector introduces certain factors into the design and operation of a national food policy. These factors are closely related to elements described above.

A multisectoral food problem faces a federal public bureaucracy organized by single sectors. Wherever this is so—as is the case of agricultural and livestock production, trade, industry, transportation, tourism, and fishing—the government apparatus operates to encourage, to support, and, wherever needed, to regulate the production of goods and services in these activities. In some cases, however, the policy tools have not been fully developed, or do not function effectively, or are managed with distinct and, at times, contradictory orientations.

During the past presidential term all parastatal agencies were "sectorized," a first step to try to overcome some of these problems. This meant organizing state enterprises and institutions by administrative sector, in an attempt to coordinate their activities under a single policy line, defined by the minister of the corresponding sector within the framework of the national development policy.

This important step in the organization of the Mexican public sector increases the effectiveness of the whole parastatal sector. However, the size and power reached by some of the parastatal entities, which for years operated with a high degree of independence, interferes with the necessary coordination, and it will surely be some time before full coordination is achieved (see Chapter 4).

Thus, one of the problems of defining a food policy is that there is still a lack of unity in the policies and operations of all public agencies that relate to the different links of the food chain. Even in those cases in which greater advances have been made, they are difficult to put into practice due to the incipient degree of development of some policy tools or due to the fact that some are not under the direct control of the sector, as is the case of finance, which is handled in an integral fashion by a single administrative sector.

Furthermore, some sectors give little attention to planning. It should be planners who coordinate sectoral participation in the various mechanisms of the National Planning System, since they have an integrated and comprehensive overview of all the policies, programs, and actions of the corresponding sector and could ensure the consistency of sectoral policy with these mechanisms.

Gradual consolidation of changes in the organization of public administration and systematic planning efforts should overcome these problems. It is now necessary for several administrative sectors to participate in the mechanisms of intersectoral coordination. Such mechanisms are provided for in the Planning Law as interministerial commissions, whereby the president is able "to deal with the activities of national planning that are to be jointly developed by several ministries or administrative departments."

The law also permits these commissions to create subcommissions for the drafting of special programs. It will be possible to incorporate agencies in the parastatal sector into those commissions or subcommissions when matters are related to their purposes.

The law also permitted the creation of the National Food Commission. At the president's request, this commission was created and began operation in October 1983. Before the formalization of the commission, a working group was formed with representatives of several ministries and parastatal agencies under the coordination of the Ministry of the Planning and Budget; this group was responsible for the design of the National Food Program.

This first document of the revised food-planning process elaborated the related concepts that are included in the National Development Plan for 1983–1988. Its architects integrated the opinions expressed at the popular consultation forums held during the presidential campaign of 1981–1982 for the formulation of the national plan, as well as some elements of SAM's studies. Materials drafted by specialized institutions and individuals were reviewed and integrated, as were the many contributions made by the agencies and entities that participated in this working group. In addition, an interinstitutional group was created to define and agree on the program's objectives, since the lack of official consultation was considered to be one of the most controversial aspects of earlier food policy making.

As a first expression of a joint effort in the field, the program may have omissions or deficiencies in dealing with some issues, but its importance lies in the fact that, first, it triggers a participatory process at the levels of the different public entities involved, second, it appeals to the organization of state coordinating mechanisms, and, finally, it proposes the creation of corresponding urban and rural coordinating entities. In addition to the federal, state, and local coordinating mechanisms, the participation of the social and private sectors is foreseen.

One of the greatest difficulties faced by recent Mexican planning has been translating propositions into regulations, criteria, and concrete modes of operation. There seems to be large gap between the grand objectives, strategies, and programs proposed, and the concrete work carried out by the multitude of governmental employees. Such a vacuum is seen between the agreements reached at the highest levels of the public, private, and social sectors, and the daily behavior of individuals or units that make up such sectors.

An important part of this gap will be filled insofar as there is greater participation by those directly responsible for the implementation of public policy, as well as members of society, in the design of plans and programs and the coordination of actions. For this reason, it is necessary to carry out food policy making in a parallel way, both at the level of steering groups and operational levels, in a process that will lead to the convergence of both tasks and to mutual adjustment. This is the only way to guarantee the consideration of global and national parameters and variables, while at the same time assessing the operational viability of the recommendations.

Another cause of this gap between design and implementation is that programs often are not made specific, nor are they reviewed and modified to fit operational needs. At times it is thought that general elements in a document automatically translate into changes throughout the whole chain of operation of the corresponding government offices. This is not the case; people at the operational level do not have easy access to these documents or find it difficult to understand the way in which their specific job is changed by the plans.

The absence of efficient communication systems in public activities that guarantee and broadcast a standard message also contributes to widen the gap between the drafting of plans and their implementation. It is not unusual for those in charge of implementing a policy to find out about changes in it through newspapers and other mass media. Perhaps they never get an official communication explaining the changes. The development of fluid and timely communication systems in all directions within the social structure is an urgent need.

Implementing PRONAL will be a challenge, but its prospects for success have been enhanced by the measures that have already been taken and by those that have been planned.

CHAPTER 16

Final Reflections

James E. Austin and Gustavo Esteva

This final chapter returns to the general food policy questions raised in the opening chapter: What is the content of, and rationale for, self-sufficiency food strategies? How are they implemented? What impact do they have? Our collective analysis of the Mexican experience has explored each of these avenues throughout the book through a series of subquestions. We now attempt to summarize and synthesize the answers, as well as evaluate, comment on, and reflect on the lessons learned.

Research Questions Revisited

Strategy: Content and Rationale

The 1979 crop crisis was the catalyst that triggered the political decision to launch a new food strategy. But the content of and rationale for SAM had roots reaching far back into Mexico's historical struggle to feed herself. From the beginning of the Spanish colonization until SAM, there were repeated failures to recognize and balance local production capacities and consumption needs. The approach to agriculture was extractive. The surpluses first fed the Spanish conquistadores and later fueled the country's industrialization. But the inevitable exhaustion set in: production stagnated, external dependency on food imports rose, and poverty and malnutrition remained pervasive among the peasantry.

SAM's peasant-oriented strategy of self-sufficiency was thus a response to both the immediate and the chronic problems. The rationale for self-sufficiency was both political and economic. The growing dependency on the giant to the north to feed the country was becoming increasingly politically sensitive, especially since the United States was using grain embargoes as a political weapon. On the economic side, the strategy collided with the law of "comparative advantage" and its advocates who had dominated the policy making prior to SAM. The argument for the shift was that, by the turn of the century, all of the oil export revenues would be absorbed just to pay for the burgeoning food imports needed to sustain the country's expanding population. Other uses of foreign exchange critical to the nation's development would have to be ignored. Furthermore, it was argued that Mexico did have the capacity to feed herself, that there was untapped productive potential in the peasantry, and that the comparative disadvantage of producing grains was much less than supposed.

Let us reexamine the concept of self-sufficiency before moving on to the focus on the peasantry. Self-sufficiency should not be seen as the final objective; it has no intrinsic merit. Rather, what is sought is food security—certainty of supply that protects the consumers' access to food. A strategy of self-sufficiency contributes to food security by increasing national control over supply and by insulating the domestic market from the political and economic instability of supply and prices in world markets. Nevertheless, two factors indicate the need to reexamine a narrow conceptualization of self-sufficiency. First, agricultural production is inherently variable; there will inevitably be surpluses and shortages over time, as the production decline in SAM's last year revealed. The imbalances imply the need to enter international markets and carefully integrate a physical and financial reserve policy into the overall food policy. Second, self-sufficiency may be attainable with large resource mobilization, but its benefits must be weighed against its costs in order to guarantee that the real objectives are pursued effectively.

To achieve self-sufficiency, SAM counted on the peasants. The traditional farmers on rain-fed land were the least productive, exposed to the greatest agronomic risks, and presumably the most resistant to change given their precarious positions. To choose them to lead the charge toward self-sufficiency would, on the surface, appear to condemn the strategy to failure. Why not go with the modern, larger farmers? The rationale of the SAM strategists, which we consider valid, was twofold. First, the peasant producers, precisely because of their low yield levels, held the greatest potential for improvement with better technology, primarily increased use of fertilizer and improved seeds. The thousands of modern commercial farmers were already far out on the technology and yield frontiers; quantum leaps were not possible for them. Yet modest and steady improvements from the millions of traditional farmers could produce major output gains. Second, the poor farmers and their families were both the subject and the object of the food policy. They were the groups suffering from inadequate food intake. By producing more they would directly improve their food supply and incomes. The key was improving their productivity, and raising productivity and capturing the corresponding benefits is "the only long-run solution to poverty."[1]

There was also a political rationale for this emphasis on the peasants. It was an attempt to revitalize the state-peasant alliance that had washed back and forth with the ebb and flow of the political tides since the Mexican revolution. The intensity of the peasant unrest had accelerated. Given the incredible capacity of the Mexican political system to shift toward and mend impending fissures in its multifaceted coalition, it is not surprising that many politicians were receptive to the proposal to place the peasants center stage as protagonists in the new strategy.

The strategy rested on eliciting a positive response from the peasants. A critical starting point in SAM's approach to this problem was its perception of the peasants as economically rational decision makers. Accordingly, the strategy aimed to change the peasants' economic equation through the pricing variable—by lowering the prices of their inputs (fertilizer, seeds, and credit) through subsidies and by raising the prices of their outputs through higher crop-support prices. State-owned enterprises were to be the primary vehicle for channeling these incentives to the peasants. The strategy also addressed, through its shared-risk insurance, the expected (and rational) reluctance of the peasants to shift from their traditional methods to the newer, debt-leveraged technology. This new mechanism was to be a concrete manifestation of the state-peasant alliance, because in case of crop failure this insurance would pay not only for the production costs but also the lost revenue. The state and the peasants were to be partners in the march up the technological curve.

Accompanying these measures was an explicit recognition that market imperfections in the form of poor information, lack of mobility, and inadequate access to buyers could result in the farmers' losing much of their new economic surplus to the off-farm agents. Attaining higher incomes was essential to increasing nutrition and generating capital for reinvestment, which could sustain and further improve productivity. Consequently, the state, through its public enterprises, was to serve as a major buyer in the marketplace. In the imperfect markets of the real world, such a governmental role is conceptually reasonable. The desirability would depend on the implementation mode and capacity, a point we reexamine subsequently.

The final and distinctive dimension of the SAM strategy was its focus on consumption needs. It started with nutritional requirements and worked backward to derive the production targets that would meet those requirements and achieve self-sufficiency. This linking of consumption

1. C. Peter Timmer, Walter Falcon, and Scott Pearson, *Food Policy Analysis* (Baltimore: The Johns Hopkins University Press for The World Bank, 1983).

needs and production capacity constituted a new and highly desirable approach to food policy formulation in Mexico and elsewhere. Traditionally, agricultural policy was formulated primarily by the Ministry of Agriculture and Water Resources and focused on production. Consumption policy, largely in the form of subsidies, was set mainly by the Ministry of Commerce. The two sides were not integrated. If anything, it was production that was driving food policy, with consumption needs being a residual. Food policy in the 1950s and 1960s did try to keep urban food prices low via low support prices for crops, but this was in order to keep wages low to stimulate industrial investment rather than to meet consumption needs. In contrast SAM employed a food systems approach that focused on all the stages from production to consumption to create an integrated strategy. This was an important conceptual contribution.

The consumption focus was used to determine not only how much should be produced but also what types of food and for whom. This established crop priorities and target groups of consumers. Such refinements gave a clearer sense of direction and purpose to the strategy. It was an important step toward achieving a position of *food sovereignty*,[2] which means self-determination of food needs and of the ways of satisfying them. With such an approach, international food trade is an expression of national development and not a fragile premise on which to hang expectations and frustrations. Trade is a way to enrich and diversify diets and strengthen the economy. In a position of food sovereignty, international trade does not imply dependence.

Food sovereignty acquires meaning only if it is translated into food security. On the supply side this means the capacity to produce or acquire from abroad. On the demand side it means ensuring access. For the poor consumer target groups this meant overcoming their low effective demand, which had traditionally prevented them from acquiring nutritionally adequate diets. The SAM strategy used subsidies as the vehicle to ensure economic access; it also planned to expand the network of state food stores to ensure physical access. The subsidies were also viewed by SAM as the means to resolve the basic food policy dilemma of the need for high prices to producers and low prices to consumers. Producer subsidies can be economically justified if they sufficiently accelerate the productivity climb and produce adequate production gains. Consumer subsidies can be socially justified if they efficiently and preferentially reach those most in need. SAM's strategy can be faulted along these lines in that the priority groups were not sufficiently disaggregated according to degree of nutritional and economic vulnerability, geography, occupation, or income level. Accordingly, subsidies were not targeted with enough preci-

2. This concept of food sovereignty was discussed in two research workshops held in conjunction with this research project; an elaboration of the idea is found in "Por una nueva politica alimentaria," a statement by the Sociedad Mexicana de Planificación, *Boletin Informativo* no. 12, June 3, 1983.

sion to discriminate by need. This design flaw carried with it two possible adverse consequences. First, the total magnitude of the subsidies and the consequent fiscal burden would be larger, and, second, the less needy might benefit more than the more needy groups.

As a final overall reflection on the SAM strategy, we conclude that it constituted a major conceptual advance in food policy for Mexico. Its systems approach, linking production capacity to consumption needs, has validity for other nations as well. The goal of self-sufficiency was understandable from the political as well as the economic perspectives, but the concept was probably defined too narrowly. A broader concept of food security and food sovereignty would have been more appropriate. Focusing on the rain-fed peasant farmers made sense both because of their potential for increasing productivity and because of their dual role as needy producers and consumers. But distinguishing more sharply between various producer and consumer target groups as well as deciding which groups most needed help would have made SAM more efficient and effective. Moreover, SAM was a strategy superimposed on existing strategies rather than one totally replacing them, and this inhibited policy coherence. Nevertheless, the SAM strategy on the whole merits praise. But many a great strategy has shriveled into insignificance because of poor implementation. Was this the fate of SAM?

Implementation: The Precarious Leap from Word to Deed

Our analysis of SAM's implementation examined its central institutions and inputs. The institutions included the state-owned enterprises (SOEs), the agricultural extension services, the private commercial enterprises, and the peasants. The inputs included credit, seeds, fertilizer, insurance, and marketing. The results? The implementation went less well than the strategists had hoped, better than the skeptics had expected, but not as well as it could have. Let us review some of the highlights.

One of the critical decisions in the strategy was to use the SOEs as the government's primary institutional conduit for SAM's public inputs. These institutions were already in place and SAM's short time frame left the strategists with few alternatives except to use existing organizations. The key doubt was whether they would respond adequately to the new strategy. Lurking behind that doubt was the question of what factors would shape their response. These concerns are of considerable relevance beyond the borders of Mexico, given the significant presence of SOEs in the food sectors of most developing countries. The SAM experience revealed a strong positive response by the SOEs. They all rapidly increased the volume of their operations to record levels and channeled their goods and services more toward peasant producers, low-income consumers, and basic staples.

There appear to have been three key factors that called forth this positive response. First, strong presidential support created a powerful incentive for compliance by the politically appointed bureaucrats and managers. Second, incremental economic resources enabled the SOEs to attend increasingly to SAM's priority products and groups without abandoning their traditional activities and clients; in effect, it was a positive-sum game (except for the treasury). Third, the SAM strategy coincided with the existing strategies of some of the SOEs; this goal congruence gave added momentum to the SOEs and produced some of the strongest responses.

Thus the SOEs proved capable of responding quickly on a large scale given the right mix of institutional and individual incentives. But all was not rosy. Some SOEs simply relabeled their traditional activities in order to comply as well as to tap into the new resources. There were also incidents of administrative corruption, although apparently not on a major scale. Problems of inefficiency arose in some instances because the SAM strategists did not involve SOE operating personnel more fully in the planning process and because the SOEs had to expand so quickly and on such short notice. Finally, there was a retrenchment by the SOEs in SAM's final year, reflecting the changing incentive structure. The economic crisis brought on budget cuts, and therefore many SOEs squeezed SAM rather than their other activities. It was no longer a positive-sum game. The political incentive evaporated as the president neared the end of his term, and the incoming president gave no clear signals of support for SAM. Overidentification with an unblessed program was a highly risky position for those managers hoping to receive new appointments in the incoming administration. SAM did show signs of shriveling as the political transition approached.

One of the key SOEs in SAM was the agricultural development bank (BANRURAL), which served as the conduit for the credit that would enable the peasants to acquire SAM's technological package. BANRURAL's response followed the general pattern of the SOEs. During SAM it greatly increased its lending to small, rain-fed farms and for maize and bean production. This was a shift away from the growing trend of the lending for livestock rather than agriculture, but there are serious doubts as to the permanence of that shift. The credit flow to the peasants was mostly working capital for basic grains. Medium-term credit appeared to go to other types of producers for other types of production investments. Such investments generate a subsequent need for working capital and so may be determinative in shaping the bank's future short-term credit portfolio. Although the bank lent to more small farmers, the amount it lent per hectare fell during SAM. This may have reduced the peasants' ability to acquire SAM's full technological package. To implement the SAM strategy, BANRURAL undertook significant organizational restructuring. It geographically decentralized personnel and facilities and delegated greater authority to these units, and so enhanced flexibility and responsiveness. However, in simultaneously greatly expanding its central bureaucracy, it introduced rigidity and lethargy as well as increased operating costs.

Liberalization in the coverage terms of the crop insurance offered by the insurance SOE, and mandatory for BANRURAL loan recipients, resulted in an improvement in BANRURAL's bad-debt portfolio but simply shifted the losses to the insurance company. The bank was unable or unwilling to implement aggressively SAM's innovative shared-risk insurance scheme, which probably should have been assigned to the insurance SOE instead. On the whole, one can conclude that enough credit flowed to the countryside to oil the farmers' acquisition process. It was not a bottleneck, but the efficiency of its operations left much room for improvement and the permanence of its strategic shift remains in doubt.

One of the primary inputs acquired with the bank credit was improved seeds. The Mexican seed industry comprised government and private-foundation research institutes, a seed-producing SOE (PRONASE), and private local and multinational seed companies. SAM had a major impact on the industry because of the huge demand it created primarily through its subsidies for improved maize and bean varieties. There was a major increase in national seed production, led by PRONASE, as well as in imports to meet the still larger demand. This meant that the seed companies' production mix shifted toward maize, which had been relatively neglected previously. Farmers of all types were enticed by the huge subsidies to buy the hybrid seeds and expand maize production. Two concerns arise: first, as the subsidies are reduced,the larger commercial farmers will quickly shift to other more profitable crops, thereby making self-sufficiency a hostage to input subsidies; second, the traditional farmers replaced their indigenous (*criollo*) seeds with hybrids, thereby losing many valuable genetic traits that had evolved over hundreds of years. The reduced genetic diversity and the increased reliance on hybrids produced by a small number of foreign firms could decrease food security. SAM shook up the seed industry, but the pieces have yet to settle. It did not foresee the possible adverse consequences of its policies on the nation's stock of germ plasm. Furthermore, the role of PRONASE relative to the private producers remains unclear. The need for greater policy and institutional coherence is the challenge confronting the seed industry.

SAM's effectiveness was critically dependent on the peasants adopting the new technology. To a great extent they did, but the experience produced several lessons. First, prescribing a standard technological package conflicts with the heterogeneity in microclimates and producer characteristics; flexibility holds greater promise for maximizing acceptance and results. Second, the production technology should be linked to the transformation and commercialization stages, and the impact of macroeconomic polices should be assessed. Third, there is a critical need to establish close links between the end-users of the technology, the peasants, and the research scientists, extension workers, and credit institutions, such that there is a feedback mechanism and an iterative process in which the peasant is an active participant rather than a passive recipient.

A stereotyped characterization of the peasants held by many is that of passivity, lethargy, disorganization, fatalism, and minimal capacity to respond to external stimuli. Fragments of these characteristics are observable among some peasants at some times, which is not surprising given the contextual constraints that envelop them. However, a more penetrating look will pierce that superficial cloak and reveal the peasants as an exceptionally dynamic force. Their organizational mode and dynamics may not fit well into the models deemed ideal by policy makers or academics, because they are the organic product of the peasants' reality rather than that of the city dwellers. SAM conceptually envisioned an active participative role for the peasants in the decision-making processes of SAM. But this never really occurred, partly because the compressed time frame of SAM prevented the integration and partly because the governmental apparatus could never shed itself of preconceived and paternalistic ideas about peasants and their organizations. There is a need to build two-way bridges to span the river of misunderstanding that separates the state and the peasants. In this process, there appears to be a need for "guides" who have lived and worked on both sides of the bridge, to assist (but not manipulate) the peasants and the bureaucrats in making the mental and operational crossing.

The peasants were the primary priority group for SAM, but the urban poor also constituted a target group. Most of these were found in Mexico City along with a quarter of the country's population. Feeding Mexico City clearly has to be an integral part of a national food strategy. The produce supply system was fraught with market imperfections: inadequate information, low access to markets, concentration, and fragmentation. The structure at the wholesaling level was oligopolistic, with the wholesalers and their network of produce assemblers holding considerable bargaining power over small producers as well as small retailers. The supply channels were lengthened by the large number of intermediaries, whose cumulative margins had the effect of raising final consumer prices and reducing the producers' share. The lack of greater competition also permitted oligopolistic profits at the expense of producers and consumers. SAM took some actions aimed at shortening the channels, improving information flows, increasing access of producers to commercialization services, of retailers to supplies on credit, and of consumers to lower-priced goods. However, these actions were focused primarily on basic grains and beans. The channels for other fresh produce remained largely untouched and unchanged by SAM. In some sense it was a blind spot of SAM. On the other hand, SAM was seeking major changes in so many other arenas that it may have been quite reasonable not to use up its political capital in a battle with the produce wholesalers. In any case, the market imperfections remain, as does the need to reduce them.

SAM chose the public enterprises as its primary institutional vehicle for implementation. However, the private commercial enterprises also play a significant role in the food system. At several stages they responded to the SAM incentives. The seed producers increased output and changed their product mix, as was discussed above. The farm-equipment and agrochemical suppliers also had boom years during SAM, although sales slumped in the final year as austerity measures reduced demand. The commercial farmers benefited significantly from the subsidized inputs; their costs were lowered, but it is not clear that their already high usage or yields increased further. The agroindustries felt the positive effects of higher throughput. The distributors or transporters did not perceive any significant effects of SAM.

From this review of the SAM actions, it is clear that the strategy of self-sufficiency was implementable and implemented. The process encountered problems and was flawed in several respects, but none of these was so severe as to abort the implementation effort. The strategy proved feasible. But what impact did it have?

Impact: The Acid Test of Results

SAM reached its self-sufficiency targets for maize and beans in its second year. Production declined in SAM's final year as the austerity measures took hold, but the average for the three-year period was the highest output level ever attained in Mexico. Imports were greatly reduced but not eliminated. Data limitations hinder any attempt to identify the sources of the growth, but some reasonable approximations reveal that productivity rather than acreage increases accounted for the output gains. Favorable weather does not appear to be the determining factor. It would appear that yield increases on irrigated land played an important role. Both commercial and small farmers contributed to the production increase; their precise relative contributions cannot be determined, but it is clear that peasants were major beneficiaries of SAM. The failure to collect production data disaggregated by producer type was a key void in SAM's information system. Nonetheless, widespread adoption of the technological package suggests that SAM was successful in moving many traditional farmers up the technological curve. Nonetheless, it would appear that weaknesses in the technology itself, particularly the improved seeds, and in the dissemination process reduced the extent and effectiveness of the adoption of new technology. The reduction in subsidies also casts doubt on the permanence of the shift in technology.

Production increased, but at what cost? SAM's price tag was large, but it was part of a generalized expansion in public spending. Oil income transfers were being made to many sectors. The food sector's share of the total federal budget changed very little and averaged less than 10%. A comparison of the incremental production costs attributable to SAM with the value of its incremental production and import substitution gains revealed a cost-benefit ratio of 1.39:1. In addition, SAM incurred similar incremental expenditures on its distribution and consumption side. The corresponding benefits were in the form of increased consumption, with average per capita maize intake rising 16%. However, the absence of nutritional status data makes it impossible to assess the level or loci of nutritional improvement. This constituted another void in SAM's information system. On the whole, SAM's average annual net costs over benefits amounted to only 1.5% of the federal budget. This is a much lower fiscal burden than many of SAM's critics had claimed. Nevertheless, it appears clear that SAM cost more than it needed to. Better administration and a sharper targeting of the subsidies would have significantly reduced costs, probably without adversely affecting benefits and possibly even enhancing them.

Although the fiscal costs did not appear overwhelming, by 1982 the negative consequences of macroeconomic decisions had so severely eroded the country's economic capacity that SAM's subsidies could not be sustained. The oil boom preceding SAM appears, ironically, to have been a significant factor contributing to agriculture's decline through a deterioration of the rural-urban terms of trade. SAM's instruments explicitly attempted to enhance those terms and create positive production incentives. In effect, the SAM strategy held potential for serving as the appropriate policy antidote to the structural effects of increases in energy prices. However, the government's failure to manage adequately its macro prices eroded SAM's and the economy's foundation. Imports skyrocketed, exports were hindered by an increasingly overvalued peso, the foreign debt-servicing burden grew, oil prices softened, the federal deficit mounted, inflation soared, and capital flight accelerated. The failure to spend prudently and to devalue sooner and decisively caused the economic distortions and consequent austerity measures that hit SAM so hard in 1982. It appears that food policy was not adequately integrated with the rest of the macroeconomic policies.

One final impact dimension to be assessed is income distribution. SAM explicitly sought to redistribute income to the poverty groups. Empirical data are lacking to provide conclusive evidence, but the results of the general-equilibrium modeling show that SAM's policy measures did produce positive redistributive effects for peasants, although the magnitude was small. However, landless rural and marginal urban workers (along with urban capitalists) suffered income and consumption declines. These first two groups constituted a significant portion of the poor and malnourished population. It appears that a design flaw in the SAM strategy was a failure to designate more clearly its target groups so that policy measures could be tailored to their specific needs. Analysis reveals that an increase in nominal wages in conjunction with the SAM measures would have improved income, consumption, and employment of these classes. The consumption benefits generated from such a measure would be over four times greater than the other SAM measures on a per-public-peso-spent basis.

So what is the overall judgment of SAM's results? Mixed. The production gains were undeniably impressive. But the excessive costliness was impressively undeniable. Self-sufficiency was approached, poverty alleviated, and consumption enhanced, but none to the degree or permanence hoped for. And the fiscal burden proved onerous when the government was faced with the imperative of austerity. Still, the vision had been created, much had been accomplished, and, most important, most of the shortcomings were correctable. The final acid test of results is whether the strategy survives and is strengthened as the future food policy emerges.

PRONAL: SAM Reincarnated?

On the evening of November 30, 1982, the lights in SAM's offices were turned off for the last time. As President Miguel de la Madrid took office, SAM's staff mourned the apparent death of the national food policy that they had so energetically nurtured and championed since its birth three years earlier. Many observers had predicted SAM's demise. Its close identification with outgoing President López Portillo made it likely that with his exit SAM would disappear from the administrative horizon. Some hoped that SAM's particular merits might save it from being another victim of Mexico's six-year political cycle. They had hoped in vain. SAM was dead. Or was it?

At the end of May 1983, when Mexico's National Development Plan was revealed, PRONAL was announced as a "Special Program" within the plan, which signaled that food policy would retain a high priority in the new administration. In October the PRONAL strategy was promulgated. A close examination of its content suggests that SAM had risen from the ashes, certainly in spirit and, to a great extent, in substance. The label had changed but the design bore a striking resemblance to the predecessor. To the PRONAL policy makers' credit, they chose to learn from SAM rather than to reject it. They modified and refined rather than reinvented. Let us compare the two.

Goals

PRONAL transformed SAM's goal of self-sufficiency into one of food sovereignty. Under this concept, self-sufficiency is viewed more flexibly. It is to be sought for some foods but not others. Autarky is out and food security is in. Selective interaction with the international markets on the export and import side is to be seen as one more tool in meeting the nation's self-determined consumption needs. We view this reconceptualization as desirable for Mexico's and other countries' food policies.

Nutritional improvement remains a priority goal in PRONAL, although divided into two time frames. The short-run goal is to protect the low income families' diets from deterioration during the immediate period of economic recession and austerity. The longer-run goal seeks intake improvements to bring consumption above the nutritional requirements. Both PRONAL's and SAM's nutritional goals suffer from lack of greater specificity. Quantitative targets for improving nutritional status of specific groups would sharpen the focus and permit measurement of progress.

Strategy

At the conceptual level, PRONAL continues SAM's systems approach. This is an important legacy of SAM and should be integral to all countries' food strategies. Consumption needs remain the starting point and primary shaper of the production targets. PRONAL added a refinement by categorizing different foodstuffs according to their nutritional priority. This gives clearer guidance to the production strategies.

PRONAL is also targeted at peasants on rain-fed land, recognizing their binomial status as producers and needy consumers. However, it expands the target producer group to include small, irrigation-dependent farmers. On the consumption side, within the poverty population, PRONAL attaches priorities to the most nutritionally vulnerable groups, in particular pregnant and lactating women and children under five years of age. This was necessary in the face of severe resource constraints. However, the new program, like SAM, does not adequately highlight the landless laborers as an acutely needy group that may require special mechanisms, nor does it put in place an impact-monitoring system.

The productivity-enhancement approach through the provision by SOEs of credit, improved seeds, and fertilizer continues under PRONAL, but the level of subsidies are significantly lower both to producers and consumers. The production response and the consumption benefits can be expected to be less. Nevertheless, the cost-effectiveness of these subsidies may improve due to PRONAL's sharper targeting and clearer priorities. But the administrative mechanisms for channeling these subsidies preferentially to the priority groups are not clearly delineated. PRONAL does not appear to have made any significant changes that would address the problems encountered by SAM due to the rigidity of its technological package and to the means of disseminating it. The organizational strategy of PRONAL is strikingly different than SAM's. Whereas SAM was run from the office of the president by advisers to the president, PRONAL established a National Food Commission chaired by the president and consisting of the ministers and directors from each of the relevant ministries and government institutions. This organizational mode has the advantage of integrating into the decision-making process the heads of the primary implementing organizations. Potentially this increases their commitment as well as achieves better coordination of the multisectoral strategy. The Ministry of Planning and Budget serves as the commission's technical staff, thus giving PRONAL better access than SAM to the budget levers. SAM was imposed on the institutions; PRONAL attempts to integrate them. Both SAM and PRONAL have purported to integrate the protagonists—the peasants and the poor consumers—into the decision-making process. The means for accomplishing this remain unclear.

Implementation

PRONAL has existed only for about a year, probably too short a period to render a reasonable judgment on how well it has been able to implement its strategy, and definitely too short to judge impact. However, it is relevant to examine the production results of 1983 and 1984 that occurred while PRONAL was just coming onto the scene.

The 1983–1984 production of basic grains remained in the higher ranges achieved during SAM. Output in 1983 was 4.2% higher than the average of the 1980–1982 SAM period, with maize and beans experiencing strong gains that offset declines in wheat and rice. In 1984, further favorable gains occurred. With SAM presumably dead and PRONAL not yet operational, these positive production results give rise to the temptation to conclude that, with favorable weather, agriculture will do fine, regardless of the apparent policy vacuum. However, our alternative interpretation is that the production gains confirm the hypothesis that SAM's name, but not its strategy, was erased.

It is a fact that government outlays for the SAM-type activities were reduced in real terms during the 1982–1984 period as part of the general austerity program. However, SAM's basic momentum could well have carried forward. The technology adopted and the resources accumulated by the producers during the SAM years, especially in rain-fed areas, may well have oriented and enabled the farmers to continue with the basics in spite of reduced government support. Furthermore, from the beginning the new administration made clear that basic grains and rain-fed farms were of high priority. This may have inhibited some institutions from abandoning in 1983 the previous SAM strategy. The broad institutional participation during 1983 in the formulation of PRONAL reinforced this continuity in strategy. In effect, there was no major deviation from the SAM strategy in the operations of the implementing organizations, even though the government expenditures were reduced.

Although PRONAL's architects clearly derived many lessons from

SAM, they seem to have fallen into a trap similar to one that caught SAM. They have kept the political and administrative commitments of the new strategy within the rigid limits of the governmental apparatus. Although some attempts have been made to gain political support, mechanisms of mechanical solidarity continue to be useless for such a purpose. In many cases, bureaucratic institutions have been happy to replace SAM's label (which might have been falsely labeled in the first place) with PRONAL's, making no substantive changes in their activities. In other cases, within the present framework of budgetary austerity, institutional inertia has reasserted itself, with the result that the organizations simply do less of the same thing. PRONAL's somewhat broader and more flexible view of self-sufficiency may have sent out signals that are subject to a broader and more ambiguous interpretation by the SOEs and other governmental institutions. A clear lesson of SAM is that committing the institutions themselves to the new strategy and trying to find a common ground for the political interests of the implementers and of the food policy makers is a necessary but not a sufficient condition for achieving effective implementation. There is also the need for the social base of producers and consumers to understand fully the food policy being adopted, to participate in the formulation of its programs, to take on its commitments and responsibilities, and to have the capacity to give feedback, formally and in reality.

PRONADRI: A New Path toward Implementation

In early 1985, President de la Madrid announced a major new piece in the administration's food and agricultural strategy: the National Program of Integral Rural Development—PRONADRI (Programa Nacional de Desarrollo Rural Integral). Was this a rupture or continuity? Had a new member of the family arrived or had a divorce occurred? Was the SAM-PRONAL policy thrust being maintained?

The Emergence of PRONADRI

The roots of PRONADRI are clearly traceable to presidential campaign statements that give priority to rural development. These were translated into a formal obligation through the constitutional amendments promoted in December 1982 by the administration, which set forth the responsibility of the state for "integral rural development" and "agrarian justice" (Article 27, items XIX and XX). These two elements—rural development and agrarian reform—were then incorporated into the National Development Plan for 1983–1988 as, in the president's words, "fundamental pillars in the national development strategy."

These "pillars" were seen as supporting the accompanying priority assigned in the National Development Plan of meeting the country's food needs and given status as a special program, which emerged in late 1983 as PRONAL. The subsequent emergence of PRONADRI constituted a logical supporting step that moved the PRONAL food policy farther down the path of implementation.

PRONAL's food goals and strategy were the points of departure for PRONADRI, which lays out in considerable detail the corresponding implementation instruments and actions on the food-production side. In this way, PRONADARI is a mechanism for operationalizing PRONAL in the rural areas. At the same time, PRONADRI constitutes a separate strategy in that it broadens the scope to address additional nonfood rural problems and needs in the areas of land reform, health, education, housing, water, electricity, and employment. In this sense, PRONADRI serves as a means to achieve greater coherence among the various sectoral development strategies for the rural areas. Thus, conceptually, PRONADRI goes beyond SAM or PRONAL in terms of scope of action in the rural areas, while at the same time serving as a vehicle for the attainment of PRONAL food production goals.

The manner in which the PRONADRI strategy is elaborated is revealing in terms of the policy-formation processes. In order for development goals or thrusts to be translated into programs of action, there generally must be a "policy entrepreneur" who is willing to take the initiative and risks to mobilize the necessary resources and institutional and political support within the bureaucratic arena. For PRONADRI this role appears to have been performed by the new secretary of agriculture and water resources, who was appointed in September 1984, and his staff.

That the entrepreneurial initiator is a line rather than a staff agency, and particularly the Ministry of Agriculture, is of particular interest in that it departs from the approaches under both SAM and PRONAL. SAM's initiator was an office of advisers to the president. One of its principal limitations was the weakness of its institutional links to the operating agencies. SAM's location in the presidency gave it essential political backing but also kept it far away from the scene of implementation. With PRONAL the institutional policy entrepreneur was the Ministry of Planning and Budget (SPP), which became the technical arm for PRONAL's interinstitutional National Food Commission nominally headed by the president. This link with SPP meant that potentially PRONAL could make its strategic orientations felt by way of budgetary allocations. The leverage over the implementers was political and economic. But PRONAL did not have a primary operating executor; all the operating agencies were responsible, which creates the risk of none feeling responsible.

With PRONADRI the food strategy potentially gains through SARH, an organizational motor directly connected to the chassis of the agricultural development vehicle. More important, it also gains a specific driver: one that will assume personal and direct responsibility for the implementation task. The fact that the driver is SARH has special meaning, given that during SAM this ministry operated somewhat independently of SAM in spite of its importance to the strategy. With SARH at the helm of PRONADRI, it is ensured that a central operating agency will be fully integrated into the food strategy implementation process.

However, there is also a risk to this approach. PRONADRI is a multisectoral strategy. The Achilles heel of all such programs is interinstitutional coordination. Thus, PRONADRI with SARH faces the same general coordinating problem as PRONAL and the National Food Commission, but it is potentially even more serious for SARH. This is because of the difficulty of one ministry trying to lead other ministries. The implied subjugation is generally unacceptable among presumed equals. For example, explicit instructions from the president were necessary in order to obtain the agreement of the Ministry of Agrarian Reform to be integrated into PRONADRI.

The challenge for SARH is to find mechanisms by which its coordinating function is not seen as a vehicle for bureaucratic domination or political manipulation. SARH and PRONADRI, like PRONAL and SAM, have strong presidential backing, and within the Mexican context this appears to be an essential ingredient for interinstitutional cooperation. An additional approach used by SARH as well as PRONAL, but not SAM, was to engage the various implementing agencies in a collective process of strategy formulation. SARH promoted and coordinated forty working groups from the different operating institutions that over an intensive six-month period produced PRONADRI strategy. Such a participative-consultative process can serve to create a sense of shared ownership of, and commitment to, the implementation of the strategy. It can also develop a set of personal relationships and trust among the technical personnel in the various institutions, which can greatly facilitate both formal and informal coordination during implementation.

The PRONADRI Strategy

The substance of the PRONADRI strategy for agricultural production falls directly on the SAM-PRONAL continuum: food sovereignty and self-sufficiency, nutritional needs dictating production goals, emphasis on rain-fed agriculture and productivity-enhancing technology. However, PRONADRI also introduced some important new dimensions.

In the definition of its target group, PRONADRI adds the landless farm laborers. Adequate attention to this critically needy group had not been given in SAM or PRONAL. In fact, under SAM this group remained the least benefited (see Chapter 14). PRONADRI emphasizes generation of employment to meet this group's needs. The implementation challenge is to avoid the creation of isolated public-works projects that are not part of an integrated strategy and or that can even cause serious distortions and disruptions in that strategy.

Another omission in SAM was agrarian reform. In fact, for many analysts, the Law of Agriculture Development that was issued at the same time as SAM was in open contradiction to SAM. Through PRONADRI the government is explicitly linking food, agriculture, and agrarian reform strategies, thereby moving closer to a coherent food policy, in contrast to the earlier series of parallel but uncoordinated policies. As mentioned initially, PRONADRI also expands the action arena to include the mix of social services.

Technical assistance plays a larger role in PRONADRI than in SAM, which is a reflection of the prominence of SARH. However, it is also a reflection of the strategic centrality of technological improvements in the production methods of the rain-fed farms. This will require SARH to make a significant shift in technical assistance away from irrigation and toward rain-fed agriculture. Internal resistance from the SARH irrigation group can be expected. Unlike SAM, PRONADRI stresses infrastructure investments. The challenge is to keep these from being excessively diverted into traditional irrigation projects and to channel them toward productivity-enhancing improvements for the rain-fed farms, for example, erosion and drainage control.

PRONADRI recognizes more explicitly than its predecessors the heterogeneity of producer situations. Consequently, it points out the need for bureaucratic structures and procedures to adapt to the requirements of the producers, rather than the opposite, which has characterized the traditional approach of the governmental apparatus. Thus the need for flexibility and adaptability of the technological package is stressed along with greater applied research to help guide this.

We consider this approach of selecting instruments capable of differentiating among producers and situations to be more effective and efficient than using standardized instruments that correspond to an unrealistic assumption of homogeneity. However, it appears that PRONADRI departs from this focus in regard to two types of instruments: support prices and input subsidies. The former is a standard price that applies equally to all farmers throughout the country; in effect, equal treatment to unequals. This approach, when combined with the methodological and data weaknesses in the price-setting process, risks producing regressive effects. In contrast, the input subsidies can be much more flexible and targeted more sharply. PRONADRI plans to use the input subsidies, as did SAM, but it plans to phase them out. Production subsidies can be highly cost-effective in a productivity-enhancing strategy. Their use should be given preference over the less cost-effective use of standard support prices.

Another area of concern with the PRONADRI strategy is its greater accent on livestock compared to SAM or PRONAL. It is not that there has been an explicit pronouncement to this effect; rather this is the impression given by the PRONADRI strategy documents as well as public comments made by officials within the administration, including SARH. For example, meat was recently relabeled as a priority food, whereas it had originally been excluded from SAM's and PRONAL's lists. One of SAM's merits was to identify the adverse implications of the trend toward livestock dominance of the sector. PRONAL translated this concern into explicit programmatic terms. For PRONADRI not to support these concerns would be a relapse and a reversal of an effort that had just begun to advance in the correct direction.

PRONADRI's goal to increase milk production is congruent with PRONAL's objectives. Nevertheless, the selection of technology remains a critical point. Capital-intensive, stable systems requiring high use of feed concentrates should be downplayed in favor of modern intensive pasture-based systems located in the available, appropriate ecological zones. Egg production might also be similarly criticized. Small-scale household production, including that in the big cities, merits attention relative to the large-scale commercial systems that rely heavily on imported inputs.

Our two final comments on PRONADRI deal with organizational issues. The first concerns producer participation. Like SAM and PRONAL, PRONADRI stresses the importance of achieving an active participation by the farmers in the design and implementation of the program. PRONADRI sees this participation as a vital part of the administration's effort toward democratic planning. There is an effort to find mechanisms to achieve this participation within the new decentralized political and planning system. The local Development Planning Committees (COPLADES) are seen as the organizational vehicles by which the farmers can have an ongoing voice in PRONADRI. The approach and attitude hold promise, but the challenge of achieving meaningful participation remains to be realized.

Participation as well as the general effectiveness of PRONADRI will be greatly affected by the second organizational aspect, namely, the internal structure of SARH and the other implementing organizations. Generally, to be effectively implemented, new strategies require new organizational structures and administrative procedures. SARH is planning to take bold steps in this direction. It plans to decentralize its personnel, from both the Federal District and the state capitals and major cities. The goal is to get technicians and extension personnel into the field to interact with and serve the farmers, rather than sitting behind desks in air-conditioned offices. Furthermore, it aims to integrate the various departments within SARH so that the traditional splintering can be converted into an integrated approach. Such reforms are long overdue and hold potential for rescuing SARH from the bureaucratic self-suffocation and interest-group fragmentation that has characterized its past. One can expect intense resistance from those being affected by this change. Strong leadership will be required to implement the change.

PRONADRI has created a new path toward implementation of Mexico's evolving food policy. It is a logical extension of SAM and PRONAL and increases the political momentum and commitment from the implementing institutions. Progress continues in the right direction, but it may be severely impeded by the country's acute macroeconomic problems.

Lessons from the Mexican Experience

Mexico's struggle for self-sufficiency continues. The effort has not produced a patentable strategy for success. Nevertheless, our exploration of the Mexican experience has revealed many lessons relevant to other countries. Undoubtedly each reader has gleaned various insights most relevant to his or her interests from each of the preceding chapters. Here we simply highlight a few.

Strategy. It is important for countries to start from the consumption side by assessing nutritional needs and then to work back to shape the requisite production approach. The consumption needs should be disaggregated by degree and priorities set accordingly.

Strategy should be holistic in two senses. It should use a systems perspective that encompasses the entire chain of food production and consumption, from input supply to final ingestion. It should also recognize the linkages between food policies and other macro policies and ensure coherence.[3] Food policies are only strands in the larger macro policy fabric; the weaving must be closely coordinated. Econometric modeling is of help.

Self-sufficiency should not be equated with autarky. Self-sufficiency in the narrow sense has no intrinsic value. What a country should be concerned about is food security. On the supply side, at the national level, this means having the ability to produce or to acquire the needed food. On the demand side, at the consumer level, it means having economic and physical access to one's food requirements. Thus participation in the international food markets, on both the export and import sides, is consistent with food-security-oriented strategy of self-sufficiency.

Subsidies can play a meaningful role in strategy, at least in the short run. Producer subsidies can be economically justified if they enhance pro-

---

3. New food policies must also be carefully integrated with prior ones, rather than simply superimposed; the latter brings the risk of multiple, parallel policies that may be contradictory.

ductivity, and consumer subsidies can be socially justified if they are targeted to the neediest groups and administered efficiently.

Implementation. State-owned enterprises can respond positively, but their response depends on appropriate political and economic incentives and a reasonable administrative capacity. One can expect that the various government bureaucracies will have varied institutional interests and agendas; incentives must be tailored to fit these differences. Involving the implementers in the planning process will enhance implementation.

New production technology will be rapidly adopted by peasant producers if the economic incentives and administrative access are adequate. The strategy's technological package should be flexible rather than standard to fit the different types of producers and ecoclimates. Technological research will be enhanced if it is closely coordinated with the producers.

Understanding the structure and competitive dynamics of the private-sector actors throughout the system is important to ensure their positive contribution to the implementation process.

Impact. Self-sufficiency is attainable, but it requires major, concentrated, and continual effort. Governments should monitor the costs of these efforts in order to avoid the "self-sufficiency-at-any-price" trap. Optimum use of scarce resources must remain a constant concern.

Impact assessment requires systematic collection and analysis of information. A monitoring and evaluation system should be given priority. Its purpose is to provide continual feedback to the policy makers on the effectiveness of their strategy, so that improvements can be made. It is important to assess how much is being produced of what, by whom, where, and why. The flow of produce from farm to consumer should be monitored to ascertain postharvest losses. The consumption effects from production increases need to determine who is eating how much of what. Nutritional-intake surveys and nutritional-status monitoring of at-risk groups are also needed, as is assessment of income-distribution effects.

Looking Forward

The fact that a policy is well intentioned is not enough; it should be designed correctly and addressed to the producer and consumer groups that need priority attention. The existence of effective leverage with government institutions does not suffice either. For a food policy to be fully effective and lasting, it must be rooted in the structures of food production and consumption. Only when producers and consumers are able to take a policy on as a political project of their own—in agreement with their interests, needs, and capacities—will they fully accept their responsibility for its implementation.

This basic principle can easily generate a general consensus, except when it comes time to put it into practice. The problem is simple. Food policies normally and naturally come from governmental offices. In practice, and especially in times of crisis, this is a mortal sin, and one for which there is no pardon, since the political markets do not have enough indulgences. This pardon will be even harder to find when the daily faults of corruption, lack of consistency, and other well-known weaknesses of bureaucratic action are added to the mortal sin.

But this contradiction is not insurmountable. When public entities try to take on productive and commercial activities themselves, and when they try to control strictly producers, traders, and consumers, they generally tend to lose sight of objectives, foster corruption, increase operating costs, and cause deterioration of social productivity. The opposite happens when governmental structures play their role of leadership and direction, encouraging actors in the system to develop orientations that are socially and nationally necessary and providing targeted resources and support for the groups that should receive priority attention.

A policy's label does not matter that much; it is always exposed to the vagaries of the economic and political interests of those who formulate and implement it. Its success does not depend so much on whether producers and consumers identify it by its name or whether they nominally agree to participate in programs conceived and implemented according to its guidelines. The key is the effective participation of these groups in the design and implementation of policy and in the policy's capacity to recognize and assume the conflicts of interests and the differences that characterize the heterogeneous population it should serve.

SAM is dead. But it has been reincarnated as PRONAL and PRONADRI, which knew how to recognize some of its most valuable contributions and how to avoid many of its weaknesses. For the new program to avoid the fate of its predecessor, it must rapidly expand beyond the administrative spheres that created it.

As was said at its time, SAM was like a eucalyptus, with a large trunk and small roots. So too it is with PRONAL-PRONADRI. A strong political or economic wind may blow it down. Like most trees, PRONAL-PRONADRI's trunk, branches, and fruit will grow to the size of its roots. It should grow administratively as much as its social and political roots permit, and only to that extent. It should be as small or as large as it is capable of mobilizing producers and consumers, providing them orientation and support. This should define its orientation and character more than anything else.

Mexico's search for self-sufficiency has been tortuous, the struggle to feed herself continues. Our hope is that this book contributes to meeting that challenge.

Glossary

ALBAMEX   Alimentos Balanceados de México (Balanced Feeds of Mexico); state-owned enterprise

AMSAC   Asociación Mexicana de Semillas A.C. (Mexican Seed Association); private industry organization

ANAGSA   Aseguradora Nacional Agrícola y Ganadera, S.A. (National Agriculture and Livestock Insurance Agency); state-owned enterprise

ANDSA   Almacenes Nacionales de Depósito, S.A. (National Warehouse Depository); state-owned enterprise

Banco de México   Bank of Mexico; the country's central bank

BANRURAL   Banco de Crédito Rural (Rural Credit Bank); state-owned enterprise

BORUCONSA   Bodegas Rurales de CONASUPO, S.A. (CONASUPO Rural Warehouses); CONASUPO warehousing subsidiary

Campesino (or *peon*)   small farmer (or laborer)

CBR   Canasto Básico Recomendado; Recommended Basic Food Basket

CIMMYT   Centro Internacional de Mejoramiento del Maíz y el Trigo (International Center for the Improvement of Maize and Wheat); private research organization

CONACYT   Consejo Nacional de Ciencia y Tecnología (National Council of Science and Technology); government entity

CONASUPO   Compañia Nacional de Subsistencias Populares (National Company of Basic Staples); state-owned enterprise

Criollo   indigenous (seeds)

D.F.   Distrito Federal (Federal District); national capital

DICONSA   Distribuidora CONASUPO, S.A. (CONASUPO Distributor); CONASUPO food retailing subsidiary

Ejido   a form of land tenure whereby an individual is given the right to farm the land but cannot sell it; farmers on an *ejido* are *ejidatarios.*

375

FERTIMEX Fertilizantes de Mexico (Fertilizers of Mexico); state-owned enterprise

FIPROR Fideicomiso para Promoción Rural; Rural Promotion Fund

FIRA Fondo de Garantía y Fomento para La Agricultura, Ganadería y Avicultura (Agriculture and Livestock Guarantee Fund); an adjunct of the Banco de México

FTA Fabrica de Tractores Agrícolas (Agricultural Tractor Factory); a joint venture between the state and the Ford Motor Company

Hacienda large ranch

ICONSA Industrias CONASUPO, S.A. (CONASUPO Industries); CONASUPO food-processing subsidiary

IMPECSA Impulsadora del Pequěno Comercio (Small Commerce Promoter); CONASUPO wholesaling subsidiary

INIA Instituto Nacional de Investigación Agrícola (National Institute of Agricultural Research); a federal institute

LICONSA Leche Industrializada CONASUPO S.A. (Conasupo Industrialized Milk); CONASUPO milk-processing subsidiary

MINSA Maíz Industrializado S.A. (Industrialized Maize); CONASUPO corn-processing subsidiary

NAFINSA Nacional Financiera, S.A. (National Finance Company); state-owned development finance company

NUTRIMEX Nutrimientos Mexicanos (Mexican Nutriments); state-owned enterprise

PACE Programa de Ayuda a la Comercialización Ejidal (Rural Commercialization Assistance Program); a small-farmer program of BORUCONSA

PEF Presupuesto de Egresos Federales (Federal Expenditure Budget)

PRONADRI Programa Nacional de Desarrollo Rural Integral (National Program of Integral Rural Development)

PRONAGRA Productora Nacional de Agricultura (National Agricultural Producer); state-owned enterprise

PRONAL Programa Nacional de Alimentación (National Food Program); food policy successor to SAM

PRONASE Productora Nacional de Semillas (National Seed Producer); state-owned enterprise

SAM Sistema Alimentario Mexicano (Mexican Food System); national food policy

SARH Secretaría de Agricultura y Recursos Hidraulicos (Ministry of Agriculture and Water Resources); federal ministry

Sexenio six-year presidential period

SOE state-owned enterprise

SPP Secretaría de Programación y Presupuesto (Ministry of Planning and Budget); federal ministry

SRA Secretaría de la Reforma Agraria (Ministry of Agrarian Reform); federal ministry

TRICONSA Trigo Industrializado S.A. (Industrialized Wheat); CONASUPO wheat-flour-milling subsidiary

Index

Library of Congress Cataloging-in-Publication Data
Food policy in Mexico.

   Includes index.
   1. Food supply—Government policy—Mexico. 2. Agriculture and state—
Mexico. I. Austin, James E. II. Esteva, Gustavo.
HD9014.M62F66   1987   338.1'9'72   86-19815
ISBN 0-8014-1962-X (alk. paper)